Algebra and Number Systems

Algebra and Number Systems

by J. Hunter and D. Monk
in association with W. T. Blackburn and D. Donald
members of the Scottish
Mathematics Group

Blackie Glasgow and London
Chambers Edinburgh and London

Blackie & Son Limited   Bishopbriggs, Glasgow
5 Fitzhardinge Street, London W1

W. & R. Chambers Limited   11 Thistle Street, Edinburgh 2
6 Dean Street, London W1

Blackie   0 216 87406 8
Chambers   0 550 75891 7

Printed in Great Britain by   Gilmour and Dean Ltd., Hamilton

Preface

THE CONTENT OF this book covers all the work required for Paper I of the Scottish Sixth Year Studies Examination in Mathematics. It provides final-year students in Scottish schools with a natural sequel to the nine books of *Modern Mathematics for Schools* (Blackie/Chambers), written by the Scottish Mathematics Group to provide Ordinary and Higher Grade courses for the Scottish Certificate of Education. In certain respects the treatment here goes beyond the minimum requirements of the Scottish Sixth Year syllabus in Mathematics. Some of the topics which are given a fuller treatment than that required for Sixth Year Studies are: truth tables in logic, the algebra of sets, number systems, number theory, complex numbers, matrices and determinants. In many places, to make the book more useful for reference purposes, more formal proofs of results have been included than the immediate object of the book dictates; in these places it should be possible by means of examples to illustrate the underlying principles without insistence on abstract proofs. It is hoped that the wider viewpoint taken will give the interested student a glimpse of part of the exciting mathematics that lies ahead and, at the same time, will make the book useful not only for study at the Advanced Level of the General Certificate of Education, but also for certain courses in Colleges of Education and Universities. In particular, the needs of students in Colleges of Education who are preparing for the degree of Bachelor of Education have been kept very much in mind.

An important aspect of the Sixth Year syllabus is consolidation. The emphasis on the basic algebraic laws which is a feature of quite elementary work at the Alternative Syllabus stage leads now to a deeper study of the concept of number. Even after the groundwork of earlier years, the amount of detail inherent in a full discussion of this theory is prohibitive, but we hope that enough has been said to indicate the problems involved and the way in which they are tackled. Inequalities, increasingly important in school work, are dealt with in this context. The work on number systems culminates in the introduction of complex numbers. Here too one aspect illustrates the theme of consolidation—the transformations on which the approach to Geometry was based turn out to be just those needed for a geometrical description of the algebraic operations on complex numbers. Closely related to the study of number systems is the topic of algebraic structures—groups, rings, integral domains and fields—which share some but not all of the familiar properties of numbers. These same basic properties provide the framework for a systematic discussion of matrices.

Prominent among the words which seem to have become canonical in the modern approach to mathematics teaching are "logic" and "sets". We believe that these must be kept in proper perspective. At this level it is just not possible to give a genuine introduction to the specialist study of mathematical logic; all one can do is to try to isolate and explain the fundamental logical techniques of elementary mathematical reasoning. The manipulation of set-theoretical identities has likewise been kept in a fairly low key. Much more important as a foundation for further work is the concept of mappings and their elementary properties, to which a separate chapter is devoted.

Traditional techniques still essential for the mathematician must not be neglected. Accordingly there are sections on mathematical induction, summation of series, applications of de Moivre's theorem and the theory of equations, all integrated as far as possible into the main framework. There is also a brief survey of determinant theory and a short collection of problems on this topic [Additional Examples I, 106 to 109].

An interesting feature of the Sixth Year Studies syllabus is the inclusion of a short section on Number Theory, dealt with here in Chapter 6. This topic has long been a source of stimulus to able mathematicians and it is hoped that the presentation given will project some of the interest of the subject even although it concentrates on the basic elementary ideas.

There is a general exercise of selected problems at the end of each chapter and there are shorter exercises scattered throughout the text. The latter have been kept short in order to preserve the continuity in development of the text; it is suggested that they should be supplemented from the collection of easier problems (Additional Examples I) given at the end of the book. An additional collection (Additional Examples II) containing some more difficult problems has also been provided at the end for challenge and revision purposes.

Contents

Introduction to Mathematical Logic

1. Introduction

What is mathematics? This is a very difficult question to answer and, in fact, an adequate answer can be appreciated only by a listener who has had an extensive training in the subject called *mathematics*. A partial and rather pompous answer might be:

"The language, techniques and results of scientific methods (logical methods based on axioms and definitions) used in studying

(i) *natural phenomena* in physics, engineering, astronomy, chemistry, biology, genetics, medicine, psychology, etc.;

(ii) *human endeavours* such as economics, actuarial work, accountancy, town planning, architecture, etc.;

(iii) *logical structures* and *special investigations* such as abstract concepts derived from intuitive notions of number and space, computer programming, linguistics, etc."

It has only fairly recently been noted that there is a similarity in the basic ideas underlying the mathematics used for all of the above subjects, and that a common education in these fundamental concepts provides a basis for any mathematical work. Some of the words and concepts that have to be mastered, to an extent depending on the particular application involved, are: sets, relations, mappings and functions, operations on a set and algebraic structure, number systems, special linear structures, and properties of functions defined on the real numbers (i.e. calculus).

We propose in this book to deal with what appear at this stage of mathematical development to be the core ideas. We shall try to avoid the danger of presenting work purely as a vehicle for problem solving, but nevertheless we feel that solid foundations can be built up only by active personal effort and that consequently problems on suitable topics are essential.

2. Mathematical Logic

In a book at this level mathematical logic is a rather difficult topic to describe since any such description must be incomplete and intuitively based and must make use of illustrations that are considered in detail later. Consequently this part of the work will be brief and informal.

Statements. In mathematics we deal with statements. For example:
  (1) In a plane, the sum of the measures in degrees of the angles of a triangle is $180°$.
  (2) Every equation of the form $x^2 = a$ with a real has a real root for x.
  (3) There exist integers x and y such that $x+y = -1$.
  (4) If x is a real number such that $x > 5$, then $x > 7$.

Statement (1) is a true statement, being a well-known result in euclidean geometry.

Statement (2) is false, since e.g. the equation $x^2 = -1$, with $a = -1$, has no real root. [Nevertheless the equation $x^2 = a$, for some real numbers a (in fact, for all $a \geqslant 0$) has a real root.]

Statement (3) is true, since e.g. $x = -2$, $y = 1$ give $x+y = -1$.

Statement (4) is false, since e.g. $6 > 5$, but $6 < 7$.

Each of the statements (1), (2), (3) and (4) is either true or false. The fundamental property of a mathematical statement is that, in each case in which it makes sense, it is either true or false, but cannot be both.

Open sentences. Consider the equation $x+3 = 5$. This equation, by itself, is not a statement since it is not possible to determine whether it is true or false until the symbol x has been replaced by some specific number. This means that we must have in the background a collection of possible replacements for x, e.g. the set of natural numbers or the set of real numbers, such that, for each possible replacement a for x, either $a+3 = 5$ is true or $a+3 = 5$ is false. For example $2+3 = 5$ is true, but $1+3 = 5$ is false.

The equation $x+3 = 5$ is called an **open sentence**.

An equation such as $x+3x = 4x$, which is true for every replacement of x, is considered to be an open sentence, since no specific value of x is indicated.

For the remainder of this section we shall, for simplicity of notation, call an open sentence a statement; then each statement is either true or false in each case in which it makes sense.

Compound statements. From given statements we can form new statements by use of the words *and, or, not*. Let p and q denote statements.

Conjunction of p, q. This is defined to be the single statement obtained by forming p *and* q; it is denoted by $p \wedge q$.

e.g.   (i) p: The integer a satisfies $a > 1$; q: The integer a satisfies $a < 3$.
Here $p \wedge q$ is: The integer a satisfies $a > 1$ and $a < 3$, i.e. in simpler form, $p \wedge q$: The integer a satisfies $1 < a < 3$.

   (ii) p: a is an integer; q: The real number a satisfies $a > 2$.
Here $p \wedge q$ is: a is an integer such that $a > 2$.

   (iii) p: $2 > 1$; q: $2 > 3$.
Here $p \wedge q$ is: $2 > 1$ and $2 > 3$. In (iii), p is true, q is false and $p \wedge q$ is false.

In the general case, $p \wedge q$ is true only when p and q are both true.

The statements p and q are called the **components** of the **compound statement** $p \wedge q$.

Disjunction of p, q. This is the compound statement obtained by forming p *or* q; it is denoted by $p \vee q$. In mathematics this *or* is almost always used in the **inclusive sense**, i.e. $p \vee q$ means: p or q *or both*. The statement $p \vee q$ is true when at least one of the statements p, q is true.

e.g.   (i) p: $2 > 1$; q: $2 > 3$.
Here $p \vee q$ is: $2 > 1$ or $2 > 3$. In this case $p \vee q$ is true since p is true.

   (ii) p: $a = 0$; q: $b = 0$ (*a* and *b* given real numbers).
Here $p \vee q$ is: $a = 0$ or $b = 0$, which can be written as $ab = 0$.

Negation of p. The statement obtained from p by prefixing the word *not* and forming *not p* is called the *negation* of p and denoted by $\sim p$; it says the opposite of p and may be worded in several equivalent ways. If p is true, then $\sim p$ is false and, if p is false, then $\sim p$ is true. The statements p and $\sim p$ cannot both be true.

e.g.   (i) p: For a plane triangle ABC, $\angle A + \angle B + \angle C = 180°$:—true;
$\sim p$: For a plane triangle ABC, $\angle A + \angle B + \angle C \neq 180°$:— false.

   (ii) p: For each pair of integers x, y, the integer $x + y$ is even:—false;
$\sim p$: There is a pair of integers x, y for which $x + y$ is odd:—true.

Truth tables. If a statement is true, we say that it has a truth value T, and, if false, that it has a truth value F. The truth value of a compound statement can be determined, when we know the truth values of the components, by using a **truth table**. We illustrate the idea by displaying the truth tables for $\sim$, $\wedge$ and $\vee$.

(1)	p	$\sim p$
	T	F
	F	T

(2)	p	q	$p \wedge q$
	T	T	T
	T	F	F
	F	T	F
	F	F	F

(3)	p	q	$p \vee q$
	T	T	T
	T	F	T
	F	T	T
	F	F	F

In (1), we list in the first column the possible truth values of p, and, in the second column, the corresponding truth values of $\sim p$.

In (2) and (3), we list in the first two columns the possible truth values of the pair p, q in that order, and, opposite each, list in the third columns the corresponding value of $p \wedge q$ in (2) and of $p \vee q$ in (3).

In a similar way we can draw up truth tables for more complicated compound statements such as $\sim (p \wedge q)$ or $(\sim p) \vee (\sim q)$, etc., or statements involving three or more components such as $(p \wedge q) \vee (\sim r)$, etc.

Implication between statements p, q. The compound statement which we obtain by forming *"If p, then q"* is called the **implication of q by p** and is denoted by $p \Rightarrow q$.

e.g.   (i) p: The integer a satisfies $a > 2$; q: The integer a satisfies $a^2 > 4$.
    Here $p \Rightarrow q$ is: If the integer a is such that $a > 2$, then $a^2 > 4$.

   (ii) p: The triangle ABC is right-angled at A;
    q: The triangle ABC has area $\frac{1}{2}AB \cdot AC$.
    Here $p \Rightarrow q$ is: If the triangle ABC is right-angled at A, then the triangle has area $\frac{1}{2}AB \cdot AC$.

  (iii) p: The integer a is odd; q: The integer a^2 is even.
    Here $p \Rightarrow q$ is: If the integer a is odd, then the integer a^2 is even.

The statement $p \Rightarrow q$ is given a truth value according to the following truth table:

p	q	$p \Rightarrow q$
T	T	T
T	F	F
F	T	T
F	F	T

The truth value is chosen so that $p \Rightarrow q$ is false when p is true and q is false, and $p \Rightarrow q$ is true in all other cases.

In example (i) above on $p \Rightarrow q$, if p has truth value T, then q has truth value T and $p \Rightarrow q$ has truth value T.

In example (ii), if p has truth value T then q has truth value T (which can be proved geometrically) and $p \Rightarrow q$ has truth value T.

In example (iii), if p has truth value T, then q has truth value F and $p \Rightarrow q$ has truth value F.

Associated with $p \Rightarrow q$ we have three other implications, namely,

$q \Rightarrow p$, called the **converse** of $p \Rightarrow q$,

$\sim q \Rightarrow \sim p$, called the **contrapositive** of $p \Rightarrow q$,

$\sim p \Rightarrow \sim q$, called the **inverse** of $p \Rightarrow q$.

Note that $\sim p \Rightarrow \sim q$ is the contrapositive of $q \Rightarrow p$, i.e. of the converse of $p \Rightarrow q$.

Logically equivalent statements. Suppose we draw up the truth tables of both $p \Rightarrow q$ and its contrapositive $\sim q \Rightarrow \sim p$.

p	q	$p \Rightarrow q$	$\sim q$	$\sim p$	$\sim q \Rightarrow \sim p$
T	T	T	F	F	T
T	F	F	T	F	F
F	T	T	F	T	T
F	F	T	T	T	T

From columns 3 and 6 it is seen that $p \Rightarrow q$ and its contrapositive $\sim q \Rightarrow \sim p$ have the same truth values. Such statements built up with the same component statements and with the same truth values are said to be **logically equivalent**. The importance from the mathematical point of view of such statements is that a proof of a true statement establishes also a proof of a logically equivalent statement.

e.g. p: a quadrilateral $ABCD$ is cyclic;

   q: a quadrilateral $ABCD$ has opposite angles supplementary.

   $p \Rightarrow q$: If a quadrilateral $ABCD$ is cyclic then its opposite angles are supplementary.

$\sim q \Rightarrow \sim p$: If the opposite angles of a quadrilateral $ABCD$ are not supplementary, then the quadrilateral is not cyclic.

In this case $p \Rightarrow q$ is true and $\sim q \Rightarrow \sim p$ is an equivalent true statement.

The converse statement $q \Rightarrow p$ in this case is also true, i.e. it is true that "If the opposite angles of a quadrilateral $ABCD$ are supplementary, then the quadrilateral is cyclic".

In a situation such as that just illustrated in which both $p \Rightarrow q$ and the converse $q \Rightarrow p$ are true, we write $p \Leftrightarrow q$. Then p is true **if and only if** q is true; in this statement the part **if** corresponds to $q \Rightarrow p$ and the part **only if** to $p \Rightarrow q$. An alternative wording for the statement $p \Leftrightarrow q$ that is often used is "A **necessary and sufficient condition** for p to be true is that q is true"; in this wording, $p \Rightarrow q$ corresponds to q true being a **necessary** condition for p to be true, and $q \Rightarrow p$ corresponds to q true being a **sufficient** condition for p to be true.

e.g. p: A positive integer a is divisible by 9.

   q: The sum of the digits of a positive integer a is divisible by 9.

It is known that $p \Leftrightarrow q$, so that a positive integer is divisible by 9 if and only if the sum of its digits is divisible by 9.

Universal and existential statements. The statements:

(1) for all real numbers x, $\cos^2 x + \sin^2 x = 1$;

(2) for all integers x, $x > 1$;

are called **universal statements** since each statement involves all the objects of a given collection of objects. The statement (1) is true, but the statement (2) is false (e.g. $0 \ngtr 1$).

The statements:

(3) there exists a rational number x such that $x^2 = 2$;

(4) there exists a real number x such that $x^2 = 2$;

are called **existential statements** since each involves the possible existence of an object with a particular property in a given collection of objects. The statement (3) is false, but the statement (4) is true.

The expressions "for all", "there exists" are called **quantifiers**. For conciseness of notation we often write:

$$\forall x \quad \text{for} \quad \text{"for all } x\text{"},$$

and $\qquad \exists x \quad \text{for} \quad \text{"there exists } x\text{"}.$

$\forall$ and $\exists$ are called respectively the **universal** and **existential symbols**.

The negation of a universal statement is an existential statement and the negation of an existential statement is a universal statement.

e.g. (i) p: $\forall$ real numbers x, $\cos x + \sin x = 1$.

$\sim p$: $\exists$ a real number x such that $\cos x + \sin x \neq 1$.

Here the statement p is false and to establish this fact we can show that $\sim p$ is true, i.e. produce a real number x such that $\cos x + \sin x \neq 1$. The particular replacement $\frac{1}{4}\pi$ for x gives $\cos \frac{1}{4}\pi + \sin \frac{1}{4}\pi = \sqrt{2} \neq 1$. We say that we have established the falsity of p by a **counterexample**, namely $x = \frac{1}{4}\pi$.

(ii) p: $\exists$ a real number x such that $x^3 = 1$.

$\sim p$: $\forall$ real numbers x, $x^3 \neq 1$.

In this illustration, the statement p is true since $1^3 = 1$.

Proof in mathematics. The mechanics of a proof in mathematics consist of establishing the truth of a statement by logical processes from the truth of a given statement or statements. At this stage it would be rather pointless to describe in detail methods of formal mathematical proof since not many such proofs have yet been met. We content ourselves with some examples illustrating the few main types of argument that we are likely to meet in early mathematical work.

(1) Examine the following list of statements:

  p: a and b are even numbers.

  $p \Rightarrow q$: If a and b are even numbers, then $a+b$ is an even number.

Conclusion: q: $a+b$ is an even number.

This example illustrates the usual simple method of **direct proof** often called the **rule of detachment** or **modus ponens**. It says:

$$\left.\begin{array}{ll} \text{From} & p \text{ true} \\ \text{and} & p \Rightarrow q \text{ true} \end{array}\right\} \text{we infer that } q \text{ is true.}$$

(2) Consider now the following statements involving an integer a:

  p: a is even

  q: a^2 is even

  r: a^2-1 is odd,

and the following table of implications:

  $p \Rightarrow q$: a is even $\Rightarrow a^2$ is even

  $q \Rightarrow r$: a^2 is even $\Rightarrow a^2-1$ is odd

Conclusion: $p \Rightarrow r$: a is even $\Rightarrow a^2-1$ is odd.

From the true statements $p \Rightarrow q$ and $q \Rightarrow r$ we infer that the implication $p \Rightarrow r$ is true. This **transitive property of implication** is another common method of direct proof.

Instead of proving results directly, as in (1) and (2), we sometimes use an **indirect proof** based on the fact that if p is a statement, then one and only one of p and $\sim p$ is true. If, by a logically correct argument, we arrive at a stage at which both p and $\sim p$ are true, then one of our assumptions must have been wrong. We often say that a proof of a result obtained by using such a process has been obtained by the **method of contradiction**. Such a method of proof often uses the fact that $p \Rightarrow q$ and its contrapositive $\sim q \Rightarrow \sim p$ are logically equivalent.

We consider the following example.

(3) Let x and y be integers and consider the statements:

  p: xy is an odd integer.

  q: x and y are both odd integers.

We show that, if p is true, then q is true, so that $p \Rightarrow q$ is true. In mathematics we could present such a result in the following formal form.

Theorem. *Let x and y be integers. If xy is an odd integer, then x and y are both odd integers.*

Proof. Let p be the statement "xy is an odd integer" and q the statement "x and y are both odd integers".

Suppose that p is true; we have to show that q is true.

Assume that q is false, so that $\sim q$ is true. It follows that at least one

of the integers x, y is even, say $x = 2z$, where z is an integer. Then
$$xy = 2zy \quad \text{and so } xy \text{ is an even integer.}$$
Thus $\sim p$ is true.

[The same conclusion holds when y is even.]

Hence p and $\sim p$ are both true.

Thus, when p is true, the assumption that q is false leads to a contradiction. Consequently, when p is true, q is also true. The proof is now complete.

Note 1. We could word the above proof slightly differently as follows:

Assume that q is false so that $\sim q$ is true. It follows that at least one of the integers x, y is even, say $x = 2z$, where z is an integer. Then
$$xy = 2zy \quad \text{and so } xy \text{ is an even integer.}$$
Thus $\sim p$ is true.

[The same conclusion holds when y is even.]

It follows that $\sim q \Rightarrow \sim p$ is true.

Hence $p \Rightarrow q$ is true, since $p \Rightarrow q$ and $\sim q \Rightarrow \sim p$ are logically equivalent. It follows that, if p is true, then q is true, and the result has been proved.

Note 2. In practice, in writing out such proofs, we do not introduce symbols for the statements involved. For example, the first version could appear as follows:

Proof. Let xy be an odd integer; we have to show that x and y are both odd integers. Suppose on the contrary that x and y are not both odd. Then at least one of x, y is even, say $x = 2z$, where z is an integer. It follows that $xy = 2zy$ and so xy is even. [The same conclusion holds when y is even.] But xy is odd; hence we have a contradiction and the proof is complete.

EXERCISE 1

1. By drawing up truth tables show that in each of (i), ..., (vii) the compound statements are logically equivalent.

$$
\begin{array}{llll}
\text{(i)} & \sim(p \wedge q) & \text{and} & (\sim p) \vee (\sim q); \\
\text{(ii)} & \sim(p \vee q) & \text{and} & (\sim p) \wedge (\sim q); \\
\text{(iii)} & \sim(\sim p) & \text{and} & p; \\
\text{(iv)} & p \Rightarrow q & \text{and} & (\sim p) \vee q; \\
\text{(v)} & \sim(p \Rightarrow q) & \text{and} & p \wedge (\sim q); \\
\text{(vi)} & p \Rightarrow (q \wedge r) & \text{and} & (p \Rightarrow q) \wedge (p \Rightarrow r); \\
\text{(vii)} & (p \vee q) \Rightarrow r & \text{and} & (p \Rightarrow r) \wedge (q \Rightarrow r).
\end{array}
$$

[Note that for (vi) and (vii) the truth tables have eight rows arising from the possible truth values of the three statements p, q and r.]

2. From your knowledge of mathematics so far obtained state the truth value of as many of the following mathematical statements as you can.

(i) The rational number $\frac{5}{8}$ is less than the rational number $\frac{16}{27}$.

(ii) ∃ real numbers x and y such that $x + y = 0$.

(iii) ∀ real numbers x and y, $x + y = y + x$.

(iv) The equation $x^3 + 8 = 0$ has no real root for x.

(v) The equation $x^2 + 2 = 0$ has no real root for x.

(vi) Similar triangles enclose equal areas.

(vii) Congruent triangles enclose equal areas.

(viii) If points P and P' are images of each other by reflection in a line l, then l is the perpendicular bisector of PP'.

(ix) If x is a real number > 2, then $-x > -2$.

(x) If x is a real number > 2, then $\dfrac{1}{x-2} > 0$.

(xi) If a, b, c, d are real numbers such that $a > b$ and $c > d$, then $ac > bd$.

(xii) If x and y are real numbers such that $\sin x < \sin y$, then $x < y$.

3. Write down the negation of each of the following mathematical statements, and determine for each given statement (i), ..., (xii) whether it is true or false.

(i) There are exactly five prime numbers p such that $2 < p < 20$.

(ii) There are exactly six integers x satisfying $-1\cdot2 < x < 5$.

(iii) The area enclosed by a circle of radius r is πr^2.

(iv) The only real numbers x for which $\sin x = 0$ are 0, π and 2π.

(v) ∀ real numbers x, $x^3 > x$.

(vi) ∃ a rational number x such that $2x^2 - x - 1 = 0$.

(vii) ∃ real numbers x and y such that $x + y = y$.

(viii) ∀ integers n, $n^2 \geqslant 1$.

(ix) ∃ real numbers x and y such that $\sin(x + y) = \sin x + \sin y$.

(x) ∀ real numbers x and y, $\cos(x + y) = \cos x \cos y - \sin x \sin y$.

(xi) ∀ rational numbers a, the equation $x^3 = a$ has a rational root for x.

(xii) ∃ an integer x such that $0 < x < 1$.

4. Use in each case a method of contradiction to prove the following mathematical results.

(i) If n^3 is odd, where n is an integer, then n is odd.

(ii) If m and n are integers such that mn^2 is even, then at least one of m, n is even.

(iii) If x and y are real numbers such that $x + y$ is an irrational number, then at least one of x, y is irrational.

(iv) If $\sin \theta \neq 0$, then $\theta \neq k\pi$ for any integer k.

5. Which of the following conditions are (a) necessary, (b) sufficient, (c) necessary and sufficient, for the natural number n to be divisible by 6?

(i) n is divisible by 3.

(ii) n is divisible by 9.

(iii) n is divisible by 12.

(iv) n^2 is divisible by 12.

(v) $n = 384$.

(vi) n is even and divisible by 3.

(vii) $n = m^3 - m$ for some natural number m.

Sets and Relations

2

1. Introduction

In Chapter 1 we noted that a mathematical statement such as "$x>2$" makes sense only when x is replaced by an object from a suitable collection of objects. In this case we could use a number system such as the integers or the rational numbers or the real numbers; but the points of a plane would not be a suitable collection of objects.

Now consider the equation $x^2-2=0$. What are the roots of this equation? A question like this makes sense only when we know the collection of replacements available for x;

e.g. $x^2-2=0$ has no root in the rational numbers;

$x^2-2=0$ has two roots, $\pm\sqrt{2}$, in the real numbers;

$x^2-2=0$ has one root, $+\sqrt{2}$, in the positive real numbers,

and so on.

In fact, in almost all mathematical work we have a collection of objects in mind from the beginning. It is useful to have available some basic notation and results concerning such collections.

2. Some basic definitions and notation for sets

A **set** is any well-defined list or collection of objects. The objects in a given set are called the **elements** or **members** of that set. Some examples of sets of numbers, with a fairly standard notation which will be used throughout this book, are:

     N: the set of natural numbers (or positive integers);

     W: the set of non-negative integers (or whole numbers);

     Z: the set of all integers;

     Q: the set of rational numbers;

     R: the set of real numbers;

     C: the set of complex numbers.

Other examples are:

        the set of all points $P(x, y)$ in the x, y-plane;

        the set of positive divisors of the integer 6;

        the set of integers x such that $0 < x < 9 \cdot 2$,

and so on.

If x is a member of a set S, we write $x \in S$, which is read as: "x belongs to S"; e.g. $x \in \mathbf{R}$ means "x is a real number". $x \notin S$ will mean that x is *not* a member of S.

A set is said to be **finite** if it has a finite number of elements; otherwise it is called **infinite**.

e.g. $\mathbf{N, W, Q, Z, R, C}$ are infinite sets;

the set of positive divisors of the integer 6 is finite.

For finite sets, whose members are easily listed, we often display the members in curly brackets, separated by commas.

e.g. $\{1, 3, 4\}$ is the set consisting of the integers 1, 3, 4;

       $\{2\}$ is the set consisting of the integer 2 alone;

       $\{1, 2, 3, \ldots, n\}$ is the set consisting of the first n positive integers;

       $\{a, e, i, o, u\}$ is the set of vowels (letters) in the alphabet.

For some infinite sets a similar notation can be used.

e.g. $\mathbf{P}$ is $\{1, 2, 3, \ldots\}$; $\mathbf{Z}$ is $\{0, \pm 1, \pm 2, \pm 3, \ldots\}$.

Specifying a set by a defining property. The type of set that occurs most frequently in mathematics is usually defined by saying that it consists of all the members x of a given set S for which some open sentence $p(x)$ is true. This set is denoted by

$$\{x : x \in S, p(x)\}, \quad \text{or} \quad \{x \in S : p(x)\}.$$

e.g. $\{x \in \mathbf{R} : x \geqslant 1\}$ is the set of real numbers $\geqslant 1$;

        $\{x \in \mathbf{R} : x^2 - 2 = 0\}$ is $\{\sqrt{2}, -\sqrt{2}\}$;

        $\{x \in \mathbf{Z} : 0 < x \leqslant 7\}$ is $\{1, 2, 3, 4, 5, 6, 7\}$.

Equality of sets. Two sets A, B are said to be *equal*, and we write $A = B$, if they consist of the same elements. Since this is often difficult to check directly, we usually use the following equivalent definition:

$$A = B \Leftrightarrow \begin{cases} \text{every member of } A \text{ is in } B \\ \text{and} \\ \text{every member of } B \text{ is in } A. \end{cases}$$

From the above examples, we have:

$$\{x \in \mathbf{R} : x^2 - 2 = 0\} = \{\sqrt{2}, -\sqrt{2}\};$$
$$\{x \in \mathbf{Z} : 0 < x \leqslant 7\} = \{1, 2, 3, 4, 5, 6, 7\}.$$

Note that $\qquad \{1, 3, 5\} = \{1, 5, 3\} = \{1, 1, 3, 3, 3, 5\}.$

This pair of equalities emphasizes that (i) the order of naming the elements of a set is ignored and (ii) the multiplicity of appearance of an element is ignored, i.e. each element is counted once only. The second of these facts means that set notation is of little value in dealing with multiplicity of roots of an equation; e.g. the equations $x(x-1) = 0$ and $x^2(x-1)^3 = 0$ have the same solution sets, viz. $\{0, 1\}$, in $\mathbf{R}$, but each root has different multiplicities for the two equations.

However in most situations in mathematics we want a notation which ignores multiplicity, e.g. the notation

$$\mathbf{Q} = \left\{ \frac{m}{n} : m \in \mathbf{Z}, n \in \mathbf{Z}, n \neq 0 \right\}$$

for the set of rational numbers enables us to count each member of $\mathbf{Q}$ once only although it appears infinitely often (e.g. $\frac{1}{2}$ appears as $\frac{2}{4}, \frac{-2}{-4}, \frac{3}{6}$, etc.).

The empty set. An empty set is a set with *no* members. If A, B are empty sets, then, $A = B$, since they have the same members, viz. none. It follows that there is *only one empty set*. We refer to it as *the* empty set and denote it by the Scandinavian symbol $\emptyset$. The empty set can be described in many ways.

e.g. $\qquad\qquad \emptyset = \{n \in \mathbf{N} : 4 < n < 5\},$
$$\emptyset = \{x \in \mathbf{R} : x^2 < 0\},$$
$$\emptyset = \{x \in \mathbf{Q} : x^2 - 2 = 0\},$$

and so on.

3. Subsets

A set A is called a **subset** of a set S if every member of A is a member of S. For example,

(i) $\{1, 2\}$ is a subset of $\{1, 2, 3\}$;

(ii) the set of all equilateral triangles in a plane is a subset of the set of all isosceles triangles in the plane.

If A is a subset of S, we shall write $A \subseteq S$, which is read as:

$\qquad$ "A **is a subset of** S" $\quad$ or $\quad$ "A **is contained in** S".

If $A \subseteq B$, we sometimes write $B \supseteq A$ and say "B **contains** A".

If $A \subseteq B$ and $A \neq B$, we sometimes write $A \subset B$ and say "*A* **is strictly contained in** *B*", or $B \supset A$ and say "*B* **strictly contains** *A*". [Compare the corresponding uses of $\leqslant$, $\geqslant$, $<$ and $>$.]

For any set S, $S \subseteq S$; also $\emptyset \subseteq S$.

Clearly, for sets A, B we have:

$$A = B \Leftrightarrow A \subseteq B \text{ and } B \subseteq A.$$

Any subset of a finite set is clearly finite.

A subset of an infinite set may be finite or infinite.

An infinite set has clearly an infinite number of subsets; we show in the following theorem that a finite set has a finite number of subsets.

Theorem 2.1. *A finite set with n elements has* 2^n *subsets.*

Proof. Let set S have n elements $a_1, a_2, \ldots, a_n$. Any subset A of S is determined when we know, for each element of S, whether it is in A or not. For a_1 there are two possibilities, viz. include it or exclude it. For each of these possibilities there are two possible ways of dealing with a_2. Thus there are $2 \cdot 2 = 2^2$ ways of dealing with a_1, a_2 together. For each of these there are two ways of dealing with a_3 and so $2^2 \cdot 2 = 2^3$ ways of dealing with a_1, a_2, a_3 together. Proceeding in this way, there are 2^n ways of dealing with $a_1, a_2, \ldots, a_n$ together. Hence S has 2^n subsets.

Example. If $S = \{a, b, c\}$, the $2^3 (=8)$ subsets of S are

$\emptyset$, the only subset with no member;

$\{a\}, \{b\}, \{c\}$, the subsets with one member each;

$\{a, b\}, \{a, c\}, \{b, c\}$, the subsets with two members each;

$\{a, b, c\}$, the only subset with all three members.

The set of all subsets of a set S is often denoted by $\mathscr{P}(S)$ and called the **power set** of S.

To end this section on subsets we note that the statements

$$a \in \{a\}, \{a\} \subseteq \{a\}, a \in \{a, b\}, \{a\} \subseteq \{a, b\}$$

are all meaningful, but that none of

$$a = \{a\}, a \subseteq \{a, b\}, \{a\} \in \{a, b\}$$

has a meaning.

4. Intervals on the real line

The set **R** of real numbers can be represented in the usual way on a line, by taking an origin O, a unit of length and a positive direction indicated by marking the point *A* corresponding to 1. Corresponding to each real number *x* there is a point *P* on the line, the measure of whose distance from *O* is *x*, and each point on the line has associated with it in this way a real number. This geometrical "representation" of the set of real numbers is called "**the real line**".

In many places in mathematics, e.g. in calculus, certain subsets of **R** called **intervals** have an important role to play. We list the general forms of these together with a shorthand notation and a method of displaying their representations on the real line.

(i) $\{x \in \mathbf{R} : a \leqslant x \leqslant b\}$, denoted by $[a, b]$.

This is called the **closed interval** $[a, b]$, the brackets [,] respectively indicating that the **end points** $x = a$ and $x = b$ are included.

(ii) $\{x \in \mathbf{R} : a < x < b\}$, denoted by (a, b).

This is called the **open interval** (a, b), the brackets (,) respectively indicating that the end points $x = a$ and $x = b$ are excluded.

(iii) $\{x \in \mathbf{R} : a \leqslant x < b\}$, denoted by $[a, b)$.

(iv) $\{x \in \mathbf{R} : a < x \leqslant b\}$, denoted by $(a, b]$.

The intervals in (iii) and (iv) are closed at one end and open at the other.

(v) $\{x \in \mathbf{R} : a \leqslant x\} = \{x \in \mathbf{R} : x \geqslant a\}$,
    denoted conventionally by $[a, \infty)$.

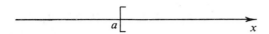

(vi) $\{x \in \mathbf{R} : a < x\} = \{x \in \mathbf{R} : x > a\}$, denoted by (a, ∞).

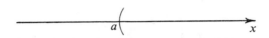

The intervals in (v) and (vi) are the subsets of $\mathbf{R}$ consisting of all real numbers $\geqslant$ and $>$, respectively, the given real number a; the symbol ∞ is read as "**infinity**"; ∞ and the word "infinity" are used in several mathematical notations and phrases which have a clearly defined meaning, but it must not be supposed that ∞ is a number.

(vii) $\{x \in \mathbf{R} : x \leqslant a\}$, denoted by $(-\infty, a]$.

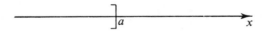

(viii) $\{x \in \mathbf{R} : x < a\}$, denoted by $(-\infty, a)$.

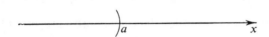

The intervals in (vii) and (viii) are the subsets of $\mathbf{R}$ consisting of all real numbers $\leqslant$ and $<$, respectively, the given real number a; the symbol $-\infty$ is read as "**minus (or negative) infinity**".

5. Operations on sets

An **operation on sets** is a rule by which, with one, two or more given sets, we associate a unique set. We give some important examples.

I. Union. The **union** of two sets A, B is defined to be the set consisting of all objects that are members of A or of B or of both [The phrase " or of both" is usually omitted but is assumed to hold.], i.e. it is the set consisting of all objects that are members of at least one of the sets A, B. The set is denoted by $A \cup B$ which is read as "A **union** B".

e.g. $\qquad\qquad\qquad \{1, 2\} \cup \{2, 3, 4\} = \{1, 2, 3, 4\}$.

Example 1. If $A = (1, 3)$ and $B = [2, 5]$, then
$A \cup B = (1, 5]$.

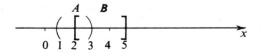

Example 2. If $A = (1, 3)$, $B = (3, 5)$, $C = [3, 5)$,
then $\qquad A \cup B = \{x \in \mathbf{R} : 1 < x < 5, x \neq 3\}$,
and $\qquad A \cup C = \{x \in \mathbf{R} : 1 < x < 5\} = (1, 5)$.

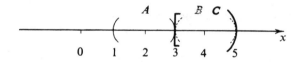

Note. It is clear that the following statements are true. For any sets S, A, B,

$$S \cup S = S, \quad S \cup \emptyset = S, \quad \emptyset \cup \emptyset = \emptyset, \quad A \cup B = B \cup A,$$

the last statement being called the **commutative property of union**.

Extension to three or more sets. $A \cup B \cup C$ is defined to be the set consisting of all the objects that are members of at least one of the sets A, B, C. Clearly the order in which the sets appear does not matter,

e.g. $\qquad\qquad A \cup B \cup C = B \cup A \cup C$, etc.

Also, $\qquad\qquad (A \cup B) \cup C = A \cup (B \cup C)$,

since each is equal to $A \cup B \cup C$. This equation is called the **associative property of union**.

Similarly the union of more than three sets can be defined, and similar properties noted. We often write

$$\bigcup_{i=1}^{n} A_i \quad \text{for} \quad A_1 \cup A_2 \cup \ldots \cup A_n.$$

II. Intersection. The **intersection** of two sets A and B is defined to be the set of all objects that are common to A and B. It is denoted by $A \cap B$ which is read as "A **intersection** B".

e.g.
$$\{2, 3, 5\} \cap \{3, 6\} = \{3\};$$
$$(1, 3) \cap [2, 5) = [2, 3);$$
$$(1, 3) \cap (3, 5) = \emptyset;$$
$$(1, 3] \cap [3, 5) = \{3\}.$$

Note. It is clear that the following statements are true. For any sets S, A, B,

$$S \cap S = S, \quad S \cap \emptyset = \emptyset, \quad \emptyset \cap \emptyset = \emptyset, \quad A \cap B = B \cap A,$$

the last statement being called the **commutative property of intersection**.

Extension to three or more sets. $A \cap B \cap C$ is defined to be the set of objects that are common to A, B and C. The order in which the sets appear does not matter. Also

$$(A \cap B) \cap C = A \cap (B \cap C)$$

[called the **associative property of intersection**], since each equals $A \cap B \cap C$.
For n sets $A_1, A_2, \ldots, A_n$, we write

$$\bigcap_{i=1}^{n} A_i \quad \text{for} \quad A_1 \cap A_2 \cap \ldots \cap A_n.$$

Note. $A \cap B \cup C$ has *no meaning*. We need brackets to indicate its meaning; it could be $(A \cap B) \cup C$ or $A \cap (B \cup C)$, and these are not equal in general.

Disjoint sets. When $A \cap B = \emptyset$, i.e. when A, B have no elements in common, we say that A, B are **disjoint** or **non-intersecting**.

III. Complements. If $A \subseteq S$, we define the **complement of A with respect to S** (or **in** S) to be the set of all elements of S that are *not* members of A. We denote this complement by S', which is read as "S dashed" or "S prime".

e.g.   if $S = \{1, 2, 3, 4\}$ and $A = \{2, 3\}$,   then $A' = \{1, 4\}$.

Example. If $S = \mathbf{R}$, $A = [1, 3)$, $B = (2, 5]$, $C = [2, 3]$, find
$$A \cap B', A \cup B', A' \cap C, A \cap C' \text{ and } A' \cup C.$$

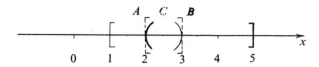

$A \cap B' = \{x \in \mathbf{R} : 1 \leqslant x \leqslant 2\} = [1, 2]$;

$A \cup B' = \{x \in \mathbf{R} : x < 3 \text{ or } x > 5\}$;

$A' \cap C = \{3\}$; $A \cap C' = \{x \in \mathbf{R} : 1 \leqslant x < 2\} = [1, 2)$;

$A' \cup C = \{x \in \mathbf{R} : x < 1 \text{ or } x \geqslant 2\}$.

Note. $S' = \emptyset$ and $\emptyset' = S$. Also, if $A \subseteq S$, then
$$A \cup A' = S, A \cap A' = \emptyset \text{ and } (A')' = A.$$

IV. Relative difference (or complement) of ordered pair of sets A, B.
The set of elements of A which are not in B is denoted by $A - B$ and
called the **relative difference** of A, B. This set is also often denoted by
$A \quad B$, which is read as "A **slash** B". The set $B - A$ (or $B \quad A$) is then the
relative difference of the pair B, A, i.e. the set of elements of B which are
not in A.

e.g.   if $A = \{1, 2, 3\}$ and $B = \{3\}$,   then
$\quad\quad A - B = \{1, 2\}$ and $B - A = \emptyset$.
Also, if $A = \{1, 2, 3\}$ and $C = \{2, 3, 6\}$,   then $A - C = \{1\}$
and $C - A = \{6\}$.

V. Symmetric difference of two sets A, B. This is defined to be the set
$(A - B) \cup (B - A)$ and is usually denoted by $A + B$.
    For the example given in **IV**, $A + B = \{1, 2\}$ and $A + C = \{1, 6\}$.
[$A + B$ is the set of elements in $A \cup B$ that are not in $A \cap B$.]

6. Venn diagrams
[named after the British logician John Venn (1834–1923).]
    These provide a useful method for displaying relations between subsets
of a given fixed set S. The fixed set S is called the **universal set** for the
subsets involved. S is represented by the interior of a rectangle and
subsets of S are represented by regions, e.g. by the interiors of circles,
inside the rectangle. Points on the boundaries of all regions involved are
ignored.

The following diagrams represent $A \cup B$, $A \cap B$, A', $A-B$, $B-A$ and $A+B$, respectively.

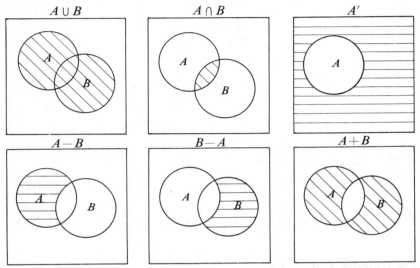

It is clear from the definition of relative difference that, when A, B are subsets of S, $A-B = A \cap B'$, $B-A = B \cap A'$,

and $$A+B = (A \cap B') \cup (B \cap A').$$

A set S is divided by a finite number of subsets and their complements into a finite number of *disjoint* subsets. Venn diagrams give a clear picture of the situation when we have one, two or three subsets of S.

One subset A of S

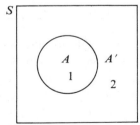

Here S is divided into two disjoint subsets represented by the regions numbered 1 and 2; we write

$$① = A \quad \text{and} \quad ② = A',$$

so that $\quad S = A \cup A' = ① \cup ②\quad$ and $\quad ① \cap ② = A \cap A' = \emptyset$.

Two subsets A, B of S

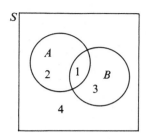

Here S is divided into four disjoint subsets represented by the regions numbered 1, 2, 3, 4; we write

$$① = A \cap B, \quad ② = A \cap B' = A - B, \quad ③ = A' \cap B = B - A, \quad ④ = A' \cap B'.$$

Then

$$S = ① \cup ② \cup ③ \cup ④; \quad A = ① \cup ② = (A \cap B) \cup (A \cap B'); \quad A \cap B = ①, \text{ etc.}$$

Three subsets A, B, C of S

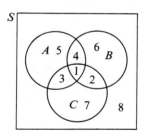

Here S is divided into eight disjoint subsets represented by the regions numbered 1, 2, 3, 4, 5, 6, 7, 8; we write

$$\begin{aligned}
① &= A \cap B \cap C, & ⑤ &= A \cap B' \cap C', \\
② &= A' \cap B \cap C, & ⑥ &= A' \cap B \cap C', \\
③ &= A \cap B' \cap C, & ⑦ &= A' \cap B' \cap C, \\
④ &= A \cap B \cap C', & ⑧ &= A' \cap B' \cap C'.
\end{aligned}$$

Then

$$\begin{aligned}
S &= ① \cup ② \cup ③ \cup ④ \cup ⑤ \cup ⑥ \cup ⑦ \cup ⑧, \\
A &= ① \cup ③ \cup ④ \cup ⑤, \\
A' &= ② \cup ⑥ \cup ⑦ \cup ⑧, \\
A \cap B &= ① \cup ④, \text{ etc.}
\end{aligned}$$

The numbered sets on the right-hand sides are called the **canonical components** of the sets on the left-hand sides.

Continuing the above process, a set S is divided by n given subsets $A_1, A_2, \ldots, A_n$ and their complements into 2^n disjoint subsets, these being of the form $A_1 \cap A_2 \cap \ldots \cap A_n$, with none, one, two, $\ldots$, or n of the A_i being replaced by the complement A_i'.

7. The algebra of sets (properties of $\cup$, $\cap$ and $'$)

The following identities for subsets of a universal set S can be proved true. We have already mentioned some of them.

Laws of the Algebra of Sets

Associative Laws

1a. $\quad A \cup (B \cup C) = (A \cup B) \cup C$ | 1b. $\quad A \cap (B \cap C) = (A \cap B) \cap C$

Commutative Laws

2a. $\quad A \cup B = B \cup A$ | 2b. $\quad A \cap B = B \cap A$

Identity Laws

3a. $\quad A \cup \emptyset = A$ | 3b. $\quad A \cap S = A$

4a. $\quad A \cup S = S$ | 4b. $\quad A \cap \emptyset = \emptyset$

Idempotent Laws

5a. $\quad A \cup A = A$ | 5b. $\quad A \cap A = A$

Distributive Laws

6a. $A \cap (B \cup C) = (A \cap B) \cup (A \cap C)$ | 6b. $A \cup (B \cap C) = (A \cup B) \cap (A \cup C)$

Complement Laws

7a. $\quad A \cup A' = S$ | 7b. $\quad A \cap A' = \emptyset$

8a. $\quad S' = \emptyset$ | 8b. $\quad \emptyset' = S$

9a. $\quad (A')' = A$ | 9b. $\quad (A')' = A$

De Morgan's Laws

10a. $\quad (A \cup B)' = A' \cap B'$ | 10b. $\quad (A \cap B)' = A' \cup B'$

For each of the integers $n = 1, 2, \ldots, 10$, the pair of identities na and nb are said to be **dual identities**. Each is obtained formally from the other by replacing $\cup$ by $\cap$, $\cap$ by $\cup$, $'$ by $'$, $\emptyset$ by S and S by $\emptyset$.

All except 6a, 6b and 10a, 10b are immediate consequences of the definitions of $\cup$, $\cap$ and $'$.

The De Morgan laws 10a and 10b can be extended as follows: If $A_1, A_2, \ldots, A_n$ are subsets of a set S, then

$$(A_1 \cup A_2 \cup \ldots \cup A_n)' = A_1' \cap A_2' \cap \ldots \cap A_n',$$

and $\qquad (A_1 \cap A_2 \cap \ldots \cap A_n)' = A_1' \cup A_2' \cup \ldots \cup A_n'.$

The identities can be proved in several ways. We first illustrate the use of Venn diagrams for proving results. Such proofs are satisfactory in the sense that they give a line of argument that can be put in analytical form, although this, in some cases, may be very complicated.

Proof by Venn diagram of Identity 10a,   i.e.

$\forall$ subsets A, B of S, $(A \cup B)' = A' \cap B'$.

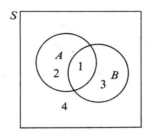

$$A \cup B = ① \cup ② \cup ③, \quad \text{and so} \quad (A \cup B)' = ④.$$
$$A' = ③ \cup ④; \quad B' = ② \cup ④, \quad \text{and so} \quad A' \cap B' = ④.$$

Hence $\qquad\qquad\qquad (A \cup B)' = A' \cap B'.$

Proof by Venn diagram of Identity 6a, i.e.

$\forall$ subsets A, B, C of S, $A \cap (B \cup C) = (A \cap B) \cup (A \cap C)$.

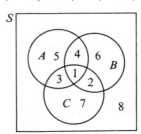

$A = ① \cup ③ \cup ④ \cup ⑤$; $B \cup C = ① \cup ② \cup ③ \cup ④ \cup ⑥ \cup ⑦$, and so

$$A \cap (B \cup C) = ① \cup ③ \cup ④.$$

Also $\qquad A \cap B = ① \cup ④, \quad A \cap C = ① \cup ③,$

and $\qquad (A \cap B) \cup (A \cap C) = ① \cup ③ \cup ④.$

Hence $\qquad A \cap (B \cup C) = (A \cap B) \cup (A \cap C).$

Exercise. Write out the similar proofs of 10b and 6b.

Deduction of 10b from 10a by the "method of complements".

Assuming that identity 10a holds for any pair of subsets of S, we deduce for the pair of subsets A', B' of S, that

$$(A' \cup B')' = (A')' \cap (B')' = A \cap B.$$

Hence, taking complements,

$$((A' \cup B')')' = (A \cap B)',$$

and so $\qquad (A \cap B)' = A' \cup B',$

proving that 10b holds.

Proof of 10a by logic. We have to show that $(A \cup B)'$ and $A' \cap B'$ have the same members, i.e. that

$$x \in (A \cup B)' \Leftrightarrow x \in A' \cap B'.$$

Let p be the statement: $x \in A$, so that $\sim p$ is: $x \in A'$;

let q be the statement: $x \in B$, so that $\sim q$ is: $x \in B'$.

Then $\qquad p \vee q$ is: $x \in A \vee x \in B$, i.e. $x \in A \cup B$,

and $\qquad p \wedge q$ is: $x \in A \wedge x \in B$, i.e. $x \in A \cap B$.

Now $x \in (A \cup B)'$ is true

$\qquad \Leftrightarrow \sim(p \vee q)$ has truth value T

$\qquad \Leftrightarrow (\sim p) \wedge (\sim q)$ has truth value T, since $\sim(p \vee q)$ and $(\sim p) \wedge (\sim q)$ are logically equivalent,

$\qquad \Leftrightarrow x \in A' \cap B'$ is true.

Hence $\qquad (A \cup B)' = A' \cap B'.$

Exercise. Write out similar proofs for 10b, 6a and 6b.

Assuming that the laws of the algebra of sets, as listed earlier, are true, we can deduce many additional properties of operations on subsets of a universal set S.

Example. By using the algebra of sets, show that

$$(A \cup B) \cap C = \{A \cup (B \cap C)\} \cap C.$$

Deduce that $(A \cup B) \cap C \subseteq A \cup (B \cap C)$, and that equality holds if and only if $A \subseteq C$.

$$\begin{aligned}
\text{Now} \quad (A \cup B) \cap C &= (A \cap C) \cup (B \cap C) \\
&= \{A \cup (B \cap C)\} \cap \{C \cup (B \cap C)\} \\
&= \{A \cup (B \cap C)\} \cap C, \quad \text{since } (B \cap C) \subseteq C.
\end{aligned}$$

This shows that $(A \cup B) \cap C$ is a subset of $A \cup (B \cap C)$,

i.e. $$(A \cup B) \cap C \subseteq A \cup (B \cap C).$$

$$\begin{aligned}
\text{Also, equality holds} &\Leftrightarrow \{A \cup (B \cap C)\} \cap C = A \cup (B \cap C) \\
&\Leftrightarrow A \cup (B \cap C) \subseteq C \\
&\Leftrightarrow A \subseteq C \text{ and } B \cap C \subseteq C \\
&\Leftrightarrow A \subseteq C, \text{ since } B \cap C \subseteq C \text{ is true } \forall A, B, C.
\end{aligned}$$

Example. By using the algebra of sets, prove the following properties of the operation $+$ on subsets of a set S:

(i) $A + B = (A \cup B) \cap (A' \cup B')$.

(ii) $A + A = \emptyset, \quad A + A' = S, \quad A + \emptyset = A, \quad A + S = A'$.

(iii) $A + B = B + A$.

(iv) $A \cap (B + C) = (A \cap B) + (A \cap C)$.

(v) $(A + B) + C = A + (B + C)$.

[**Note.** Prove some of them by Venn diagrams as an exercise on the use of these diagrams.]

Note. For some of the properties (i), . . . , (v), it is useful to have the following two results available:

(a) $(A \cup B) \cap (C \cup D) = (A \cap C) \cup (A \cap D) \cup (B \cap C) \cup (B \cap D)$;

(b) $(A \cap B) \cup (C \cap D) = (A \cup C) \cap (A \cup D) \cap (B \cup C) \cap (B \cup D)$.

Each is easily proved by two applications of a distributive law.

Proof of (iv).
$$A \cap (B+C) = A \cap \{(B \cap C') \cup (B' \cap C)\} = (A \cap B \cap C') \cup (A \cap B' \cap C).$$

Also $(A \cap B) + (A \cap C)$
$$= \{(A \cap B) \cap (A \cap C)'\} \cup \{(A \cap C) \cap (A \cap B)'\}$$
$$= \{(A \cap B) \cap (A' \cup C')\} \cup \{(A \cap C) \cap (A' \cup B')\}$$
$$= (A \cap B \cap A') \cup (A \cap B \cap C') \cup (A \cap C \cap A') \cup (A \cap B' \cap C)$$
$$= \emptyset \cup (A \cap B \cap C') \cup \emptyset \cup (A \cap B' \cap C)$$
$$= (A \cap B \cap C') \cup (A \cap B' \cap C).$$

Hence $$A \cap (B+C) = (A \cap B) + (A \cap C).$$

Sketch of proof of (v):

Show that:
$$(A+B)+C = (A \cap B \cap C) \cup (A \cap B' \cap C') \cup (A' \cap B \cap C') \cup (A' \cap B' \cap C).$$
Note that, by symmetry, $(B+C)+A$ equals the same union of subsets. Use the fact that, by (iii), $(B+C)+A = A+(B+C)$.

Prove the remaining properties of $+$.

8. Enumerative results for finite sets

If A is a finite set, we shall denote by $n(A)$ the number of elements in A. If A and B are *disjoint* sets, so that $A \cap B = \emptyset$, then it is clear that

$$n(A \cup B) = n(A) + n(B).$$

For any pair of sets A, B we show that the following result holds:

$$n(A \cup B) = n(A) + n(B) - n(A \cap B).$$

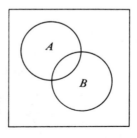

Proof. Clearly $A = (A \cap B) \cup (A - B)$, and so

$$n(A) = n(A \cap B) + n(A - B),$$

since $A \cap B$ and $A - B$ are disjoint.

Also $\qquad A \cup B = B \cup (A - B)$, and so
$$n(A \cup B) = n(B) + n(A - B),$$

since B and $A - B$ are disjoint.

From these two equations involving numbers of elements in sets we deduce, by subtraction, that

$$n(A) - n(A \cup B) = n(A \cap B) - n(B),$$
and so $\qquad n(A \cup B) = n(A) + n(B) - n(A \cap B).$

This result can be extended to the case of three sets A, B, C when we have:

$$n(A \cup B \cup C) = n(A) + n(B) + n(C)$$
$$- n(A \cap B) - n(A \cap C) - n(B \cap C) + n(A \cap B \cap C).$$

Proof.

$$n(A \cup B \cup C) = n((A \cup B) \cup C)$$
$$= n(A \cup B) + n(C) - n((A \cup B) \cap C)$$
$$= n(A) + n(B) - n(A \cap B) + n(C) - n((A \cap C) \cup (B \cap C))$$
$$= n(A) + n(B) + n(C)$$
$$- n(A \cap B) - [n(A \cap C) + n(B \cap C) - n(A \cap B \cap C)]$$
$$= n(A) + n(B) + n(C)$$
$$- n(A \cap B) - n(A \cap C) - n(B \cap C) + n(A \cap B \cap C).$$

Example. Each student in a class of 40 studied at least one of the subjects English, French and Philosophy. 16 studied English, 22 Philosophy and 26 French; 5 studied English and Philosophy, 14 French and Philosophy, and 2 English, French and Philosophy. Determine how many students studied English and French but not Philosophy.

We use a Venn diagram in which E, F, P denote the sets of students studying English, French and Philosophy, respectively. We indicate the numbers studying various known combinations of subjects as shown; the integer 2, e.g., equals $n(E \cap F \cap P)$; $3 = n(E \cap F' \cap P)$, $12 = n(E' \cap F \cap P)$, $5 = n(E' \cap F' \cap P)$ and $0 = n(E' \cap F' \cap P')$ since each student studies at least one subject.

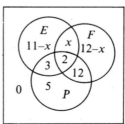

If $x = n(E \cap F \cap P')$, then we can complete the diagram with

$$11 - x = n(E \cap F' \cap P') \quad \text{(since } n(E) = 16),$$

and $\qquad 12 - x = n(E' \cap F \cap P') \quad \text{(since } n(F) = 26).$

Since $n(E \cup F \cup P) = 40$, it follows that

$$16 + 17 + (12 - x) = 40, \quad \text{and so } x = 5.$$

Hence the number of students studying English and French but not Philosophy is 5.

9. Cartesian product of non-empty sets A, B

We denote by $A \times B$ the set of all ordered pairs (a, b) where $a \in A$ and $b \in B$, i.e.

$$A \times B = \{(a, b) : a \in A, b \in B\}.$$

The set $A \times B$ is called the **cartesian product** of the sets A, B in the order A, B.

e.g. $\qquad$ if $A = \{1, 2, 3\}$ and $B = \{a, b\}$,

then $\qquad A \times B = \{(1, a), (1, b), (2, a), (2, b), (3, a), (3, b)\}$.

Also $\qquad B \times A = \{(a, 1), (b, 1), (a, 2), (b, 2), (a, 3), (b, 3)\}$.

Clearly, in general, $\qquad A \times B \neq B \times A$.

Also, if A and B are finite, then $n(A \times B) = n(A) \cdot n(B) = n(B \times A)$.

Similarly we can define $A \times B \times C$, the cartesian product of three sets A, B, C by:

$$A \times B \times C = \{(a, b, c) : a \in A, b \in B, c \in C\},$$

and we can go on to define $A_1 \times A_2 \times \ldots \times A_n$, the cartesian product of n sets.

The special cartesian products $A \times A$, $A \times A \times A$, $\ldots$ are usually denoted by A^2, A^3, $\ldots$. In particular, if $\mathbf{R}$ is the set of real numbers, then

$$\mathbf{R}^2 = \{(x, y) : x \in \mathbf{R}, y \in \mathbf{R}\}.$$

$\mathbf{R}^2$ can be represented by the set of all points in a plane by using a coordinate system.

We often, in fact, use a coordinate pictorial representation for $A^2 = A \times A$, where A is any given non-empty set, in which we imagine the elements of A as represented by points on two rectangular axes (or half-axes) intersecting as shown.

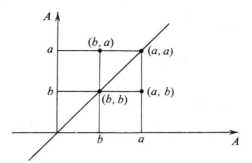

Note. As indicated in the diagram, if a, b are distinct elements of a set A, then the pairs (a, b) and (b, a) are distinct elements of $A \times A$—in an "ordered pair" the order of the elements matters.

The subset $\{(a, a) : a \in A\}$ of A^2 is called the **diagonal** of A^2.

e.g.   if $A = \{1, 2, 3, 4\}$,

then the diagonal of A^2 is $\{(1, 1), (2, 2), (3, 3), (4, 4)\}$.

10. Partitions of a set, equivalence relations

A **partition** of a nonempty set S is a collection of disjoint nonempty subsets of S such that S is the union of these subsets, e.g.

If $S = \{1, 2, 3, 4, 5, 6, 7, 8, 9, 10\}$,   and if
$$B_1 = \{1, 2, 4\}, \quad B_2 = \{3, 6, 9\}, \quad B_3 = \{5\}, \quad B_4 = \{7, 8, 10\},$$
then $S = B_1 \cup B_2 \cup B_3 \cup B_4$ and $B_i \cap B_j = \emptyset$ when $i \neq j$.

Hence $\{B_1, B_2, B_3, B_4\}$ provides a partition of S.

An important technique in mathematics and its applications is that of partitioning a set into subsets such that all the elements in the same subset have some property in common or are related in some definite way. e.g. take the set $\mathbf{N}$ of natural numbers; if $a \in \mathbf{N}$ and $b \in \mathbf{N}$ let us write $a \, R \, b$, and say that b is **related to** a, when the positive integer b has the same *parity* as the positive integer a, that is when a and b are either both odd or both even.

If B_1 denotes the subset of elements of $\mathbf{N}$ related to the integer 1, and B_2 denotes the subset of elements of $\mathbf{N}$ related to the integer 2, then

$$a \in B_1 \Leftrightarrow a \text{ is odd}, \qquad \text{so } B_1 = \{1, 3, 5, 7, \ldots\};$$
$$a \in B_2 \Leftrightarrow a \text{ is even}, \quad \text{therefore } B_2 = \{2, 4, 6, 8, \ldots\}.$$
$$\mathbf{N} = B_1 \cup B_2 \text{ and } B_1 \cap B_2 = \emptyset.$$

Thus $\{B_1, B_2\}$ gives a partition of $\mathbf{N}$.

As a second example let us again take $\mathbf{N}$, the set of natural numbers, and in this case write $a R b$ when b has the same least non-negative remainder as a on division by 3. We note that the only possible remainders are 0, 1 and 2.

If B_i denotes the subset of those members of $\mathbf{N}$ with remainder i ($i = 0, 1, 2$) on division by 3, then clearly

$$B_0 = \{3, 6, 9, 12, \ldots\} = \{3k : k \in \mathbf{N}\};$$
$$B_1 = \{1, 4, 7, 10, \ldots\} = \{3k - 2 : k \in \mathbf{N}\};$$
$$B_2 = \{2, 5, 8, 11, \ldots\} = \{3k - 1 : k \in \mathbf{N}\}.$$

Then $\mathbf{N} = B_0 \cup B_1 \cup B_2$ and $B_i \cap B_j = \emptyset$ $(i \neq j)$.

Thus $\{B_0, B_1, B_2\}$ gives a partition of $\mathbf{N}$.

An important method of partitioning a set S arises from a certain type of relation defined on S. We first explain what we mean by a **binary relation** R or, simply, a **relation** R on a set S. It is simply an association which assigns to each ordered pair (a, b) in $S^2 = S \times S$ exactly one of the following statements:

(i) "b is related to a", written $a R b$;

(ii) "b is not related to a", written $a \mathrlap{\,/}R b$.

Examples (1) The order symbol $<$ is a relation on the set of integers $\mathbf{Z}$.

(2) The inclusion symbol $\subseteq$ is a relation on the set of all subsets of a given set S.

(3) The identity relation $=$ on any set S; each element of S is related only to itself.

The subset $\{(a, b) \in S^2 : a R b\}$ of S^2 is called the **graph** of the relation R. Conversely, any subset T of S^2 defines a relation R on S as follows:

$$a R b \text{ if and only if } (a, b) \in T.$$

Because of the correspondence between relations R on S and subsets of S^2, a relation on S is often defined by saying:

A relation R on S is a subset of S^2.

Example. Take $S = \{1, 2, 3, 4\}$ and define a relation R on S by writing for $a, b \in S$, $a R b$ if and only if $b = a + 1$.

The graph of R is $\{(1, 2), (2, 3), (3, 4)\}$.

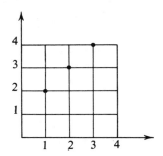

Some special properties that a relation R on S may possess:

I. Reflexive property. A relation R on a set S is called **reflexive** if, $\forall a \in S$, $a\,R\,a$, i.e. if $\{(a, a) : a \in S\} \subseteq$ graph of R, that is, if the graph of R contains the diagonal of S^2.

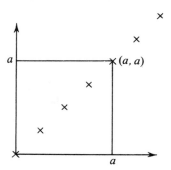

Examples (1) Let R be the relation of *congruence* on the set of triangles in a plane; then R is reflexive since every triangle is congruent to itself.

         (2) Let R be the relation of order $<$ on the set of real numbers; then R is not reflexive since $a \nless a$ for any real number a.

II. Symmetric property. A relation R on a set S is called **symmetric** if whenever $a\,R\,b$ then $b\,R\,a$, that is,

$$\text{if } a\,R\,b \Rightarrow b\,R\,a.$$

The graph of a symmetric relation R is symmetrical about the diagonal of S^2.

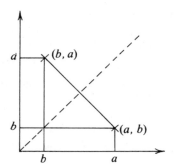

Examples (1) Let R be the relation of *similarity* on the set of triangles in a plane; then R is clearly symmetric.

(2) Let R be the relation defined on the set $S = \{1, 2, 3, 4\}$ by the subset $\{(1, 2),\ (1, 3),\ (2, 1),\ (2, 2),\ (2, 3)\}$ of S^2. R is not symmetric since for example $1\,R\,3$ but $3\,\cancel{R}\,1$.

III. Transitive property. A relation R on a set S is called **transitive** if whenever $a\,R\,b$ and $b\,R\,c$ then $a\,R\,c$, that is, if

$$a\,R\,b \wedge b\,R\,c \Rightarrow a\,R\,c.$$

The graph of a transitive relation has a sort of "rectangular property" as indicated in the diagram shown, namely that, if the points (a, b) and (b, c) are in the graph, then so also is the vertex (a, c) of the rectangle, with sides parallel to the coordinate axes, determined by the points (a, b) and (b, c).

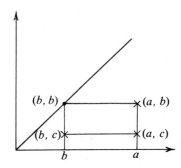

Examples (1) Let R be the relation $\leqslant$ on the set of real numbers; then R is transitive since $a\leqslant b \wedge b\leqslant c \Rightarrow a\leqslant c$.

(2) Let R be the relation $\perp$, that is *perpendicular to*, on the set of lines in a plane; then R is not transitive, since if a, b and c are lines such that $a \perp b \wedge b \perp c$, then a is not perpendicular to c. In fact a is parallel to c.

A relation R on a set is called an **equivalence relation** if it possesses all three of the properties **I, II** and **III**, that is, if R is reflexive, symmetric and transitive. Instead of R we often use for an equivalence relation the notation $\sim$, writing $a \sim b$ when b is related to a and $a \nsim b$ when b is not related to a. When $a \sim b$ we say that b is **equivalent to** a.

The equivalence relation defined by a partition of a set S. We can define a relation $\sim$ on S by writing $a \sim b$ when a and b belong to the same subset in the partition of S. Then clearly

(i) $a \sim a \ \forall a \in S$ and
(ii) $a \sim b \Rightarrow b \sim a$.
Also (iii) $a \sim b \wedge b \sim c \Rightarrow a, b, c \in$ the same subset in the partition of S
$\Rightarrow a \sim c$.

Hence $\sim$ is an equivalence relation on S.

For example, if $S = \{1, 2, 3, 4, 5, 6\}$ then the partition

$$S = \{1\} \cup \{2, 3, 4\} \cup \{5, 6\}$$

of S defines as described above an equivalence relation $\sim$ on S such that the graph of $\sim$ is
$\{(1, 1), (2, 2), (2, 3), (2, 4), (3, 2), (3, 3), (3, 4), (4, 2), (4, 3), (4, 4), (5, 5), (5, 6), (6, 5), (6, 6)\}$.

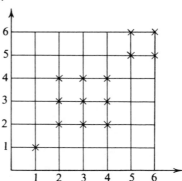

The graph illustrates the geometrical properties associated with the reflexive, symmetric and transitive properties of an equivalence relation.

The partition of a set S by an equivalence relation on S. We now show that an equivalence relation R on S provides a partition of S. Let $[a]$, called the **equivalence class of** a, be the set of elements related to a, that is,

$$[a] = \{x \in S : a \, R \, x\}.$$

The set of equivalence classes is often denoted by S/R and called the **quotient** of S by R, so that $S/R = \{[a] : a \in S\}$.

The fact that S/R forms a partition of S follows from the following results, part (2) of which is rather difficult to prove, and a proof is included for completeness only.

(1) $a \in [a] \quad \forall a \in S$;
(2) $\forall a, b \in S$ either $[a] = [b]$ or $[a] \cap [b] = \emptyset$.

Proof of (1). $\forall a \in S$, $a \, R \, a$, and so $a \in [a]$.

Proof of (2). We have to show that

$$[a] \cap [b] \neq \emptyset \Rightarrow [a] = [b].$$

Suppose that $[a] \cap [b]$ is non-empty, containing an element c. Then $c \in [a]$ and $c \in [b]$, so $a \, R \, c$ and $b \, R \, c$. By the symmetry property $c \, R \, b$, and the transitive property then shows that $a \, R \, b$.

Now let d be *any* element of $[b]$. Then $b \, R \, d$ and by transitivity again $a \, R \, d$. Hence $d \in [a]$. This shows that $[b] \subseteq [a]$.

We also have $b \, R \, a$ by symmetry and a similar argument shows that $[a] \subseteq [b]$. Hence $[a] = [b]$.

Note. It is clear from (1) and (2) that $\{[a] : a \in S\}$ forms a partition of S. As described earlier in this section a partition of S defines an equivalence relation on S, and it is not difficult to check that the equivalence relation on S defined by the partition $\{[a] : a \in S\}$ is simply the given equivalence relation R.

Example. A relation $\sim$ is defined on the set V of all points $P(x, y)$ in the x, y-plane by writing: $(x, y) \sim (z, t) \Leftrightarrow x + t = y + z$. Show that $\sim$ is an equivalence relation on V and find the equivalence class determined by the point $A(1, 2)$.

(i) $\forall (x, y) \in V$, $(x, y) \sim (x, y)$ since $x + y = y + x$; thus $\sim$ is reflexive.
(ii) $(x, y) \sim (z, t) \Rightarrow x + t = y + z$
$\Rightarrow z + y = t + x \Rightarrow (z, t) = (x, y)$;
thus $\sim$ is symmetric.

(iii) $\left.\begin{array}{l}(x, y) \sim (z, t) \Rightarrow x+t = y+z \\ (z, t) \sim (u, v) \Rightarrow z+v = t+u\end{array}\right\} \Rightarrow x+t+z+v = y+z+t+u$

$$\Rightarrow x+v = y+u \Rightarrow (x, y) \sim (u, v);$$

thus $\sim$ is transitive.

From (i), (ii) and (iii), $\sim$ is an equivalence relation on V.

Now $(1, 2) \sim (x, y) \Leftrightarrow 1+y = 2+x \Leftrightarrow y-x = 1$.

Thus the equivalence class determined by the point A $(1, 2)$ is the set of points $\{(x, y): y = x+1\}$.

Geometrically, this is the line in the x, y-plane through the point A $(1, 2)$ of gradient 1.

11. General relations

If S, T are sets, a **binary relation** or, simply, a **relation** R from S to T assigns to each pair $(a, b) \in S \times T$ exactly one of the statements

(i) "b is related to a", written $a \, R \, b$;

(ii) "b is not related to a", written $a \, \mathcal{R} \, b$.

For the relations discussed earlier, the sets S and T were equal.

If R is a relation S to T, then the graph of R is the subset

$$\{(a, b) \in S \times T : a \, R \, b\} \text{ of } S \times T.$$

In fact, just as for a relation from S to S, a relation from S to T is often defined as a subset of $S \times T$.

Although these general relations are important in certain later parts of mathematics the only type that is really important at this stage of a mathematical training is that in which there is exactly one element of T related to each element of S. Such a relation from S to T is called a **mapping from S to T** and we discuss in the next chapter some basic ideas associated with such mappings.

EXERCISE 2

1. If S is the set of all positive integers $\leqslant 8$ and A, B and C are the subsets $\{1, 5, 6\}$, $\{2, 3, 5, 7, 8\}$ and $\{1, 3, 6, 8\}$, respectively, of S, find the subsets
$$A \cap B' \cap C \text{ and } (A \cup C) \cap (B \cup C').$$

2. Determine which of the following sets are equal to the empty set:
   (i) $\{x \in \mathbf{R} : x^2 = 9 \wedge 2x = 4\}$,   (ii) $\{x \in \mathbf{R} : x \neq x\}$,   (iii) $\{x \in \mathbf{R} : x+2 = 2\}$,
   (iv) $\{x \in \mathbf{R} : 1 < x < 2\}$,   (v) $\{x \in \mathbf{Z} : 1 < x < 2\}$.

3. If A and B are the sets defined by
$$A = \{1, 2, 3, \{1, 2, 3\}\} \quad \text{and} \quad B = \{1, 2, \{1, 2\}\},$$
find $A \cap B$ and $A \cup B$.

4. If A and B are the intervals $(0, 3]$ and $[3, 5)$, respectively, find $A \cap B$, $A \cap B'$ and $A \cup B'$.

5. If $A = (1, 3]$, $B = [3, 6)$, $C = (4, \infty)$, find $A \cup B$, $A \cap C$, $B \cap C'$ and $(A' \cup B) \cap C'$.

6. Let $S = [-1, 1]$ and let $A = [-1, 0)$, $B = (0, \frac{1}{2})$, $C = (-\frac{1}{2}, \frac{1}{4})$. Find $(A' \cap B') \cup (A \cap C)$, where A', B' are the complements of A, B in S.

7. If S is the x, y-plane and if A, B, C are the subsets of S defined by
$$A = \{(x, y): x^2 + y^2 \leqslant 2\},$$
$$B = \{(x, y): x = 1\},$$
$$C = \{(x, y): y < 1\},$$
indicate clearly in diagrams the sets $A \cap B$, $A \cap C$, $B \cap C'$ and $A' \cap B \cap C$.

8. If A, B, C are subsets of a set S, show by means of Venn diagrams that
   (i) $A \subseteq B \Rightarrow B' \subseteq A'$,
   (ii) $A \cap B \subseteq C'$ and $A \cup C \subseteq B \Rightarrow A \cap C = \emptyset$,
   (iii) $A \subseteq (B \cup C)'$ or $B \subseteq (A \cup C)' \Rightarrow A \cap B = \emptyset$,
   (iv) $A \subseteq B \Leftrightarrow A \cap B' = \emptyset \Leftrightarrow A' \cup B = S$,
   (v) $A \cap C = B \cap C$ and $A \cup C = B \cup C \Rightarrow A = B$,
   (vi) $A \cap (B \cup C) \subseteq (A \cap B) \cup C$ with equality $\Leftrightarrow C \subseteq A$,
   (vii) $(A \cap B) \cup (B' \cap C) \cup (C' \cap A) = A \Leftrightarrow A' \cap B' \cap C = \emptyset$.

9. If A, B, C are subsets of a set S, show (a) by means of Venn diagrams, (b) by the algebra of sets, that
   (i) $(A - B) + B = A \cup B$,
   (ii) $A + B = \emptyset \Leftrightarrow A = B$,
   (iii) $(A + B) - C \subseteq A + (B - C)$,
   (iv) $(A - B) \cup (A - C) = A - (B \cap C)$,
   (v) $A + C \subseteq (A + B) \cup (B + C)$.

10. Prove, by using the algebra of sets, that
   (i) $(A \cap B') \cup (B \cap A') = (A \cup B) \cap (A \cap B)'$,
   (ii) $A \cup (A \cap B) = A$,
   (iii) $(A \cap C') \cap (B' \cup C) = A \cap B' \cap C'$,
   (iv) $(A \cap B \cap C) \cup (A' \cap C) \cup (B' \cap C) = C$,
   (v) $A \cap (B + C) = (A' \cup B) + (A' \cup C)$,
   (vi) $A + (A \cup B) = B - (A \cap B)$,
   (vii) $A - (B - (C - D)) = (A - B) \cup ((A \cap C) - D)$.

11. A, B are subsets of a set S, such that $A \cap B = A$. Show, without using a Venn diagram, that the simultaneous equations
$$X \cup A = B \quad \text{and} \quad X \cap A = \emptyset$$
for subsets X of S have one and only one solution, namely, $X = A' \cap B$.

12. If P, Q are given subsets of a set S, show that the equation
$$P + X = Q$$
for subsets X of S has one and only one solution, namely, $X = P + Q$.

13. In a group of 70 students, each of whom studied at least one of the subjects economics, history and geography, 53 studied economics, 3 studied only history, 4 only geography, 35 economics and history and 34 history and geography. How many students studied both economics and history but not geography?

14. Prove that $A \times (B \cap C) = (A \times B) \cap (A \times C)$.

15. Which of the relations on $\mathbf{N}$ defined by the following statements are equivalence relations on $\mathbf{N}$?
   (i) $a \sim b$ when $a + b$ is an even integer;
   (ii) $a \sim b$ when a and b are both squares of integers;
   (iii) $a \sim b$ when $|a - b| \leqslant 3$, (where $|a - b|$ is the absolute value of $a - b$);
   (iv) $a \sim b$ when $ab = n^2$ for some integer n, i.e. when ab is the square of an integer.

16. Which of the following relations on $\mathbf{Z}$ defined by the following statements are equivalence relations on $\mathbf{Z}$?

   (i) $a\,R\,b$ when $|a-b|$ is of the form $5k$ for some integer k;

   (ii) $a\,R\,b$ when $a+2b$ is divisible by 3;

   (iii) $a\,R\,b$ when $a \leqslant b$.

17. A relation R is defined on the set V of all points $P(x, y)$ in the x, y-plane by writing

$$(x, y)\,R\,(z, t) \quad \text{if and only if} \quad x^2 + y^2 = z^2 + t^2.$$

Show that R is an equivalence relation on V and describe geometrically the partition of V into equivalence classes. What is the equivalence class determined by the point $A(1, 1)$?

18. Give an example of a relation on a set which is symmetric and transitive but not reflexive.

19. If R_1 and R_2 are equivalence relations on a set S and if R_1 and R_2 are regarded as subsets of $S^2 = S \times S$, show that $R_1 \cap R_2$ is also an equivalence relation on S.

20. Let R be a relation from set S to set T. The **inverse relation** of R, denoted by R^{-1}, is the relation from T to S defined by

$$b\,R^{-1}\,a \quad \text{when} \quad a\,R\,b.$$

The graph of R^{-1} is the subset $\{(b, a): a\,R\,b\}$ of $T \times S$.

   If $T = S$, so that R is a relation in S, show that R is symmetric if and only if $R = R^{-1}$. [**Hint.** Regard R and R^{-1} as subsets of $S^2 = S \times S$.]

Mappings

1. Basic definitions and some examples

The idea of a mapping is one of the most important concepts in mathematics. As mentioned at the end of the main text of Chapter 2, the idea is associated with a special type of relation between sets.

Consider first the following examples.

Example 1. Let S be the set of points P in a plane and l the set of points on a given line in the plane S.

With each point $P \in S$ we can associate a unique point $P' \in l$ by letting P' be the orthogonal projection of P on l; P' coincides with P when $P \in l$.

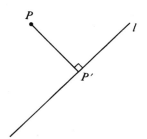

Example 2. The **absolute value** (or **modulus**) $|x|$ of a real number x is defined by

$$|x| = \begin{cases} x, & x \geqslant 0, \\ -x, & x < 0. \end{cases}$$

[For example, $|3| = 3$, $|-\pi| = \pi$.]

In this case we have a rule which associates with each $x \in \mathbf{R}$ a unique member, namely $|x|$, of $\mathbf{R}^+$, the set of non-negative real numbers.

37

Example 3. The table

1	2	3	4	5
5	3	1	1	4

associates with each member of the set $\{1, 2, 3, 4, 5\}$ a unique member of the same set, the number corresponding to each member of the set being listed in the second row immediately below that member.

Example 4. With each $x \in \mathbf{R}$ we associate the real number 0.

These four examples all display the same pattern, namely a rule which associates with each member of one set a unique member of another set. We cover relations of this type by the following definition.

Definition. Let S and T be nonempty sets. A **mapping from S to T** is a rule which associates with each member of S a *unique* member of T. S is called the **domain** of the mapping and T the **codomain** of the mapping. In this definition the sets S and T may be equal.

If we denote the mapping by f we write $f: S \rightarrow T$ to indicate that f is a mapping **from S to T**. If $s \in S$, we denote by $f(s)$ the unique member of T which f associates with s; $f(s)$ is called the **image of s under f**, or the **value of f at s**. We can represent the mapping diagrammatically as follows:

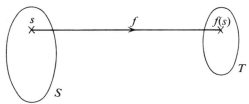

The subset of T consisting of the members of T that are images under f of elements of S, i.e. the subset $\{t \in T : t = f(s) \text{ for some } s \in S\}$, is called the **image of S under f**, or, simply, the **image of f**. It is often denoted by $f(S)$; it can be represented in a diagram as shown. [The word **range** is often used instead of **image**.]

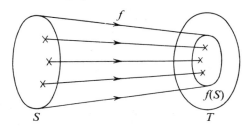

In Examples 1, 2, 3, 4, we can define mappings whose S, T and $f(S)$ are as follows:

Example number	S	T	$f(S)$
1.	Plane S	line l	line l
2.	$\mathbf{R}$	$\mathbf{R}^+$	$\mathbf{R}^+$
3.	$\{1, 2, 3, 4, 5\}$	$\{1, 2, 3, 4, 5\}$	$\{1, 3, 4, 5\}$
4.	$\mathbf{R}$	$\mathbf{R}$	$\{0\}$

It is clear that the image of a mapping need not equal the codomain.

An important application of mappings and a good source of examples is provided by the functions which appear in calculus. Here a function is simply a mapping from $\mathbf{R}$ to $\mathbf{R}$ or from a subset of $\mathbf{R}$ to $\mathbf{R}$. We list some examples.

Example 5. A mapping $f:\mathbf{R}\to\mathbf{R}$ is defined by the formula

$$f(x) = x^2+1 \quad \forall x \in \mathbf{R}.$$

The image of f is the set

$$\begin{aligned}\{f(x):x\in\mathbf{R}\} &= \{x^2+1:x\in\mathbf{R}\} \\ &= \{y\in\mathbf{R}:y = x^2+1 \text{ for some } x\in\mathbf{R}\} \\ &= \{y\in\mathbf{R}:y\geqslant 1\}.\end{aligned}$$

Example 6. A mapping $g:\mathbf{R}-\{1\}\to\mathbf{R}$ is defined by the formula

$$g(x) = \frac{1}{x-1} \ \forall x \in \mathbf{R}-\{1\}.$$

The image of g is

$$\left\{y\in\mathbf{R}:y = \frac{1}{x-1} \text{ for some } x\in\mathbf{R}-\{1\}\right\},$$

and it can be shown that this set is $\mathbf{R}-\{0\}$.

Graph of a mapping. Let $f:S\to T$ be a mapping. The subset

$$\{(s, f(s)):s \in S\}$$

of $S\times T$ is called the **graph of** f. As we noted for relations in Chapter 2, it is often convenient to identify a mapping with its graph.

If S and T are subsets of $\mathbf{R}$ the graph gives a picture of the mapping if we regard $(s, f(s))$ as the coordinates of a point relative to a coordinate system.

The graphs of the mappings in examples 2, 3, 4, 5 and 6 are:

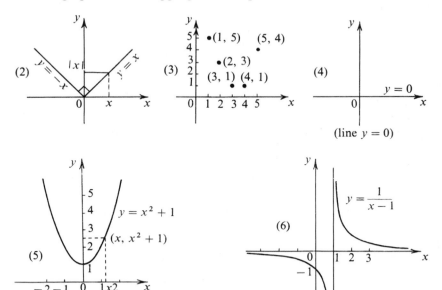

Equality of mappings. We first note that a mapping involves three things, namely, (i) its domain, (ii) its codomain, and (iii) its rule. Consequently our definition of equality is as follows:

Two mappings $f: S \to T$ and $g: S_1 \to T_1$ are said to be **equal** (or the same mapping), and we write then $f = g$, if

$$(1) \ S = S_1, \qquad (2) \ T = T_1, \qquad (3) \ f(s) = g(s) \ \forall s \in S.$$

Note. *All* the above conditions must be satisfied for the mappings to be equal. It follows that the mappings

$$f: \mathbf{R} \to \mathbf{R} \quad \text{defined by } f(x) = x^2 \ \forall x \in \mathbf{R}$$

and
$$g: \mathbf{R} \to \mathbf{R}^+ \quad \text{defined by } g(x) = x^2 \ \forall x \in \mathbf{R}$$

are not equal since they have different codomains, even although they have the same image.

Geometrical examples of mappings. The standard geometrical transformations of a plane, such as reflections, rotations, translations and dilatations, provide examples of mappings from a plane to itself.

Example 7. Reflection R_l in a line l maps the plane to itself by mapping a point P to its mirror image P' in l. This transformation can be represented in terms of coordinates. For example, taking l as the x-axis, the point $P(x, y)$ is mapped to the point $P'(x', y')$, where $x' = x$ and $y' = -y$.

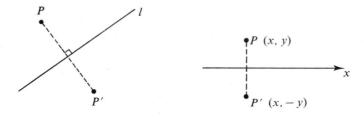

2. Injections

Example 8. Consider the mapping $f: \mathbf{R} \rightarrow \mathbf{R}$ defined by

$$f(x) = \begin{cases} 2x & \text{when } x \geqslant 0, \\ x+1 & \text{when } x < 0. \end{cases}$$

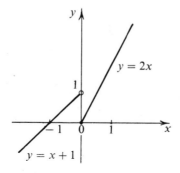

The graph of f consists of two half lines as shown in the diagram. The small circle surrounding the point $(0, 1)$ indicates that this point is not part of the graph. $f(-1) = f(0) = 0$, so that -1 and 0 have the same image by f, namely 0.

Give other pairs of numbers with the same image under f.

On the other hand, consider the following example.

Example 9. Let $g: \mathbf{R} \rightarrow \mathbf{R}$ be the mapping defined by

$$g(x) = 3x - 2.$$

The graph of g is the straight line $y = 3x - 2$, if we identify the graph with its geometrical representation. For this mapping,

$$x_1, x_2 \text{ have the same image under } g$$
$$\Rightarrow \quad g(x_1) = g(x_2)$$
$$\Rightarrow 3x_1 - 2 = 3x_2 - 2$$
$$\Rightarrow \quad x_1 = x_2.$$

Thus distinct members of the domain of g have distinct images under g. It is useful to have a name for mappings with this property.

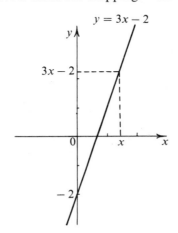

Definition. A mapping $f: S \rightarrow T$ is called **injective** or **one-to-one** if, whenever s_1, s_2 are distinct members of S, their images $f(s_1), f(s_2)$ are distinct members of T. Equivalently, f is injective if

$$f(s_1) = f(s_2) \Rightarrow s_1 = s_2. \tag{2.1}$$

An injective mapping is called an **injection**.

The mapping in Example 9 is injective, but that in Example 8 is not.

For mappings from **R** (or a subset of **R**) to **R** (or a subset of **R**) we can give a geometrical condition involving the graph of the mapping which is equivalent to the condition (2.1) for an injection. Such a mapping f is injective if and only if the line through each point on the y-axis, in the codomain of f, parallel to the x-axis meets the graph of f in **at most** one point. Check this for Examples 8 and 9.

Example. Let $S = \mathbf{R} - \{3\}$ and let $f: S \rightarrow \mathbf{R}$ be defined by

$$f(x) = \frac{1}{x-3}.$$

Show that f is injective.

Here, $f(x_1) = f(x_2)$

$$\Rightarrow \frac{1}{x_1-3} = \frac{1}{x_2-3} \Rightarrow x_1-3 = x_2-3 \Rightarrow x_1 = x_2,$$

and so f is injective.

Exercise. Determine those mappings in Examples 1, 2, . . . , 7 which are injective.

3. Surjections

Example 10. Let $f:\mathbf{R}\rightarrow\mathbf{R}$ be the mapping defined by

$$f(x) = \begin{cases} x+1 & \text{when } x\geqslant0, \\ x & \text{when } x<0. \end{cases}$$

Here there are real numbers, e.g. $\frac{1}{2}$, which are not images of any member of the domain.

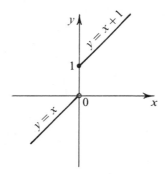

On the other hand consider the following example.

Example 11. Let $g:\mathbf{R}\rightarrow\mathbf{R}$ be the mapping defined by

$$g(x) = \begin{cases} x^2 & \text{when } x\geqslant0, \\ x & \text{when } x<0. \end{cases}$$

If y is any member of the codomain, i.e. any real number in this case, then there is a member x of the domain, i.e. a real number in this case, such that $y = g(x)$. If $y\geqslant0$, then $x = \sqrt{y}$, and, if $y<0$, then $x = y$.

In this case every member of the codomain is the image of some member (a unique member in this case) of the domain.

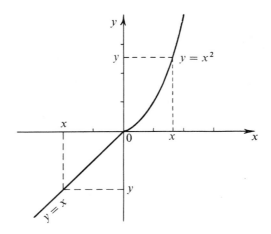

A glance at the graph of the mapping in Example 8 shows that for that mapping we have: every member of the codomain is the image of at least one member of the domain.

It is useful to have a name for mappings with this property.

Definition. A mapping $f: S \to T$ is called **surjective** or **onto** if every member of the codomain T is the image under f of **at least** one member of the domain S, i.e. if $f(S) = T$, i.e. if the image of f is the whole of T.

A surjective mapping is called a **surjection**.

The mappings in Examples 8 and 11 are surjective, but that in Example 10 is not.

For mappings from $\mathbf{R}$ (or a subset of $\mathbf{R}$) to $\mathbf{R}$ (or a subset of $\mathbf{R}$) a geometrical condition for a surjection is: such a mapping f is surjective if and only if the line through each point on the y-axis, in the codomain of f, parallel to the x-axis meets the graph of f in **at least** one point. Check this for Examples 8, 10 and 11.

Example 12. Show that the mapping $f: \mathbf{R} \to \mathbf{R}^+$ defined by $f(x) = (x-1)^2$ is surjective.

Let $y \in \mathbf{R}^+$. Then

$$y = (x-1)^2 \Leftrightarrow \pm\sqrt{y} = x-1 \Leftrightarrow x = 1 \pm \sqrt{y}.$$

Since $y \geqslant 0$, $1 \pm \sqrt{y} \in \mathbf{R}$ and so

$$y = f(1 + \sqrt{y}) = f(1 - \sqrt{y}).$$

Thus y is the image of at least one member of the domain of f, and so f is surjective.

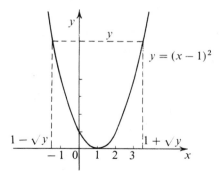

Exercise. (i) Determine those mappings in Examples $1, 2, \ldots, 7, 9$ which are surjective.

(ii) A mapping $f: \mathbf{R} \rightarrow \mathbf{R}$ is defined by

$$f(x) = \begin{cases} x^2 & \text{when } x \geqslant 0, \\ x+2 & \text{when } x < 0. \end{cases}$$

Determine geometrically whether f is injective and whether it is surjective.

4. Bijections, inverse mappings

A mapping which is both injective and surjective is called **bijective**; a bijective mapping is called a **bijection**.

Check that the mappings in Examples 1, 7, 9 and 11 are bijections.

For mappings of $\mathbf{R}$ (or a subset of $\mathbf{R}$) to $\mathbf{R}$ (or a subset of $\mathbf{R}$) a geometrical condition for a bijection is: such a mapping f is bijective if and only if the line through each point on the y-axis, in the codomain of f, parallel to the x-axis meets the graph of f in **exactly** one point. Check this for Examples 9 and 11.

Inverse mappings. We now show that, *if $f: S \rightarrow T$ is a bijection*, there is a special mapping $g: T \rightarrow S$ which can be associated with f in the following manner.

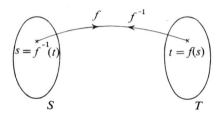

Consider any member t of T. Since f is *surjective*, $\exists s \in S$ such that $f(s) = t$. Since f is *injective*, this s is unique; for, if $f(s_1) = t$ and $f(s_2) = t$, where $s_1 \in S$ and $s_2 \in S$, then $f(s_1) = f(s_2)$ and so $s_1 = s_2$.

Thus with each $t \in T$ there is associated a unique $s \in S$ such that $f(s) = t$. This ensures that we can define a mapping $g : T \to S$ by the rule:

$$g(t) = s \Leftrightarrow f(s) = t.$$

This mapping g is called the **inverse mapping of the bijection f and is denoted by f^{-1}**. Hence

$$f^{-1}(t) = s \Leftrightarrow f(s) = t.$$

f^{-1} is a bijection: $T \to S$.

Example 13. Show that the mapping $f : \mathbf{R} \to \mathbf{R}$ defined by

$$f(x) = \begin{cases} x^2 + 1 & \text{when } x \geqslant 0, \\ x + 1 & \text{when } x < 0 \end{cases}$$

is a bijection and find f^{-1}.

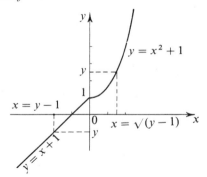

From the graph of f it is clear that the line through any point on the y-axis parallel to the x-axis meets the graph in exactly one point, and so f is bijective. We can of course write out analytic statements proving this without using geometrical thinking.

If $y \geqslant 1$, then $y = x^2 + 1$ with $x \geqslant 0 \Leftrightarrow x = \sqrt{(y-1)}$.
If $y < 1$, then $y = x + 1$ with $x < 0 \Leftrightarrow x = y - 1$.

Hence the inverse mapping $f^{-1} : \mathbf{R} \to \mathbf{R}$ is given by the rule

$$f^{-1}(y) = \begin{cases} \sqrt{(y-1)} & \text{when } y \geqslant 1, \\ y - 1 & \text{when } y < 1. \end{cases}$$

Exercise. Find formulae for the inverse mappings of the bijections in Examples 9 and 11.

5. Composition of mappings

Let $f: S \to T$ and $g: T \to U$ be mappings. With each $s \in S$, the mapping f associates a unique member $f(s)$ of T. Since the domain of g is T, g associates with $f(s)$ a unique member $g(f(s))$ of U. Thus we have a rule for associating with each $s \in S$ a unique member of U, i.e. we have a mapping from S to U. This mapping is called the **composition of f and g** and denoted by $g \circ f$; this is read as "g **circle** f". [It should be noted that f operates first and then g.] Thus $g \circ f: S \to U$ is defined by the rule

$$(g \circ f)(s) = g(f(s)) \quad \forall s \in S.$$

The mapping $g \circ f$ exists if and only if the domain of g equals the codomain of f.

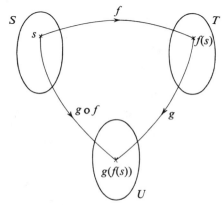

Example 14. Let $f: \mathbf{R} \to \mathbf{R}$ and $g: \mathbf{R} \to \mathbf{R}$ be defined by $f(x) = x+1$, $g(x) = -x$. Find the numbers $(g \circ f)(0)$ and $(f \circ g)(-1)$, and find formulae for mappings $g \circ f$ and $f \circ g$.

$$(g \circ f)(0) = \quad g(f(0)) = g(1) = -1.$$
$$(f \circ g)(-1) = f(g(-1)) = f(1) = 2.$$

$g \circ f$ is a mapping from $\mathbf{R}$ to $\mathbf{R}$ given by

$$(g \circ f)(x) = g(f(x)) = g(x+1) = -(x+1) = -x-1 \quad \forall x \in \mathbf{R}.$$

$f \circ g$ is a mapping from $\mathbf{R}$ to $\mathbf{R}$ given by

$$(f \circ g)(x) = f(g(x)) = f(-x) = -x+1 \quad \forall x \in \mathbf{R}.$$

Clearly $(g \circ f)(x) \neq (f \circ g)(x) \quad \forall x \in \mathbf{R}$, and so $g \circ f \neq f \circ g$.

Note. This example shows that, for a given pair of mappings f, g, the composite mappings $g \circ f$ and $f \circ g$ need not be equal even when both exist and have equal domains and equal codomains. We say that **composition of mappings is not commutative**.

Example 15. Let $f: \mathbf{R} \to \mathbf{R}$ be defined by

$$f(x) = \begin{cases} x^2 & \text{when } x \geqslant 0, \\ x-1 & \text{when } x < 0, \end{cases}$$

and $g: \mathbf{R} \to \mathbf{R}$ be defined by

$$g(x) = \begin{cases} x+1 & \text{when } x \geqslant 1, \\ 2x & \text{when } x < 1. \end{cases}$$

Find formulae for the mappings $g \circ f: \mathbf{R} \to \mathbf{R}$ and $f \circ g: \mathbf{R} \to \mathbf{R}$.

$$(g \circ f)(x) = \begin{cases} g(x^2), x \geqslant 0 \\ g(x-1), x < 0 \end{cases} = \begin{cases} x^2+1, x \geqslant 1 \\ 2x^2, 0 \leqslant x < 1 \\ 2x-2, x < 0. \end{cases}$$

$$(f \circ g)(x) = \begin{cases} f(x+1), x \geqslant 1 \\ f(2x), x < 1 \end{cases} = \begin{cases} (x+1)^2, x \geqslant 1 \\ 4x^2, 0 \leqslant x < 1 \\ 2x-1, x < 0. \end{cases}$$

This rather more complicated example shows again that normally $g \circ f \neq f \circ g$ when both exist and have the same domain and the same codomain.

We now show that composition of three suitable mappings satisfies a basic algebraic property.

Theorem 3.1. *Let $f: S \to T$, $g: T \to U$ and $h: U \to V$ be mappings. Then*

$$h \circ (g \circ f) = (h \circ g) \circ f.$$

Proof.

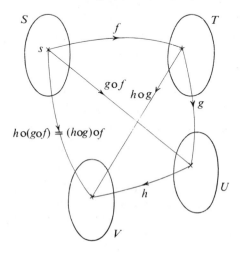

$$g \circ f \text{ is a mapping: } S \to U,$$

and so $$h \circ (g \circ f) \text{ is a mapping: } S \to V.$$

Also $$h \circ g \text{ is a mapping: } T \to V,$$

and so $$(h \circ g) \circ f \text{ is a mapping: } S \to V.$$

Thus $h \circ (g \circ f)$ and $(h \circ g) \circ f$ have equal domains and equal codomains. Also, $\forall s \in S$,

$$(h \circ (g \circ f))(s) = h((g \circ f)(s)) = h(g(f(s))),$$

and $$((h \circ g) \circ f)(s) = (h \circ g)(f(s)) = h(g(f(s))).$$

Hence $$h \circ (g \circ f) = (h \circ g) \circ f.$$

Note. This property of mappings is called the **associative property**. We omit brackets and write $h \circ g \circ f$ for both $h \circ (g \circ f)$ and $(h \circ g) \circ f$.

We complete this section on composition of mappings with two properties of a bijection and its inverse mapping. We note first that the **identity mapping** on a nonempty set S is the mapping

$$i_S : S \to S \quad \text{defined by: } i_S(s) = s \quad \forall s \in S,$$

i.e. it is the mapping which maps each element of S to itself.

Theorem 3.2. *If $f : S \to T$ is a bijection and $f^{-1} : T \to S$ is the inverse of f, then*

$$f \circ f^{-1} = i_T \text{ and } f^{-1} \circ f = i_S.$$

Proof. $f \circ f^{-1}$ and i_T are both mappings from T to T. Also, if $t \in T$,

$$\begin{aligned}(f \circ f^{-1})(t) &= f(f^{-1}(t)) \\ &= f(s), \quad \text{where } f(s) = t, \\ &= t = i_T(t).\end{aligned}$$

Hence, $\forall t \in T$, $(f \circ f^{-1})(t) = i_T(t)$, and so $f \circ f^{-1} = i_T$.

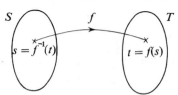

Similarly, $f^{-1} \circ f$ and i_S are both mappings from S to S. Also, if $s \in S$,

$$(f^{-1} \circ f)(s) = f^{-1}(f(s)) = f^{-1}(t) = s = i_S(s).$$

Hence, $\forall s \in S$, $(f^{-1} \circ f)(s) = i_S(s)$, and so $f^{-1} \circ f = i_S$.

Exercise. Verify these results for the bijections given in Examples 9, 11 and 13.

6. Restriction of mappings

It is sometimes important in dealing with a mapping which is not a bijection to be able to take a subset of the domain or the codomain (or of both) and, with the same rule as that which defines the mapping, produce a new mapping which is a bijection. The formal definition underlying this "cutting down in size" process is the following:

Definition. A mapping $g:S_1 \to T_1$ is called a **restriction** of a mapping $f:S \to T$ if the following conditions hold:

(1) $S_1 \subseteq S$, i.e. the domain of g is a subset of the domain of f,
(2) $T_1 \subseteq T$, i.e. the codomain of g is a subset of the codomain of f,
(3) $g(s_1) = f(s_1) \; \forall s_1 \in S_1$, i.e. the values of f and g on S_1 are equal.

Example 16. Consider the mapping $f:\mathbf{R} \to \mathbf{R}$ defined by

$$f(x) = x^2 \quad \forall x \in \mathbf{R}.$$

f is not injective and not surjective.
The image of f is $\mathbf{R}^+ = \{y \in \mathbf{R} : y \geqslant 0\}$.

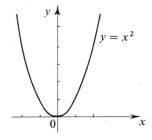
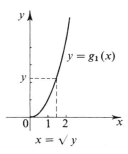

We show that we can easily find two restrictions of f which have the same image as f and which are bijections.

Let $g_1:\mathbf{R}^+ \to \mathbf{R}^+$ be defined by $g_1(x) = x^2 \; \forall x \in \mathbf{R}^+$. From the graph of g_1, it is clear that g_1 is a bijection and has an inverse $g_1^{-1}:\mathbf{R}^+ \to \mathbf{R}^+$ defined by

$$g_1^{-1}(y) = \sqrt{y} \quad \forall y \in \mathbf{R}^+.$$

The graph of g_1 is the "right-hand half" of the graph of f, and clearly g_1 is a restriction of f.

Now let $g_2: \mathbf{R}^- \to \mathbf{R}^+$ be defined by $g_2(x) = x^2 \quad \forall x \in \mathbf{R}^-$,

where $\mathbf{R}^- = \{x \in \mathbf{R} : x \leqslant 0\}$.

From the graph of g_2, it is clear that g_2 is a bijection and has an inverse $g_2^{-1}: \mathbf{R}^+ \to \mathbf{R}^-$ defined by $g_2^{-1}(y) = -\sqrt{y} \quad \forall y \in \mathbf{R}^+$.

The graph of g_2 is the "left-hand half" of the graph of f, and clearly g_2 is a restriction of f.

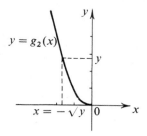

Example 17. Consider the mapping $f: \mathbf{R} \to \mathbf{R}$ defined by

$$f(x) = \sin x.$$

f is neither injective nor surjective; the image of f is the closed interval $[-1, 1]$.

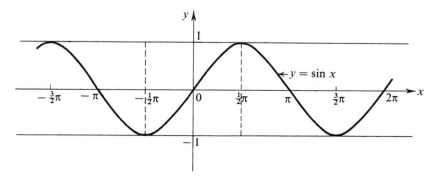

From the graph of f it is clear that one restriction of f with the same image as f is the mapping $g: [-\tfrac{1}{2}\pi, \tfrac{1}{2}\pi] \to [-1, 1]$ given by

$$g(x) = \sin x \quad \forall x \in [-\tfrac{1}{2}\pi, \tfrac{1}{2}\pi].$$

Also from the graph it is clear that g is a bijection. The inverse of g is called the **inverse sine function** and denoted by $\sin^{-1}$ or arcsin.

7. Real functions

To complete this chapter on mappings we mention briefly the class of mappings which appears in elementary calculus. This class is normally described in more detail in a book on calculus.

A formula such as $f(x) = \sqrt{(1-x^2)}$, used in connection with real functions, does not by itself define a unique mapping. We must also specify the domain and the codomain; the domain will be some subset S of $\mathbf{R}$ such that $f(x) \in \mathbf{R} \; \forall x \in S$. In this case some possible domains for mappings defined by the formula $f(x) = \sqrt{(1-x^2)}$ are the intervals $[0, \frac{1}{2}]$, $(0, 1]$, $[-\frac{1}{2}, \frac{1}{2}]$, $[-1, 1]$ and so on. In calculus however it is inconvenient to be constantly concerned with domains and codomains and we make the convention that the mapping $f : S \to \mathbf{R}$ defined by a given formula, e.g. by $f(x) = \sqrt{(1-x^2)}$, has for domain S the *largest subset of* $\mathbf{R}$ *such that* $f(x) \in \mathbf{R} \; \forall x \in S$. This particular mapping is called **the real function** f defined by the formula for $f(x)$.

For example, the real function defined by the formula $f(x) = \sqrt{(1-x^2)}$ has domain $[-1, 1]$ since this is the maximal subset S of $\mathbf{R}$ such that $f(x)$ is a real number $\forall x \in S$.

It is often rather difficult to determine the maximal domain for the real function defined by a given formula. We consider two examples.

Example 18. Find the maximal domains for the real functions defined by the following formulae:

(i) $$f(x) = \sin^{-1}(x^2 - 2);$$

(ii) $$f(x) = \frac{1}{1 + 2\cos x}.$$

(i) From Example 17, $\sin^{-1} y \in \mathbf{R} \Leftrightarrow -1 \leqslant y \leqslant 1$.
Thus the required domain $= \{x \in \mathbf{R} : -1 \leqslant x^2 - 2 \leqslant 1\}$
$$= \{x \in \mathbf{R} : 1 \leqslant x^2 \leqslant 3\}$$
$$= [-\sqrt{3}, -1] \cup [1, \sqrt{3}].$$

(ii) In this case, the required domain $= \mathbf{R} - \{x \in \mathbf{R} : \cos x = -\frac{1}{2}\}$
$$= \mathbf{R} - \{2n\pi \pm \tfrac{2}{3}\pi : n \in \mathbf{Z}\}.$$

EXERCISE 3

[**Note.** For some of the following problems a knowledge of some simple graphs such as $y = ax + b$ with a, b constants, $y = \pm x^2$, $y = \pm x^2 + k$ with k a constant, $y = x^3$, etc., will be assumed.]

1. (i) For the mapping $f : \mathbf{R} \to \mathbf{R}$ defined by $f(x) = x^2 - 1$, find $f(0), f(2), f(-3)$, sketch the graph of f, and state the image of f.
   (ii) Sketch the graph of the mapping $g : [-3, 2] \to \mathbf{R}$ defined by $g(x) = 3x + 4$, and state the image of g.

(iii) For the mapping $h:\mathbf{R}\to\mathbf{R}$ defined by $h(x) = \frac{1}{2}(x+|x|)$, where $|x|$ is the absolute value of x, evaluate $h(3)$, $h(-3)$, $h(0)$; sketch the graph of h and state the image of h.

(iv) Sketch the graph of the mapping $k:[0,\pi]\to\mathbf{R}$ defined by $k(x) = \cos x$, and state the image of k.

2. For each of the following mappings, sketch its graph and state whether it is (a) injective, (b) surjective. If it is not surjective, state its image; if it is not injective, specify an element of the codomain that is the image of more than one element of the domain.

(i) $f:\mathbf{R}^+\to\mathbf{R}^+$ defined by $f(x) = x^2 + 1$.

(ii) $g:\mathbf{R}\to\mathbf{R}$ defined by $g(x) = \begin{cases} 2x+1 & \text{when } x \geqslant 0, \\ x-1 & \text{when } x < 0. \end{cases}$

(iii) $h:\mathbf{R}\to\mathbf{R}$ defined by $h(x) = \begin{cases} x^2 & \text{when } x \geqslant 0, \\ 2x+3 & \text{when } x < 0. \end{cases}$

(iv) $k:\mathbf{R}\to\mathbf{R}^+$ defined by $k(x) = \begin{cases} x^2+1 & \text{when } x \geqslant 0, \\ 1-x & \text{when } x < 0. \end{cases}$

3. For each of the following mappings determine whether it is (a) injective, (b) surjective. Find the inverse mapping of any mapping which is bijective.

(i) $f:\mathbf{R}\to\mathbf{R}$ defined by $f(x) = x+2$.

(ii) $g:\mathbf{R}\to\mathbf{R}^+$ defined by $g(x) = x^2$.

(iii) $h:\mathbf{R}\to\mathbf{R}$ defined by $h(x) = \begin{cases} x^2-1 & \text{when } x \geqslant 0, \\ -x^2-1 & \text{when } x < 0. \end{cases}$

(iv) $k:\mathbf{R}\to\mathbf{R}$ defined by $k(x) = \begin{cases} 2x & \text{when } x \text{ is an integer}, \\ x & \text{when } x \text{ is not an integer}. \end{cases}$

(v) $u:\mathbf{Z}\to\mathbf{Z}$ defined by $u(n) = n^3$.

(vi) $v:\mathbf{Z}\to\mathbf{Z}$ defined by $v(n) = \begin{cases} n+2 & \text{when } n \text{ is even}, \\ n & \text{when } n \text{ is odd}. \end{cases}$

4. Let $S = \{1,3,5,\ldots\}$, the odd positive integers, and $T = \{2,4,6,\ldots\}$, the even positive integers. Give examples of mappings from S to T such that

(i) f_1 is bijective,

(ii) f_2 is neither injective nor surjective,

(iii) f_3 is injective but not surjective,

(iv) f_4 is surjective but not injective.

5. Mappings $f:\mathbf{R}\to\mathbf{R}$ and $g:\mathbf{R}\to\mathbf{R}$ are defined by the formulae

$$f(x) = x^2, \quad g(x) = \begin{cases} x-1 & \text{when } x \geqslant 0, \\ -x & \text{when } x < 0. \end{cases}$$

Find formulae for the mappings $f \circ f$, $f \circ g$, $g \circ f$ and $g \circ g$.

6. Mappings $f:\mathbf{R}\to\mathbf{R}$ and $g:\mathbf{R}\to\mathbf{R}$ are defined by

$$f(x) = \begin{cases} 2x & \text{when } x \geqslant 0, \\ x+2 & \text{when } x < 0 \end{cases} \quad \text{and} \quad g(x) = \begin{cases} 2x+1 & \text{when } x \geqslant 0, \\ x & \text{when } x < 0. \end{cases}$$

Find formulae for $f \circ g$ and $g \circ f$, show that one of them is bijective and the other is not, and find the inverse mapping of the one that is bijective.

7. Mappings $f:\mathbf{R}\to[-1,1]$ and $g:[-1,1]\to\mathbf{R}$ are defined by $f(x) = \sin x$ and $g(x) = \sqrt{(1-x^2)}$. Find formulae for $g \circ f$ and $f \circ g$ specifying in each case the domain and codomain. Find also the images of $g \circ f$ and $f \circ g$.

8. Mappings $f: \mathbf{R} \to \mathbf{R}$ and $g: \mathbf{R} \to \mathbf{R}$ are defined by

$$f(x) = \begin{cases} x+1 & \text{when } x \geqslant 0, \\ x^2 & \text{when } x < 0 \end{cases} \quad \text{and} \quad g(x) = \begin{cases} 2x-3 & \text{when } x \geqslant 1, \\ 1-x & \text{when } x < 1. \end{cases}$$

Find formulae for $g \circ f$ and $f \circ g$.

9. Show that, for any mapping $f: S \to T$,

$$f \circ i_S = f \quad \text{and} \quad i_T \circ f = f.$$

10. Let $f: S \to T$ and $g: T \to S$ be mappings such that

$$g \circ f = i_S \quad \text{and} \quad f \circ g = i_T.$$

Show that f is bijective and that $g = f^{-1}$.

11. If $f: S \to T$ and $g: T \to U$ are bijections, show that $g \circ f$ is also a bijection and that

$$(g \circ f)^{-1} = f^{-1} \circ g^{-1}.$$

12. Mappings $f: S \to T$, $g: S \to T$ and $h: T \to U$ are such that

$$h \circ g = h \circ f.$$

If h is injective (but not necessarily bijective), show that $g = f$.

13. For the mapping $f: \mathbf{R} \to \mathbf{R}$ defined by $f(x) = x^2 + 3$,
  (i) sketch the graph and state the image;
  (ii) show that f is neither injective nor surjective;
  (iii) find a restriction g of f which is bijective and has the same image as f;
  (iv) find g^{-1}.

14. Show that the mapping $f: \mathbf{R} \to [-1, 1]$ defined by $f(x) = \cos x$ is surjective but not injective. Find a restriction of f which is bijective and has the same image as f.
    [Note that the usual inverse cosine function $\cos^{-1}$ is the inverse mapping of the restriction g of f given by

$$g: [0, \pi] \to [-1, 1], \ g(x) = \cos x.$$

15. What is the (maximal) domain of the real function (if any) defined by each of the following formulae?

  (i) $f(x) = \sqrt{x}$,

  (ii) $f(x) = \dfrac{1}{1 + \cos x}$,

  (iii) $f(x) = (-x^2)^{1/4}$,

  (iv) $f(x) = \sin \sqrt{(4 - x^2)}$,

  (v) $f(x) = \sqrt{\left(\dfrac{1-x}{1+x} \right)}$,

  (vi) $f(x) = \sqrt{(2 \sin x - 3)}$,

  (vii) $f(x) = \cos (x^{1/3})$,

  (viii) $f(x) = \sin^4 x$.

Number Systems

1. Peano's axioms for the set N of natural numbers or positive integers

Man has for many centuries intuitively pictured the counting or natural numbers as objects obtained in a sequence by adding at each step one additional stroke or circle or cross, etc., depending on the recording symbol being used, e.g.

(i) $\{/, //, ///, ////, /////, //////, ///////, ////////, \ldots\}$;
(ii) $\{o, oo, ooo, oooo, ooooo, oooooo, ooooooo, oooooooo, \ldots\}$;
(iii) $\{x, xx, xxx, xxxx, xxxxx, xxxxxx, xxxxxxx, xxxxxxxx, \ldots\}$,

and so on. This notation was eventually simplified to the usual decimal shorthand notation:

$$\{1, 2, 3, 4, 5, 6, 7, 8, \ldots\}.$$

It was only in the nineteenth century that a successful attempt was made by the Italian mathematician Peano (1858–1932) and others to describe the set N in a way which, while taking advantage of intuition, brought out in a mathematical way the nature of the basic properties of N. The obvious ideas that we must take into account in such an attempt are:

(a) the existence of a "first" element;
(b) the existence of a "successor" element for each element;
(c) the fact that the whole set is obtained from the first element and its successive successor elements;
(d) the elements are all distinct.

We can formalize this mathematically in the following way.

The set N is a non-empty set for which $\exists$ a mapping from N to itself called the **successor mapping** on N; we denote the successor of $x \in N$ by x'. The set and mapping have the following properties, called **Peano's axioms**:

(1) $\exists$ an element $1 \in N$ which is not the successor of any member of N.

(2) $x' = y' \Rightarrow x = y$.

(3) If A is a subset of $\mathbf{N}$ such that $1 \in A$ and

$$x \in A \Rightarrow x' \in A,$$

then $A = \mathbf{N}$.

If we denote $(1')'$ by $1''$, $(1'')'$ by $1'''$, and so on, and if

$$A = \{1, 1', 1'', 1''', 1'''', 1''''', \ldots\},$$

i.e. the set of all successors of elements arising by succession from 1, then, by using (3), we can show that $A = \mathbf{N}$.

The properties (1) and (2) ensure that the elements in this sequence of successors are all distinct. We can now simplify the notation to $\mathbf{N} = \{1, 2, 3, 4, \ldots\}$ in the usual form.

Property (3) is the basis of the so-called **"Principle of Induction"** method for proving results involving integers n.

Suppose that $P(n)$ is a statement and that the universal set for n is $\mathbf{N}$. Let

$$A = \{n \in \mathbf{N} : P(n) \text{ has truth value } T\}.$$

If $1 \in A$ and if $k \in A \Rightarrow k' \in A$, then, by axiom (3), $A = \mathbf{N}$, i.e.

$$P(n) \text{ is true } \forall n \in \mathbf{N}.$$

This can be stated as follows:

"If $P(1)$ is true and if $P(k)$ true $\Rightarrow P(k')$ true, then $P(n)$ is true $\forall n \in \mathbf{N}$."

We shall illustrate the use of this "Principle of Induction" in Section 1 of Chapter 5 and in many places later in the text.

Having obtained a suitable definition for $\mathbf{N}$, our next task is to introduce addition, multiplication and ordering, i.e. inequality, in $\mathbf{N}$.

Addition on N. We define addition on $\mathbf{N}$ in the following inductive manner:

$$\begin{cases} x + 1 = x' \\ x + y' = (x + y)', \end{cases}$$

i.e. we first define the adding on of 1 and then, assuming that we can add on y, define the adding on of y'.

Multiplication on N. We define multiplication on $\mathbf{N}$ in a similar inductive manner:

$$\begin{cases} x1 = x \\ xy' = xy + x, \end{cases}$$

i.e. we first define multiplying by 1 and then, assuming that we can multiply by y, define multiplying by y'. [The formula for xy' is informally suggested by thinking of xy' as $x(y+1)$.]

Order on N. If $x, y \in \mathbf{N}$ we say that y **is less than** x and write $y < x$ if $\exists t \in \mathbf{N}$ such that $x = y + t$, i.e. if we have to add a positive integer to y to obtain x.

We also write $x > y$ and say that x **is greater than** y.

If $y < x$ or $y = x$, we write $y \leqslant x$; similarly $x \geqslant y$ means that either $x > y$ or $x = y$.

We are now in a position to prove a large number of properties of addition, multiplication and order on $\mathbf{N}$. The most important are:

Addition.
$$x + (y + z) = (x + y) + z \ \forall x, y, z \in \mathbf{N} : \text{associative property}$$
$$x + y = y + x \ \forall x, y \in \mathbf{N} \qquad : \text{commutative property}$$
$$x + y = x + z \Rightarrow y = z \qquad : \text{cancellation property}$$

Multiplication.
$$x(yz) = (xy)z \ \forall x, y, z \in \mathbf{N} \qquad : \text{associative property}$$
$$xy = yx \ \forall x, y \in \mathbf{N} \qquad : \text{commutative property}$$
$$xy = xz \Rightarrow y = z \qquad : \text{cancellation property}$$

Addition and multiplication together.
$$x(y + z) = xy + xz \ \forall x, y, z \in \mathbf{N} \quad : \text{distributive property}$$

Order.

(i) If $x, y \in \mathbf{N}$, then one and only one of the following three statements is true:
$$x < y, \ x = y, \ x > y.$$

(ii) $\qquad x < y \wedge y < z \Rightarrow x < z.$

(iii) $\qquad x < y \Rightarrow x + z < y + z.$

(iv) $\qquad x < y \Rightarrow xz < yz.$

Also the so-called well-ordering property holds:

Every non-empty subset of $\mathbf{N}$ *contains a least member.*

All of the above statements are proved from the axioms and the definitions of addition, multiplication and ordering. The well-ordering property is intuitively obvious when we think about the real number line.

If S is a non-empty subset of $\mathbf{N}$, then $\exists n_1 \in S$. The smallest of the finite set of integers $\{n \in \mathbf{N} : 1 \leqslant n \leqslant n_1\} \cap S$ is clearly the least member of S.

We do not believe in stressing at this stage formal proofs of the properties of **N** listed, but we shall include a proof of the associative property of addition to stress the importance of the idea of induction in these foundations.

Property. $\forall x, y, z \in \mathbf{N}, \; x+(y+z) = (x+y)+z.$

Proof. Let $A = \{z \in \mathbf{N} : x+(y+z) = (x+y)+z, \; \forall x, y \in \mathbf{N}\}.$
We first show that $1 \in A$.

$$\forall x, y \in \mathbf{N}, \quad x+(y+1) = x+y' = (x+y)' = (x+y)+1.$$

Thus $1 \in A$.

Suppose now that $z \in A$, so that $x+(y+z) = (x+y)+z \; \forall x, y \in \mathbf{N}$.
We show that $z' \in A$. We have:

$$\begin{aligned}
\forall x, y \in \mathbf{N}, \; x+(y+z') &= x+(y+z)' \\
&= (x+(y+z))' \\
&= ((x+y)+z)' \\
&= (x+y)+z'.
\end{aligned}$$

Thus $z \in A \Rightarrow z' \in A$.

It follows, by the induction axiom (3), that $A = \mathbf{N}$, and so the result is proved.

Subtraction in N. We first note that, if $y < x$ and if $x = y+t$, then t is uniquely determined by x and y; for, if $x = y+t_1$ and $x = y+t_2$, then $y+t_1 = y+t_2$, and so, by the cancellation property of addition, $t_1 = t_2$.

This unique t is often denoted by $x-y$, and the process of forming $x-y$ for $x > y$ is called **subtraction**.

Exercise. Try to prove some of the other properties of addition, multiplication and order listed.

2. The extension of N to $\mathbf{Z} = \{0, \pm 1, \pm 2, \ldots\}$, the set of all integers

We start with the observation that subtraction is not always possible *within* **N**, i.e., given $a, b \in \mathbf{N}$, the equation

$$x+b = a$$

is not always solvable for $x \in \mathbf{N}$.

For example, the equation $x+5 = 2$ has no solution in **N**.

On the other hand the equation $x+2 = 5$ has solution $x = 3$, and $x = 3$ is also the solution of each equation of the form

$$x+b = a,$$

with $a, b \in N$ and $a-b = 3$. The set of all such equations is given by the pairs

$$(a, b) = (4, 1), (5, 2), (6, 3), (7, 4), (8, 5), (9, 6), \ldots .$$

Similarly the integer $k \in N$ is the solution of all the equations

$$x+b = a$$

given by the set of pairs $\{(a, b) \in N \times N : a-b = k\}$,

i.e. $\{(k+1, 1), (k+2, 2), (k+3, 3), (k+4, 4), \ldots\}.$

Thus, corresponding to each $k \in N$ we have a subset C_k of $N \times N$. These subsets C_k $(k = 1, 2, 3, \ldots)$ partition the subset A of $N \times N$ consisting of all the pairs $(a, b) \in N \times N$ with $a > b$. From this partition we can define an equivalence relation $\sim$ on A in which

$$(a, b) \sim (c, d) \Leftrightarrow (a, b) \text{ and } (c, d) \in \text{ same } C_k$$
$$\Leftrightarrow a-b = c-d$$
$$\Leftrightarrow a+d = b+c$$

[see Section 10 of Chapter 2].

The last statement gives us a clue to a method for extending N to a set in which each equation of the form $x+b = a$ has a solution in that set. We take $N \times N$ and define an equivalence relation $\sim$ on $N \times N$ by

$$(a, b) \sim (c, d) \Leftrightarrow a+d = b+c.$$

We can easily check that this is an equivalence relation [see worked example in Section 10 of Chapter 2].

The set of resulting equivalence classes gives us our required set when we have introduced suitable definitions of addition, multiplication and inequality. In defining addition, multiplication and inequality for pairs, we remember that we are intuitively associating $a-b$ with the pair (a, b); thus

$(a, b) = (c, d)$ should mean $a-b = c-d,$ i.e. $a+d = b+c$;

$(a, b)+(c, d)$ should mean $(a-b)+(c-d),$

       i.e. $(a+c)-(b+d)$, i.e. $(a+c, b+d)$;

$(a, b)(c, d)$ should mean $(a-b)(c-d),$

       i.e. $(ac+bd)-(ad+bc)$, i.e. $(ac+bd, ad+bc)$;

$(a, b) < (c, d)$ should mean $a-b < c-d,$ i.e. $a+d < b+c.$

For simplicity of notation we shall write $=$ in place of $\sim$. We can now make the formal definitions of equivalence, addition, multiplication and inequality on $\mathbf{N} \times \mathbf{N}$ as follows:

$$(a, b) = (c, d) \Leftrightarrow a+d = b+c$$
$$(a, b)+(c, d) = (a+c, b+d)$$
$$(a, b)(c, d) = (ac+bd, ad+bc)$$
$$(a, b) < (c, d) \Leftrightarrow a+d < b+c.$$

Before proceeding further, we have first to check that the definitions of addition, multiplication and inequality are not affected when a pair is replaced by an equivalent pair, in other words to check that we can define addition, multiplication and inequality for equivalence classes by dealing with any one representative for each class.

The set of all equivalence classes obtained in this way from $\mathbf{N} \times \mathbf{N}$ is denoted by $\mathbf{Z}$ and called the **set of all integers**. The integer $k \in \mathbf{N}$ can be identified with the equivalence class determined by the pair $(k+1, 1)$ or any of its equivalents [note that $(k+1)-1 = k$]. In this way we can regard $\mathbf{N}$ as a subset of $\mathbf{Z}$ strictly contained in $\mathbf{Z}$ and call $\mathbf{Z}$ an **extension of N**. The class determined by the pair $(1, 1)$ or any of its equivalents is denoted by 0 and called **zero**, and the class determined by the pair $(1, 1+k)$ or any of its equivalents is denoted by $-k$ and called **negative k** or **minus** k. In this simplified notation, $\mathbf{Z} = \{0, \pm 1, \pm 2, \ldots\}$; the elements $-1, -2, -3, \ldots$ are called the **negative integers** and all are <0.

Some properties of Z that can now be proved are:
[We shall omit the quantifier $\forall$ and the names of the properties.]

$$x+(y+z) = (x+y)+z \quad ; \quad x(yz) = (xy)z$$
$$x+y = y+x \qquad\qquad xy = yx$$
$$x+0 = x \qquad\qquad x1 = x.$$

For each $x \in \mathbf{Z}$, $\exists -x \in \mathbf{Z}$ such that $x+(-x) = 0$;

$\qquad\qquad\qquad -x$ is called the **additive inverse** of x.

$\quad x(y+z) = xy+xz$
$\qquad\quad xy = 0 \Rightarrow$ either $x = 0$ or $y = 0$ (or both).

Also,

$$\text{either } x < y \text{ or } x = y \text{ or } x > y,$$
$$x < y \wedge y < z \Rightarrow x < z.$$
$$x < y \Rightarrow x+z < y+z.$$
$$\begin{cases} x < y \wedge z > 0 \Rightarrow xz < yz, \\ x < y \wedge z < 0 \Rightarrow xz > yz. \end{cases}$$

We say that **Z** forms an **integral domain** under addition and multiplication; it is called an **ordered integral domain** since it has the ordering relation $<$.

Z also possesses the following ordering property: every non-empty subset A of **Z** which is "**bounded below**" has a least member (in **Z**). ["Bounded below" means that $\exists k \in \mathbf{Z}$ such that every member of A is $\geqslant k$.] We say that **Z** has the **lower bound ordering property**. Similarly **Z** has the **upper bound ordering property**, namely that every non-empty subset B of **Z** which is "**bounded above**" has a greatest member (in **Z**). ["Bounded above" means that $\exists k \in \mathbf{Z}$ such that every member of B is $\leqslant k$.]

Notes. (1) $-(-x) = x \quad \forall x \in \mathbf{Z}$.

(2) If $a, b \in \mathbf{Z}$, the equation $x + b = a$ has solution

$$x = a + (-b) \text{ in } \mathbf{Z}.$$

This is denoted by $a - b$, and defines **subtraction in Z**.

Example. *Show that* $\forall x \in \mathbf{Z}$, $x0 = 0$.

Let x be represented by the pair $(a, b) \in \mathbf{N} \times \mathbf{N}$ and take representative $(1, 1)$ for 0. Then the result follows from the fact that

$$(a, b)(1, 1) = (a + b, a + b) = (1, 1).$$

Exercise. Try to prove some of the above properties.

3. The extension of Z to $\mathbf{Q} = \left\{ \dfrac{m}{n} : m \in \mathbf{Z}, n \in \mathbf{Z}, n \neq 0 \right\}$, the set of all rational numbers

We start with the observation that division is not always possible *within* **Z**, i.e. given $a, b \in \mathbf{Z}$, the equation

$$bx = a, \text{ with } b \neq 0,$$

is not always solvable for $x \in \mathbf{Z}$.

For example, the equation $5x = 2$ has no solution in **Z**.

We aim at an extension of **Z** to a set in which the equation $bx = a$ with $b \neq 0$ can always be solved and such that the equations $bx = a$ and $kbx = ka$ for any $k \neq 0$ are equivalent.

Following the same sort of pattern as in Section **2** we are led to a consideration of the subset B of $\mathbf{Z} \times \mathbf{Z}$ consisting of all pairs (a, b) of integers a, b with $b \neq 0$. Here the definitions of equivalence, which we shall denote by the equals sign $=$, addition, multiplication and inequality are suggested by intuitively thinking of the pair (a, b) as the fraction $\dfrac{a}{b}$.

$$\left[\frac{a}{b} = \frac{c}{d} \text{ means } ad = bc; \; \frac{a}{b} + \frac{c}{d} = \frac{ad + bc}{bd}; \; \frac{a}{b}\frac{c}{d} = \frac{ac}{bd}; \right.$$

$$\left. \frac{a}{b} < \frac{c}{d} \text{ means } \frac{a}{b}b^2d^2 < \frac{c}{d}b^2d^2, \text{ i.e. } abd^2 < cdb^2. \right]$$

Consequently we make the following definitions on B:

(1) $(a, b) = (c, d) \Leftrightarrow ad = bc$
(2) $(a, b) + (c, d) = (ad + bc, bd)$
(3) $(a, b)(c, d) = (ac, bd)$
(4) $(a, b) < (c, d) \Leftrightarrow abd^2 < cdb^2$.

Note that, for (2) and (3), $bd \neq 0$ since $b \neq 0$ and $d \neq 0$.

Exercise. Show that equality is an equivalence relation on B.

The equivalence class which contains the pair (a, b) is denoted by $\dfrac{a}{b}$ and called the **rational number** $\dfrac{a}{b}$. The set of all classes $\dfrac{a}{b}$ is denoted by $\mathbf{Q}$. Addition, multiplication and inequality on $\mathbf{Q}$ are defined by:

$$\frac{a}{b} + \frac{c}{d} = \frac{ad + bc}{bd}; \; \frac{a}{b}\frac{c}{d} = \frac{ac}{bd}; \; \frac{a}{b} < \frac{c}{d} \Leftrightarrow abd^2 < cdb^2.$$

It can be shown that these definitions are independent of the representative pairs chosen for the equivalence classes.

The integer $a \in \mathbf{Z}$ can be identified with the equivalence class $\dfrac{a}{1}$ determined by the pair $(a, 1)$ or any of its equivalents; in particular we identify 0 with $\dfrac{0}{1}$ and 1 with $\dfrac{1}{1}$. In this way we can regard $\mathbf{Z}$ as a subset of $\mathbf{Q}$, strictly contained in $\mathbf{Q}$, and call $\mathbf{Q}$ an extension of $\mathbf{Z}$.

If $x = \dfrac{a}{b} \in \mathbf{Q}$ and $\dfrac{a}{b} \neq \dfrac{0}{1}$, then $a \neq 0$ and $\dfrac{b}{a}$ exists. This is called the **multiplicative inverse** of x and denoted by x^{-1} or $\dfrac{1}{x}$; we also write $-x$ for $\dfrac{(-a)}{b}$.

The following list contains some of the properties of the rational numbers which can now be proved.

$$x+(y+z) = (x+y)+z \quad ; \quad x(yz) = (xy)z$$
$$x+y = y+x \qquad\qquad xy = yx$$
$$x+0 = x \qquad\qquad x1 = x$$
$$x+(-x) = 0 \qquad\qquad xx^{-1} = 1 \ (x \neq 0)$$

$$x(y+z) = xy+xz.$$
Either $x<y$ or $x = y$ or $x>y$
$$x<y \wedge y<z \Rightarrow x<z$$
$$x<y \Rightarrow x+z<y+z$$
$$\begin{cases} x<y \wedge z>0 \Rightarrow xz<yz \\ x<y \wedge z<0 \Rightarrow xz>yz. \end{cases}$$

We say that **Q** forms a **field** under addition and multiplication; it is called an **ordered field** since it possesses the ordering relation $<$.

Exercise. Try to prove some of the above properties.

We have now extended **N** to a set **Q** in which we can solve all equations of the form $x+b = a$ and of the form $bx = a$ $(b \neq 0)$, but we have paid a price, namely, that the lower bound and upper bound ordering properties no longer hold, i.e. we can have a non-empty subset of **Q** which is bounded below but does not have a least member (in **Q**) and similarly we can have a non-empty subset of **Q** which is bounded above but which does not have a greatest member (in **Q**). If we assume for the moment that the number $\sqrt{2} \notin \mathbf{Q}$ and take the decimal representation for $\sqrt{2}$, $\sqrt{2} = 1\cdot4142135\ldots$, we have the following situation:

Let S be the set of rational numbers,
$$S = \{1, 1\cdot4, 1\cdot41, 1\cdot414, 1\cdot4142, 1\cdot41421, 1\cdot414213, 1.4142135, \ldots\}, \quad (3.1)$$
obtained by taking the decimal expansion of $\sqrt{2}$ to "no place", one place, two places, three places, etc., of decimals. Then S is certainly bounded above, e.g. by $1\cdot5$, but does not have a greatest member; it increases to the non-rational "limit" $\sqrt{2}$.

4. The extension of Q to R, the set of all real numbers

The set **Q** of rational numbers is adequate for practical purposes where measurement to any given degree of accuracy involves only rational approximation. However **Q** is quite inadequate for mathematical purposes for a variety of reasons; e.g. calculus is not possible over **Q** and so Physics cannot be described in terms of **Q**; also simple equations such as $x^2 = 2$ have no solution in **Q**. We now prove this result.

Theorem 4.1. *There is no rational number x satisfying $x^2 = 2$.*

Proof. Suppose that the statement is not true; then $\exists x = m/n$ with $x^2 = 2$ where $m, n \in \mathbb{N}$ and where m, n are not both even (in fact, we can arrange that m, n have no common factor > 1). Then

$$m^2 = 2n^2.$$

Hence m^2 is even and so m is even. For, either m is even or m is odd, and if m is odd, $= 2k+1$ say, then $m^2 = 2(2k^2+2k)+1$ is odd. Putting $m = 2m_1$ where $m_1 \in \mathbb{N}$, we deduce that

$$n^2 = 2m_1^2.$$

The same argument as above shows that n is also even, and so m and n are both even. But this contradicts our assumption that m and n are not both even.

Hence the equation $x^2 = 2$ has no rational solution.

This result is usually expressed in the form: "$\sqrt{2}$ is irrational". This form of words, however, begs the question of the existence of a number system in which 2 has a square root. Such systems do indeed exist. Most important is the existence of a field, the field **R** of **real numbers**, which contains solutions of $x^2 = 2$ and many other (but not all) such algebraic (i.e. polynomial) equations. It also contains a multitude of so-called **transcendental numbers** like π. These numbers do *not* satisfy any polynomial equation with rational coefficients, i.e. any equation of the form $a_n x^n + a_{n-1} x^{n-1} + \ldots + a_1 x + a_0 = 0$ with $a_i \in \mathbb{Q}$ and $a_n \neq 0$. Real numbers which satisfy such polynomial equations are called **algebraic**; every rational number m/n is algebraic since it satisfies the equation $nx - m = 0$. Methods of extending **Q** to **R** are more complicated than the previous extensions **N** to **Z** and **Z** to **Q** where we considered ordered pairs.

A clue to one possible approach to a method of extending **Q** to **R** is obtained by considering the sequence of rational approximations for $\sqrt{2}$ given in (3.1). This suggests a method of extending **Q** by considering suitable sequences of rational numbers. In carrying out such a programme there are several problems to be faced:

    1. Select an admissible set of sequences $\{a_n\}$ of rational numbers.

    2. Define a suitable equivalence relation on this set.

    3. Define addition, multiplication and inequality on the set **R** of equivalence classes.

    4. Check that the system is an ordered field, i.e. that **R** satisfies the properties listed for **Q**.

5. Check that **R** contains a subset which can be identified with **Q**; **R**−**Q** $\neq \emptyset$ and any $x \in$ **R**−**Q** is called an **irrational number**.

6. Check that the well-ordering property holds in **R**.

7. Show that **Q** is **dense** in **R**, i.e. given any $a \in$ **R**, $\exists x \in$ **Q** as close to a as we please.

This is a formidable programme and here we cannot do more than give an assurance that it can indeed be carried out successfully to yield the **field of real numbers R**, and that **R** has important properties other than those listed. For example, for each real number $a > 0$, $\exists$ a unique real number $b > 0$ such that $b^2 = a$. This number b is denoted by $\sqrt{a}$ and called the **square root of** a. If $a < 0$, then a does *not* have a square root. Similarly we can define $\sqrt[n]{a}$ in suitable circumstances.

The procedure for establishing **R** outlined above was originated by the German mathematician Georg Cantor in 1872. Another theory, on different lines, was evolved by Richard Dedekind about 1858, though not published until 1872.

For many mathematical purposes it is enough to have in mind a picture of the **real number line** as described in Section **4** of Chapter **2** when considering **R**. Also it is important to note that each real number x can be represented uniquely as a decimal expansion and that x is the "limit" of the sequence of rational numbers obtained by successive approximation from the expansion: e.g. $\frac{1}{2} = 0 \cdot 50000 \ldots$, $2 = 2 \cdot 00000 \ldots$, $\sqrt{2} = 1 \cdot 4142135 \ldots$, $\pi = 3 \cdot 1415926 \ldots$. Conversely every decimal expansion gives a unique real number.

5. Order properties of R, inequalities

We recall that in Example **2**, Section **1** of Chapter **3**, the **absolute value** (or **modulus** or **numerical value**) $|x|$ of a real number x was defined by

$$|x| = \begin{cases} x & \text{when } x \geq 0, \\ -x & \text{when } x < 0. \end{cases}$$

We first list some basic properties of the absolute value function.

(i) $|a| = \sqrt{(a^2)}$.

(ii) $|ab| = |a||b|$.

(iii) $|a+b| \leq |a|+|b|$, **(the triangle inequality)**.

(iv) $|x| \leq a \quad (a > 0) \Leftrightarrow -a \leq x \leq a$.

Proof of (i). It is clear that $|a|^2 = a^2$. Since $\sqrt{(a^2)}$ is defined to be non-negative, it follows that $|a| = \sqrt{(a^2)}$; e.g. $\sqrt{(-1)^2} = 1$, $\sqrt{(\cos^2 x)} = |\cos x|$, and so on.

Proof of (ii). This follows from a consideration of the four possible cases:

(1) $a \geqslant 0, b \geqslant 0$: here $|a||b| = ab$ and $|ab| = ab$;

(2) $a \geqslant 0, b < 0$: here $|a||b| = a(-b)$ and $|ab| = -(ab) = a(-b)$;

(3) $a < 0, b \geqslant 0$: here $|a||b| = (-a)b$ and $|ab| = -(ab) = (-a)b$;

(4) $a < 0, b < 0$: here $|a||b| = (-a)(-b)$ and $|ab| = ab = (-a)(-b)$.

Proof of (iii). We first observe that either $ab > 0$ or $ab < 0$ or $ab = 0$. In any case, we have:

$$ab \leqslant |ab| = |a||b|.$$

Thus $$2ab \leqslant 2|a||b|,$$

and so $$a^2 + 2ab + b^2 \leqslant a^2 + 2|a||b| + b^2,$$

$$(a+b)^2 \leqslant (|a|+|b|)^2 \text{ [since } |a|^2 = a^2 \text{ and } |b|^2 = b^2],$$
$$\sqrt{(a+b)^2} \leqslant \sqrt{(|a|+|b|)^2},$$
$$|a+b| \leqslant |a|+|b|.$$

Proof of (iv). We shall be content with noting that the result is obvious geometrically:

Exercise. Try to write out a non-geometrical proof of (iv), separating the cases $x \geqslant 0$ and $x < 0$.

Example 1. Express the set $\{x \in \mathbf{R}: |x+1| < 2\}$ as an interval and the set $\{x \in \mathbf{R}: |x+1| > 2\}$ as a union of intervals.

$$|x+1| < 2 \Leftrightarrow -2 < x+1 < 2 \quad \text{(by property (iv))}$$
$$\Leftrightarrow -3 < x < 1.$$

Thus $$\{x \in \mathbf{R}: |x+1| < 2\} = (-3, 1).$$

Also, $|x+1| > 2 \Leftrightarrow \begin{cases} \text{either} & x+1 > 2 \quad \text{(when } x+1 \geqslant 0) \\ \text{or} & x+1 < -2 \quad \text{(when } x+1 < 0) \end{cases}$

$\Leftrightarrow$ either $x > 1$ or $x < -3$.

Hence $\{x \in \mathbf{R}: |x+1| > 2\} = \{x \in \mathbf{R}: x > 1\} \cup \{x \in \mathbf{R}: x < -3\}$
$= (1, \infty) \cup (-\infty, -3).$

Example 2. Express the set

$$S = \left\{x \in \mathbf{R}: 0 < x - \frac{4}{x} \leqslant 3\right\}$$

as a union of intervals.

Inequalities (or inequations) of this type involving rational functions are probably best dealt with by tables of sign.

$$S = \left\{ x \in \mathbf{R} : 0 < x - \frac{4}{x} \right\} \cap \left\{ x \in \mathbf{R} : x - \frac{4}{x} \leqslant 3 \right\},$$
$$= U \cap V, \text{ say.}$$

For U:　$x - \dfrac{4}{x} = \dfrac{x^2-4}{x} = \dfrac{(x+2)(x-2)}{x}$; the "critical" values here are -2, 0, 2, the zeros of numerator and denominator in increasing order. We draw up a table of sign as follows:

$x:$	$\rightarrow$	-2	$\rightarrow$	0	$\rightarrow$	2	$\rightarrow$
$x+2:$	$-$	0	$+$	$+$	$+$	$+$	$+$
$x-2:$	$-$	$-$	$-$	$-$	$-$	0	$+$
$x:$	$-$	$-$	$-$	0	$+$	$+$	$+$
$\dfrac{x^2-4}{x}:$	$-$	0	$+$	∞	$-$	0	$+$

The symbol ∞ is placed under $x = 0$ merely to indicate that $\dfrac{x^2-4}{x}$ is not defined at $x = 0$ and $\left|\dfrac{x^2-4}{x}\right|$ becomes unbounded as $x \to 0$.

Here $\dfrac{x^2-4}{x} > 0$ holds when $-2 < x < 0$ or $x > 2$, and so

$$U = \{x \in \mathbf{R} : -2 < x < 0 \text{ or } x > 2\}.$$

For V: We have to consider $x - \dfrac{4}{x} - 3 \leqslant 0$.

Since　　　　$x - \dfrac{4}{x} - 3 = \dfrac{x^2 - 3x - 4}{x} = \dfrac{(x+1)(x-4)}{x},$

we have to consider　$\dfrac{(x+1)(x-4)}{x} \leqslant 0.$

Here the critical values in increasing order are -1, 0, 4, and the corresponding table of sign is

$x:$	$\rightarrow$	-1	$\rightarrow$	0	$\rightarrow$	4	$\rightarrow$
$x+1:$	$-$	0	$+$	$+$	$+$	$+$	$+$
$x-4:$	$-$	$-$	$-$	$-$	$-$	0	$+$
$x:$	$-$	$-$	$-$	0	$+$	$+$	$+$
$\dfrac{(x+1)(x-4)}{x}:$	$-$	0	$+$	∞	$-$	0	$+$

We see that $\dfrac{(x+1)(x-4)}{x} \leqslant 0$ holds when $x \leqslant -1$ or $0 < x \leqslant 4$.

Hence $V = \{x \in \mathbf{R} : x \leqslant -1 \text{ or } 0 < x \leqslant 4\}.$

The set $S = U \cap V$ is best shown graphically with U and V displayed in different colours or as indicated.

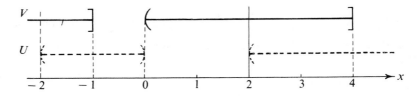

$$S = \{x \in \mathbf{R} : -2 < x \leqslant -1 \text{ or } 2 < x \leqslant 4\} = (-2, -1] \cup (2, 4].$$

Exercise. Express the set

$$S = \left\{x \in \mathbf{R} : 1 \leqslant \frac{3x}{x^2 - 4}\right\}$$

as a union of intervals and determine $S \cap [-1, 3]$.

EXERCISE 4

1. Show that the following are irrational:
   (i) $\sqrt[3]{2}$,   (ii) $\sqrt{3}$,   (iii) $\sqrt{5}$,   (iv) $\sqrt{6}$,   (v) $\sqrt{2} + \sqrt{3}$.
   [For (ii), assume that $\sqrt{3} = m/n$ where m, n are positive integers with no common factor > 1; similarly for (iii) and (iv).]
2. Prove that there is no rational number a such that $10^a = 2$, so that $\log_{10} 2$ is irrational.
3. Prove that, if a is irrational, then $1/a$ is irrational; prove also that, if b is rational, then $a + b$ is irrational.
4. Prove that, if a is irrational and $x + ya = z + ta$, where x, y, z, t are rational, then $x = z$ and $y = t$.
5. Prove that, if a is algebraic (i.e. is a root of a polynomial equation with rational co-efficients), then $-a$ and $a + 1$ are also algebraic.
6. For each of the following statements about real numbers, state whether it is true or false. If it is true, prove it; if it is false, give a counterexample.
   (i) If a and b are irrational, then $a + b$ is irrational.
   (ii) If a is irrational and $a + b$ is rational, then b is irrational.
   (iii) If a and b are transcendental, then ab is transcendental.
   (iv) No rational number is a transcendental number.
   (v) If a^2 is algebraic, then a is algebraic.
7. Show that the number $\sqrt{2} + \sqrt{3}$ is algebraic. [**Hint.** Put $x = \sqrt{2} + \sqrt{3}$.]
8. Determine whether the number $\sqrt{(3 + 2\sqrt{2})} - \sqrt{(3 - 2\sqrt{2})}$ is rational or irrational.

9. Express each of the following subsets of **R** in terms of intervals:

(i) $\left\{x \in \mathbf{R} : \dfrac{1-x}{x} \geqslant 0\right\}$,

(ii) $\left\{x \in \mathbf{R} : \dfrac{x+1}{3x-7} < 2\right\}$,

(iii) $\left\{x \in \mathbf{R} : x-1 > \dfrac{6}{x-2}\right\}$,

(iv) $\left\{x \in \mathbf{R} : \dfrac{x-2}{x} > \dfrac{x}{x+2}\right\}$,

(v) $\left\{x \in \mathbf{R} : \dfrac{x}{x-2} > \dfrac{3x-4}{x}\right\}$,

(vi) $\left\{x \in \mathbf{R} : -1 \leqslant \dfrac{x^2}{x-2} \leqslant 8\right\}$,

(vii) $\{x \in \mathbf{R} : |x+1| > 3x-7\}$,

(viii) $\{x \in \mathbf{R} : |x+1| \geqslant |x+2| + |x+3|\}$.

10. Show that, if a, b are positive real numbers, then
$$\sqrt{(ab)} \leqslant \tfrac{1}{2}(a+b).$$
[This says: the geometric mean of a, b is $\leqslant$ the arithmetic mean.]

11. Show, by using the triangle inequality, that, if a, b are real numbers, then

(i) $|a-b| \leqslant |a| + |b|$,   (ii) $|a+b| \geqslant |a| - |b|$,

(iii) $|a-b| \geqslant |a| - |b|$,   (iv) $||a| - |b|| \leqslant |a-b|$.

12. Show that, $\forall n \in \mathbf{N}$,
$$\sqrt{\left(\frac{2n-1}{2n+1}\right)} > \frac{2n-1}{2n} > \sqrt{\left(\frac{n-1}{n}\right)}.$$

13. Show that, if $p > 1$, then, $\forall n \in \mathbf{N}$,
$$\frac{1+p^2+p^4+\ldots+p^{2n}}{p+p^3+p^5+\ldots+p^{2n-1}} > \frac{n+1}{np}.$$

14. If a, b, c are real numbers and b, c are positive, show that $\dfrac{a+c}{b+c} \lesseqgtr \dfrac{a}{b}$ according as $b \lesseqgtr a$.

Deduce that $\dfrac{a+c}{b+c}$ lies between 1 and $\dfrac{a}{b}$.

15. Use Problem 14 to show that, if a, b are rational numbers with $a < b$, then there exists an irrational number x satisfying $a < x < b$. [In fact, there are infinitely many such irrational numbers.]

Induction
Finite Summations
Permutations
and Selections

1. Use of induction

We recall that in Section 1 of Chapter 4 we described how the Principle of Induction arises from Peano's axioms for the set $\mathbf{N}$ of positive integers. A slightly modified form of the principle can be stated as follows:

Suppose that $P(n)$ is a statement and that the universal set for n is $\mathbf{W}$, the set of non-negative integers. If

(i) $P(n_0)$ *is true for some* $n_0 \in \mathbf{W}$,

and

(ii) *for any* $k \geqslant n_0$, $P(k)$ *true* $\Rightarrow$ $P(k+1)$ *true, then* $P(n)$ *is true* $\forall n \geqslant n_0$.

[Our intuitive thinking about this situation is as follows:

By (i), $P(n)$ is true for $n = n_0$; and so,
by (ii), $P(n)$ is true for $n = n_0+1$; and so,
by (ii), $P(n)$ is true for $n = n_0+2$; and so,
by (ii), $P(n)$ is true for $n = n_0+3$, and so on.]

Usually n_0 will be 0 or a small positive integer.
In the examples that follow we shall use the notation:

$$\sum_{r=1}^{n} a_r \text{ for } a_1 + a_2 + \ldots + a_n, \quad \text{and} \quad \prod_{r=1}^{n} a_r \text{ for } a_1 a_2 \ldots a_n;$$

e.g.
$$1^2 + 2^2 + 3^2 + \ldots + n^2 = \sum_{r=1}^{n} r^2,$$

$$1.2 + 2.3 + 3.4 + \ldots + (n-1).n = \sum_{r=1}^{n-1} r(r+1),$$

$$1.3.5. \ldots .(2n-1) = \prod_{r=1}^{n} (2r-1).$$

70

Example 1. Prove each of the following results by induction.

(i) $\forall n \in \mathbf{N},\ \displaystyle\sum_{r=1}^{n} r = \tfrac{1}{2}n(n+1).$

(ii) $\forall n \in \mathbf{N},\ \displaystyle\sum_{r=1}^{n} r^2 = \tfrac{1}{6}n(n+1)(2n+1).$

(iii) $\forall n \in \mathbf{N},\ \displaystyle\sum_{r=1}^{n} r(r+1) = \tfrac{1}{3}n(n+1)(n+2).$

(iv) $\forall n \in \mathbf{N},\ \displaystyle\sum_{r=1}^{n} r^3 = \tfrac{1}{4}n^2(n+1)^2 = \left(\displaystyle\sum_{r=1}^{n} r\right)^2.$

(v) $\forall n \in \mathbf{N},\ \displaystyle\sum_{r=1}^{n} (2r-1) = n^2.$

(vi) $\forall n \in \mathbf{N},\ \displaystyle\prod_{r=1}^{n} (1+a^{2^{r-1}}) = \dfrac{1-a^{2^n}}{1-a} \quad (a \neq 1).$

Proof of (i), i.e. that, $\quad \forall n \geqslant 1,\ \displaystyle\sum_{r=1}^{n} r = \tfrac{1}{2}n(n+1).$

The statement involved in (i) is:

$$\sum_{r=1}^{n} r = \tfrac{1}{2}n(n+1). \qquad (1.1)$$

Consider $n = 1$: Each side of (1.1) has then the value 1, and so the statement (1.1) is true for $n = 1$.

Take any $k \geqslant 1$ and suppose that (1.1) is true for $n = k$, so that

$$\sum_{r=1}^{k} r = \tfrac{1}{2}k(k+1).$$

Consider $n = k+1$: $\displaystyle\sum_{r=1}^{k+1} r = \sum_{r=1}^{k} r + (k+1)$

$$= \tfrac{1}{2}k(k+1) + (k+1)$$

$$= \tfrac{1}{2}(k+1)(k+2)$$

$$= \tfrac{1}{2}(k+1)\{(k+1)+1\}.$$

Thus, if the statement (1.1) is true for any $k \geqslant 1$, it is also true for $n = k+1$. Hence, by induction, the result (i) is established.

Proof of (iii). The statement involved in (iii) is:

$$\sum_{r=1}^{n} r(r+1) = \tfrac{1}{3}n(n+1)(n+2). \tag{1.2}$$

Consider $n = 1$: Each side of (1.2) has then the value 2, and so the statement (1.2) is true for $n = 1$.

Take any $k \geqslant 1$ and suppose that (1.2) is true for $n = k$, so that

$$\sum_{r=1}^{k} r(r+1) = \tfrac{1}{3}k(k+1)(k+2).$$

Consider $n = k+1$:
$$\sum_{r=1}^{k+1} r(r+1) = \sum_{r=1}^{k} r(r+1) + (k+1)(k+2)$$
$$= \tfrac{1}{3}k(k+1)(k+2) + (k+1)(k+2)$$
$$= \tfrac{1}{3}(k+1)(k+2)(k+3)$$
$$= \tfrac{1}{3}(k+1)\{(k+1)+1\}\{(k+1)+2\}.$$

Thus, if the statement (1.2) is true for any $k \geqslant 1$, it is also true for $n = k+1$. Hence, by induction, (1.2) is true $\forall n \geqslant 1$, and so the result (iii) is established.

Exercise. Prove the results (ii), (iv), (v) and (vi) similarly.

Example 2. Prove that, if $x > 0$, then $\forall n \geqslant 2$,

$$(1+x)^n > 1 + nx, \tag{1.3}$$

and deduce that, $\forall n \geqslant 2, \left(1 + \dfrac{1}{n}\right)^n > 2.$

The statement involved in this example is:

$$(1+x)^n > 1 + nx. \tag{1.4}$$

Consider $n = 2$: $(1+x)^2 = 1 + 2x + x^2 > 1 + 2x$ since $x^2 > 0$, and so the statement (1.4) is true for $n = 2$.

Take any $k \geqslant 2$ and suppose that (1.4) is true for $n = k$, so that

$$(1+x)^k > 1 + kx.$$

Consider $n = k+1$: $(1+x)^{k+1} = (1+x)^k(1+x)$
$$> (1 + kx)(1+x), \text{ since } 1 + x > 0,$$
$$= 1 + (k+1)x + kx^2$$
$$> 1 + (k+1)x.$$

Thus, if the statement (1.4) is true for any $k \geqslant 2$, it is also true for $n = k+1$. Hence, by induction, (1.4) is true $\forall n \geqslant 2$, and so the result is established.

The last part of the example is obtained from (1.3) by putting

$$x = \frac{1}{n}.$$

2. The binomial coefficients and binomial theorem

For each integer $n \geqslant 1$ we shall, as usual, write $n!$ for the product

$$\prod_{r=1}^{n} r = 1.2.3.\ldots.n,$$

and read this as "factorial n" or "n factorial". Also we shall define $0!$ to be 1.

Now take an integer $n \geqslant 0$; for each r satisfying $0 \leqslant r \leqslant n$, we write

$$\binom{n}{r} = \frac{n!}{r!(n-r)!}.$$

The number $\binom{n}{r}$ is called the **binomial coefficient** for n and r. It is sometimes denoted by ^nC_r or c_r when n is understood.

When $n \geqslant 1$, we have:

$$\binom{n}{r} = \frac{n(n-1)\ldots(n-r+1).(n-r)!}{r!(n-r)!} = \frac{n(n-1)\ldots(n-r+1)}{r!}.$$

e.g. $\binom{5}{2} = \frac{5.4}{1.2} = 10; \quad \binom{9}{3} = \frac{9.8.7}{1.2.3} = 84; \quad \binom{11}{5} = \frac{11.10.9.8.7}{1.2.3.4.5} = 462.$

Note that, for each integer $n \geqslant 0$, $\binom{n}{0} = \binom{n}{n} = 1$.

It is not at first glance obvious that $\binom{n}{r}$ is an integer. We can prove this by induction on n using the following result:

If $n \geqslant 1$ and $r \geqslant 1$, then

$$\binom{n}{r-1} + \binom{n}{r} = \binom{n+1}{r}. \qquad (2.1)$$

Proof. $\displaystyle \binom{n}{r-1} + \binom{n}{r} = \frac{n!}{(r-1)!(n-r+1)!} + \frac{n!}{r!(n-r)!}$

$$= \frac{n!\{r+n-r+1\}}{r!(n-r+1)!}$$

$$= \frac{(n+1)!}{r!\{(n+1)-r\}!} = \binom{n+1}{r}.$$

Proof that $\binom{n}{r}$ is an integer. Consider the following statement:

If n is a non-negative integer, then each of the binomial coefficients $\binom{n}{r}$ given by $r = 0, 1, \ldots, n$ is an integer. $\qquad (2.2)$

For $n = 0$ we have one binomial coefficient $\binom{0}{0}$ which has the integral value 1. Thus (2.2) is true for $n = 0$.

For $n = 1$ we have two binomial coefficients $\binom{1}{0}$ and $\binom{1}{1}$ each of which has the integral value 1. Thus the statement (2.2) is also true for $n = 1$. We show by using induction that it is true for each $n \geqslant 1$.

Suppose that it is true for $n = k \geqslant 1$ so that $\binom{k}{r}$ is an integer for each $r = 0, 1, \ldots, k$. Consider the coefficients

$$\binom{k+1}{r}, \quad r = 0, 1, \ldots, k+1.$$

Each of $\binom{k+1}{0}$, $\binom{k+1}{k+1}$ is the integer 1.

For $r = 1, 2, \ldots, k$, we have, from (2.1),

$$\binom{k+1}{r} = \binom{k}{r-1} + \binom{k}{r}$$

$$= \text{(an integer)} + \text{(an integer)}, \text{ by the inductive hypothesis,}$$

$$= \text{an integer.}$$

Hence, if the statement (2.2) is true for any $k \geqslant 1$, then it is also true for $n = k+1$. It follows by induction that the statement is true for all positive integers and so for all non-negative integers since it is true also for $n = 0$.

For small values of n the binomial coefficients are conveniently displayed in a triangular array, called **Pascal's triangle** as follows:

n	Values of $\binom{n}{r}$, $0 \leqslant r \leqslant n$.

n														
0							1							
1						1		1						
2					1		2		1					
3				1		3		3		1				
4			1		4		6		4		1			
5		1		5		10		10		5		1		
6	1		6		15		20		15		6		1	
7	1	7	21	35	35	21	7	1						

The additive property (2.1) is illustrated by the fact that any integer in a row, from $n = 2$ onwards, which is not 1, is the sum of the two nearest integers in the row above. Also it is clear that the coefficients, for a given n, have a symmetric pattern in that they read the same from left to right as from right to left. This fact can be expressed in symbols as:

$$\binom{n}{r} = \binom{n}{n-r}, \tag{2.3}$$

and is easily proved; for,

$$\binom{n}{r} = \frac{n!}{r!(n-r)!} = \frac{n!}{(n-r)!\{n-(n-r)\}!} = \binom{n}{n-r}.$$

The binomial theorem for a positive integer n. We begin by examining the following algebraic identities:

$$x+y = x+y;$$

$$(x+y)^2 = (x+y)(x+y) = x^2+2xy+y^2;$$

$$(x+y)^3 = (x+y)(x+y)^2 = (x+y)(x^2+2xy+y^2)$$
$$= x^3+3x^2y+3xy^2+y^3;$$

$$(x+y)^4 = (x+y)(x+y)^3 = (x+y)(x^3+3x^2y+3xy^2+y^3)$$
$$= x^4+4x^3y+6x^2y^2+4xy^3+y^4;$$

$$(x+y)^5 = (x+y)(x+y)^4 = (x+y)(x^4+4x^3y+6x^2y^2+4xy^3+y^4)$$
$$= x^5+5x^4y+10x^3y^2+10x^2y^3+5xy^4+y^5,$$

and so on.

It will be noted that the coefficients in these expansions of $(x+y)^n$ for $n = 1, 2, 3, 4, 5$ are the binomial coefficients in Pascal's triangle corresponding to $n = 1, 2, 3, 4, 5$.

The binomial theorem is simply the statement that the corresponding result is true for any positive integer n.

Theorem 5.1 (The binomial theorem). *If n is a positive integer,*

$$(x+y)^n = \sum_{r=0}^{n} \binom{n}{r} x^{n-r} y^r \tag{2.4}$$

$$= \binom{n}{0} x^n + \binom{n}{1} x^{n-1} y + \binom{n}{2} x^{n-2} y^2 + \ldots$$

$$+ \binom{n}{r} x^{n-r} y^r + \ldots + \binom{n}{n-1} x y^{n-1} + \binom{n}{n} y^n$$

$$= x^n + n x^{n-1} y + \frac{n(n-1)}{2!} x^{n-2} y^2 + \ldots$$

$$+ \frac{n(n-1)\ldots(n-r+1)}{r!} x^{n-r} y^r + \ldots + n x y^{n-1} + y^n.$$

Proof. We use induction.

For $n = 1$, each side of (2.4) is $x+y$, and so the statement is true for $n = 1$.

Suppose now that the result is true for $n = k(\geqslant 1)$, so that

$$(x+y)^k = \binom{k}{0} x^k + \binom{k}{1} x^{k-1} y + \ldots + \binom{k}{r-1} x^{k-r+1} y^{r-1} +$$

$$+ \binom{k}{r} x^{k-r} y^r + \ldots + \binom{k}{k} y^k.$$

Consider $n = k+1$.

$$(x+y)^{k+1} = (x+y)(x+y)^k$$

$$= (x+y) \left\{ \binom{k}{0} x^k + \binom{k}{1} x^{k-1} y + \ldots + \binom{k}{r-1} x^{k-r+1} y^{r-1} + \right.$$

$$\left. + \binom{k}{r} x^{k-r} y^r + \ldots + \binom{k}{k} y^k \right\}$$

$$= x \left\{ \binom{k}{0} x^k + \binom{k}{1} x^{k-1} y + \ldots + \binom{k}{r} x^{k-r} y^r + \ldots + \binom{k}{k} y^k \right\}$$

$$+ y \left\{ \binom{k}{0} x^k + \ldots + \binom{k}{r-1} x^{k-r+1} y^{r-1} + \ldots + \binom{k}{k-1} x y^{k-1} + \binom{k}{k} y^k \right\}$$

$$= \binom{k}{0} x^{k+1} + \left\{ \binom{k}{0} + \binom{k}{1} \right\} x^k y + \left\{ \binom{k}{1} + \binom{k}{2} \right\} x^{k-1} y^2 + \ldots$$

$$+ \left\{ \binom{k}{r-1} + \binom{k}{r} \right\} x^{k+1-r} y^r + \ldots$$

$$+ \left\{ \binom{k}{k-1} + \binom{k}{k} \right\} xy^k + \binom{k}{k} y^{k+1}$$

$$= \binom{k+1}{0} x^{k+1} + \binom{k+1}{1} x^k y + \binom{k+1}{2} x^{k-1} y^2 + \ldots$$

$$+ \binom{k+1}{r} x^{(k+1)-r} y^r + \ldots + \binom{k+1}{k} xy^k + \binom{k+1}{k+1} y^{k+1},$$

using (2.1) and the facts that $\forall n \geqslant 0$, $\binom{n}{0} = 1$ and $\binom{n}{n} = 1$. Thus

$$(x+y)^{k+1} = \sum_{r=0}^{k+1} \binom{k+1}{r} x^{(k+1)-r} y^r.$$

Hence, if the result is true for any integer $k \geqslant 1$, then it is also true for $n = k+1$. It follows by induction that the result is true for all positive integers n.

Notes.

1. Since $x+y = y+x$, it follows that $\forall n \in \mathbf{N}$,

$$(x+y)^n = (y+x)^n \text{ and so } \sum_{r=0}^{n} \binom{n}{r} x^{n-r} y^r = \sum_{r=0}^{n} \binom{n}{r} x^r y^{n-r}.$$

2. As a special case of (2.4) we have:

$$\forall n \in \mathbf{N}, \ (1+x)^n = \sum_{r=0}^{n} \binom{n}{r} x^r \left(= \sum_{r=0}^{n} \binom{n}{r} x^{n-r} \right). \tag{2.5}$$

Example 1. Show that, $\forall n \geqslant 1$,

$$\sum_{r=0}^{n} \binom{n}{r} = 2^n \text{ and } \sum_{r=0}^{n} (-1)^r \binom{n}{r} = 0.$$

These results follow at once from (2.5) by putting $x = 1$ and $x = -1$, respectively.

Example 2. Find the term independent of x in the binomial expansion of

$$\left(\frac{4}{3}x^2 - \frac{3}{2x}\right)^9.$$

We have:

$$\left(\frac{4}{3}x^2 - \frac{3}{2x}\right)^9 = \sum_{r=0}^{9}\binom{9}{r}\left(\frac{4}{3}x^2\right)^{9-r}\left(\frac{-3}{2x}\right)^r$$

$$= \sum_{r=0}^{9}\binom{9}{r}\frac{2^{18-2r}}{3^{9-r}}\cdot(-1)^r\cdot\frac{3^r}{2^r}x^{18-3r}.$$

The term independent of x is given by $18 - 3r = 0$, i.e. by $r = 6$; it is

$$(-1)^6\binom{9}{6}\cdot 3^3 = \binom{9}{3}\cdot 27 = \frac{9.8.7}{1.2.3}\cdot 27 = 2268.$$

Example 3. By writing $(1+x)^{m+n}$, where m and n are positive integers, as a product of two factors, or otherwise, prove that, when $0 \leqslant r \leqslant m$ and $0 \leqslant r \leqslant n$, then

$$\binom{m+n}{r} = \sum_{s=0}^{r}\binom{m}{s}\binom{n}{r-s}.$$

We have:

$$(1+x)^{m+n} = (1+x)^m(1+x)^n.$$

Thus,

$$\sum_{r=0}^{m+n}\binom{m+n}{r}x^r = \left\{\sum_{s=0}^{m}\binom{m}{s}x^s\right\}\left\{\sum_{t=0}^{n}\binom{n}{t}x^t\right\}$$

$$= \left\{\binom{m}{0}+\binom{m}{1}x+\ldots+\binom{m}{r}x^r+\ldots\right\}$$

$$\times\left\{\binom{n}{0}+\binom{n}{1}x+\ldots+\binom{n}{r}x^r+\ldots\right\}$$

If we now compare coefficients of x^r in this identity, we have:

$$\binom{m+n}{r} = \binom{m}{0}\binom{n}{r}+\binom{m}{1}\binom{n}{r-1}+\binom{m}{2}\binom{n}{r-2}+\ldots+\binom{m}{r}\binom{n}{0},$$

and so

$$\binom{m+n}{r} = \sum_{s=0}^{r}\binom{m}{s}\binom{n}{r-s}.$$

3. Applications of the binomial coefficients to permutations and selections (or combinations).

It is convenient to introduce the symbol ^nP_r for

$$n(n-1)(n-2)\ldots(n-r+1),$$

where $n \in \mathbf{N}$ and $1 \leqslant r \leqslant n$.

Thus $^5P_3 = 5.4.3 = 60,$ $^9P_4 = 9.8.7.6 = 3024,$ etc.

Clearly $^nP_r = {}^nC_r.r!,$

since $^nC_r = \dfrac{n(n-1)\ldots(n-r+1)}{r!}.$

We now prove the following two important "combinatorial" results:

I. ^nP_r *is the number of distinct permutations of n different objects taken r at a time.*

II. ^nC_r *is the number of distinct selections of n different objects taken r at a time (i.e. order does not matter in this case).*

Proof of I: The first place can be filled in n ways.

For each of these choices there are $(n-1)$ ways of filling the second place. Thus there are $n(n-1)$ ways of filling the first two places.

For each of these $n(n-1)$ ways of filling the first two places there are $(n-2)$ ways of filling the third place. Thus there are $n(n-1)(n-2)$ ways of filling the first three places.

Proceeding in this way, and noting that the number of factors is always the same as the number of places filled, we see that the number of ways in which the r places may be filled is

$$n(n-1)(n-2)\ldots(n-r+1), \text{ i.e. } {}^nP_r.$$

Note. In particular,

$$^nP_n = n(n-1)(n-2)\ldots 1 = n!,$$

i.e. there are $n!$ permutations of n distinct objects n at a time.

For example, when $n = 3$, there are $3! = 6$ permutations. If we number the objects $1, 2, 3$, the 6 permutations are 123, 231, 312, 213, 321 and 132.

Proof of II. Let T_r denote the number of different selections of r objects from n different objects without regard to order.

In each of these selections the r objects can be arranged among themselves in ^rP_r, i.e. $r!$, ways.

Thus the number of different permutations of the n different objects taken r at a time is $T_r . r!$.

But this number is, by **I**, ^nP_r.

Hence,
$$T_r = \frac{1}{r!} {}^nP_r = \frac{n(n-1)\ldots(n-r+1)}{r!} = {}^nC_r = \binom{n}{r}.$$

For example, if $n = 4$ and if the objects are denoted by a, b, c, d, the $\binom{4}{3} = 4$ selections of these objects 3 at a time are $\{b, c, d\}$, $\{a, c, d\}$, $\{a, b, d\}$ and $\{a, b, c\}$.

There are $^4P_3 = 4.3.2 = 24$ permutations of these 4 objects taken 3 at a time, 6 permutations arising from each of the 4 selections.

Example 1. In how many ways may a committee of 7 be chosen from 8 men and 6 women so that it contains (i) 2 women, (ii) at most 2 women?

(i) 2 women can be selected from 6 in $\binom{6}{2}$ ways. The other 5 members of the committee have to be chosen from the 8 men, which can be done in $\binom{8}{5}$ ways. Each selection of men can be associated in turn with each of the $\binom{6}{2}$ selections of women. Hence the required number is

$$\binom{6}{2}.\binom{8}{5} = \frac{6.5}{1.2}.\frac{8.7.6}{1.2.3} = 840.$$

(ii) In this case the committee may contain 0, 1 or 2 women. Dealing, as in (i), with each of these cases in turn, we deduce that the required number is

$$\binom{6}{0}.\binom{8}{7}+\binom{6}{1}.\binom{8}{6}+\binom{6}{2}.\binom{8}{5} = 8+168+840 = 1016.$$

Example 2. How many words of 5 letters having 2 vowels can be made from 14 consonants and 5 vowels if no letter is used more than once in the same word? [Here a "word" is simply an ordered set of letters.]

The consonants can be chosen in $\binom{14}{3}$ ways and the vowels in $\binom{5}{2}$ ways. Thus the 5 letters to form a word can be selected in $\binom{14}{3}.\binom{5}{2}$

ways. In each selection the 5 letters can be permuted in $^5P_5 = 5!$ ways, giving 5! different words. Hence the total number of words is

$$\binom{14}{3} \cdot \binom{5}{2} \cdot 5! = \frac{14.13.12}{1.2.3} \cdot \frac{5.4}{1.2} \cdot 120 = 436,800.$$

Note. In problems, such as Example 2, which involve selection and arrangement, we usually deal with selection first and then with the arrangements of the selections.

So far we have been concerned with sets of *distinct* objects. We now consider **permutations of objects not all different**.

Theorem 5.2. *The number of distinct permutations of n objects when p are alike of one kind, q are alike of another kind, and the remaining $n-p-q$ are all different is*

$$\frac{n!}{p!q!}.$$

Proof. Let N denote the required number.

Now let the p like objects be replaced by p objects different from each other and from all the rest of the n objects. By permuting these p distinct objects, keeping the other objects fixed in position, we obtain from *each* of the N arrangements, $p!$ different arrangements. We now have $N \times p!$ arrangements.

Similarly, if we *now* replace the q like objects by q objects different from each other and from all the rest of the n objects, and permute these q different objects among themselves, we obtain $q!$ arrangements from *each* of the $N \times p!$ arrangements above, giving in all $N \times p! \times q!$ permutations.

But the objects are now all different, and so can be permuted in $n!$ distinct ways. Thus

$$N \times p! \times q! = n!, \text{ i.e. } N = \frac{n!}{p!q!}.$$

Extension. *The number of distinct permutations of n objects, p_1 alike of one kind, p_2 alike of a second kind, ..., p_r alike of an rth kind, and the remaining $n-(p_1+p_2 \ldots +p_r)$ all different is*

$$\frac{n!}{p_1!p_2! \ldots p_r!}.$$

The proof is obtained by an obvious extension of the argument used to prove Theorem 5.2.

Note. The number of permutations of n distinct objects r at a time, *when repetition is allowed*, is n^r, since there are always n ways of filling each place.

Example 3. How many different permutations are there of all the letters of the word *ALABAMA*? Of these permutations,

(a) in how many do vowels and consonants alternate,
(b) how many contain the word *LAMB*,
(c) in how many are the consonants in alphabetical order?

We can list the letters as follows:

$$
\begin{array}{llll}
A & B & L & M \\
A \\
A \\
A
\end{array}
$$

The total number of permutations is $\dfrac{7!}{4!} = 7.6.5 = 210$.

(a) The vowels must go in the odd places $A \times A \times A \times A$.

Thus, the required number is $3! = 6$, arising from the permutations of the set $\{B, L, M\}$.

(b) Here *LAMB* acts as one object, and we have to find the number of distinct permutations of the objects

$$
\begin{array}{ll}
A & (LAMB), \\
A \\
A
\end{array}
$$

i.e. a collection of 4 objects, of which 3 are alike. This number is

$$
\frac{4!}{3!} = 4.
$$

(c) The alphabetical order of the consonants is B, L, M. The four places for the As can be selected in $\dbinom{7}{4}$ ways; then B, L, M can be placed in one way only if the order B, L, M is preserved. Hence the required number is $\dbinom{7}{4} = \dbinom{7}{3} = 35$.

EXERCISE 5

1. Prove the following results by induction:

(i) $\forall n \in \mathbf{N}, \dfrac{1}{1.2} + \dfrac{1}{2.3} + \ldots + \dfrac{1}{n(n+1)} = \dfrac{n}{n+1}$;

(ii) $\forall n \in \mathbf{N}, \displaystyle\sum_{r=1}^{n} r(r+1)(r+2) = \tfrac{1}{4}n(n+1)(n+2)(n+3)$;

(iii) $\forall n \in \mathbf{N}, \displaystyle\sum_{r=0}^{n-1} ax^r = a\dfrac{x^n - 1}{x - 1}, \ (x \neq 1)$;

(iv) $\forall n \in \mathbf{N}, \displaystyle\sum_{r=1}^{n} (-1)^{r-1} r^2 = \tfrac{1}{2}(-1)^{n-1} n(n+1)$;

(v) $\forall n \in \mathbf{N}, \displaystyle\sum_{r=1}^{n} r^5 = \tfrac{1}{12}n^2(n+1)^2(2n^2 + 2n - 1)$;

(vi) $\forall n \in \mathbf{N}, \ 1.2^2 + 2.3^2 + \ldots + n.(n+1)^2 = \tfrac{1}{12}n(n+1)(n+2)(3n+5)$;

(vii) $\forall n \in \mathbf{N}, \displaystyle\sum_{r=1}^{n} \dfrac{(r+1)^2}{r(r+2)} = \dfrac{n(4n^2 + 15n + 13)}{4(n+1)(n+2)}$;

(viii) $\forall n \in \mathbf{N}, \displaystyle\sum_{r=1}^{2n} (-1)^r r^3 = n^2(4n+3)$;

(ix) $\forall n \in \mathbf{N}, \displaystyle\sum_{r=1}^{n} \dfrac{3r+2}{r(r+1)(r+2)}(-2)^{r-1} = \dfrac{1}{2} - \dfrac{(-2)^n}{(n+1)(n+2)}$;

(x) If $f_r(x) = \dfrac{x(x+1)\ldots(x+r-1)}{r!}$, then, $\forall n \in \mathbf{N}$,

$$\sum_{r=1}^{n} f_r(x) = f_n(x+1) - 1;$$

(xi) $\forall n \in \mathbf{N}, \displaystyle\sum_{r=0}^{n} x^r(1+x)^{n-r} = (1+x)^{n+1} - x^{n+1}$;

(xii) $\forall n \geqslant 4, \ 3^n > n^3$;

(xiii) $\forall n \geqslant 5, \ \dbinom{2n}{n} < 2^{2n-2}$;

(xiv) $\forall n \geqslant 2, \ \sqrt[n]{n} < 2 - \dfrac{1}{n}$;

(xv) $\forall n \geqslant 2, \ \dfrac{1}{\sqrt{(2n+1)}} > \dfrac{1.3.5.\ldots.(2n-1)}{2.4.6.\ldots.(2n)} > \dfrac{1}{2\sqrt{n}}$.

2. Prove that, for all integers $r \geqslant 3$,

$$\binom{r+2}{3} - \binom{r}{3} = r^2,$$

and hence show that

$$\forall n \geqslant 1, \ \sum_{r=1}^{n} r^2 = \tfrac{1}{6}n(n+1)(2n+1).$$

3. By using suitable binomial expansions, find approximations for (i) $(1 \cdot 995)^6$, (ii) $(1 \cdot 04)^{15}$ and (iii) $(1 \cdot 002)^{10}$.

4. (i) Show that

$$\{1+\sqrt{(1-x^2)}\}^5 + \{1-\sqrt{(1-x^2)}\}^5 = 32 - 40x^2 + 10x^4.$$

(ii) Show that $(2+\sqrt{3})^7 + (2-\sqrt{3})^7$ is rational, and evaluate it in simplest form.

5. Find
  (i) the coefficient of x^3 in the expansion of $(3x-2)^{12}$;

  (ii) the term independent of x in the expansion of $\left(x^2 + \dfrac{1}{x}\right)^9$;

  (iii) the middle term in the expansion of $\left(\dfrac{x}{y} - \dfrac{y}{x}\right)^8$;

  (iv) the coefficient of x^9 and the term independent of x in the expansion of $\left(\dfrac{1}{x^2} - x\right)^{18}$;

  (v) the largest coefficient in the expansion of $(1+2x)^8$;
  (vi) the expansion as far as the term in x^3 of $(1+2x-x^2)^6$;
  (vii) the coefficient of x^{3n} $(n \in \mathbf{N})$ in the expansion of

$$(1+x)^6(1-x)^2(1-x^3)^{2n}.$$

6. (i) In the expansion of $(3+\tfrac{1}{2}x)^n$, $(n \in \mathbf{N}, n \geqslant 7)$, the coefficients of x^7 and x^8 are equal; find n.
  (ii) Find the consecutive terms which have equal numerical coefficients in the expansion of

$$\left(\tfrac{1}{2}x^3 + \dfrac{1}{x}\right)^{20}.$$

  (iii) If the coefficients of three successive terms in the expansion of $(1+x)^n$, where n is a positive integer, are in the ratios $1:2:3$, find the value of n.
  (iv) Show that, when n is even, the largest coefficient in the expansion of $(1+x)^n$ occurs in the term containing $x^{n/2}$.
    Find the corresponding result when n is odd.

7. Show that, if $n \in \mathbf{N}$,

  (i) $$\binom{n}{0} + \binom{n}{2} + \binom{n}{4} + \ldots = 2^{n-1};$$

  (ii) $$\sum_{r=1}^{n} \binom{2n+1}{r} = 2^{2n} - 1;$$

  (iii) $$\sum_{r=0}^{n} \binom{n}{r}^2 = \dfrac{(2n)!}{(n!)^2};$$

(iv) $\qquad \sum_{r=0}^{n} (-1)^r \binom{n+1}{n-r} = 1;$

(v) $\qquad \binom{n}{1} + \binom{n}{3} + \ldots + \binom{n}{n} = 2^{n-1}, \quad$ when n is odd;

(vi) $\qquad \binom{n}{1} + \binom{n}{3} + \ldots + \binom{n}{n-1} = 2^{n-1}, \quad$ when n is even;

(vii) $\qquad \binom{n}{1} + 2\binom{n}{2} + \ldots + n\binom{n}{n} = n.2^{n-1};$

$$\left[\text{Differentiate the identity } (1+x)^n = \sum_{r=0}^{n} \binom{n}{r} x^r. \right]$$

(viii) $\binom{n}{1} - 2\binom{n}{2} + \ldots + (-1)^{n-1} n\binom{n}{n} = 0 \quad (n \geqslant 2);$

(ix) $\binom{n}{0} + \frac{1}{2}\binom{n}{1} + \frac{1}{3}\binom{n}{2} + \ldots + \frac{1}{n+1}\binom{n}{n} = \frac{1}{n+1}(2^{n+1} - 1).$

8. A staff-student committee of 6 is to be chosen from 6 staff and 12 students so that there is a majority of staff. In how many ways can this be done?

9. How many seven-digit numbers (beginning with a non-zero digit) can be formed by permuting the digits 0, 0, 2, 2, 2, 3, 5? How many of these are divisible by 5?

10. How many permutations are there of all the letters of the word *DIFFERENTIATE*? In how many of these does the word *TEN* appear?

11. How many different permutations are there of all the letters of the word *ABSTEMIOUS*? How many of these have the vowels in alphabetical order?

12. How many different permutations are there of all the letters of the word *RECEIVER*? In how many of these do vowels and consonants appear alternately? In how many do at least two *Es* come together? How many of the permutations begin and end with the same letter?

13. On each of $n (\geqslant 3)$ given straight lines $m (\geqslant 2)$ points are taken such that no other straight line can be drawn through any three of the mn points. How many triangles can be formed with these points as vertices?

14. How many permutations are there of all of the first 5 letters of the alphabet, no repetition of a letter being allowed? In how many of these do the 2 vowels not occur together?
    How many permutations are there of 4 letters from the first 5 letters of the alphabet, repetition of letters being allowed? How many of these contain exactly 2 vowels, not necessarily different, these vowels not occurring together?

15. State the number of different permutations of $p+q+r$ objects of which p are alike of one kind, q are alike of a second kind and r are alike of a third kind.
    Show that if $p = q+r+2$ there must in any permutation be at least 2 objects of the first kind adjacent. Find the number of permutations in which exactly 2 objects of the first kind are adjacent in the case when $p = 9$, $q = 3$, $r = 4$.

Elementary Number Theory

1. Introduction

Elementary number theory deals with basic properties of the set **N** of natural numbers or positive integers and includes ideas such as those concerned with the expression of an integer in essentially one way as a product of one or more prime numbers, the existence and determination of the least common multiple and greatest common divisor of a set of integers, the expressions for an integer in various scales and the congruence relation between integers. This work provides an introduction to basic results for the integers that are in constant mathematical use and gives motivation for many important later generalizations in mathematics. The subject has many unsolved problems that are easy to state but have so far not yielded to attacks by many great mathematicians. For example:

(i) Is the set of twin primes, i.e. prime numbers differing by 2, infinite?

(ii) Is the set of primes of the form $x^2 + 1$ for some integer x infinite?

(iii) Is Fermat's "Last Theorem" true, namely "The equation $x^n + y^n = z^n$, when n is an integer $\geqslant 3$, has no solution for integers x, y, z with none of these zero"?

2. The division identity

What is the remainder when 367 is divided by 13? If we write out the long-division procedure in this case we have a scheme of working as follows:

```
            28
       ┌──────
   13  │ 367
       │ 26
       ├──────
       │ 107
       │ 104
       ├──────
       │   3
       └──────
```

which shows that the required remainder is 3 and that the quotient in the division is 28. In terms of multiplication and addition of integers we have

$$367 = 28.13 + 3.$$

This equation implies that the largest number of 13s whose sum does not exceed 367 is 28 and that 367 exceeds this sum by 3.

In a similar way, we have

$$483 = 37.13 + 2,$$
$$585 = 45.13 + 0,$$

and so on.

The general result of which these equations are particular cases is the basic result of Elementary Number Theory. It can be stated as follows.

Theorem 2.1. *If b is a positive integer and a a non-negative integer, then there are unique non-negative integers q and r such that*

and
$$a = qb + r \qquad\qquad (2.1)$$
$$0 \leqslant r < b. \qquad\qquad (2.2)$$

Proof. On the real number line, the integers

$$a, a-b, a-2b, a-3b, \ldots$$

form a decreasing sequence of integers.

Since only finitely many of these are $\geqslant 0$, there is a unique integer $q \geqslant 0$ for which

and so
$$a - (q+1)b < 0 \leqslant a - qb,$$
$$0 \leqslant a - qb < b.$$

If we write $r = a - qb$, then
$$a = qb + r \quad \text{with} \quad 0 \leqslant r < b.$$

Thus we have found non-negative integers q, r for which (2.1) and (2.2) hold.

To show that q and r are unique, suppose that

$$a = q_1 b + r_1 \quad \text{with} \quad 0 \leqslant r_1 < b.$$

Then $r_1 = a - q_1 b$ and $0 \leqslant a - q_1 b < b$. It follows that

$$a - (q_1 + 1)b < 0 \leqslant a - q_1 b,$$

so that q_1 is the integer q determined above and

$$r_1 = a - qb = r.$$

We have thus proved the theorem.

Notes. 1. Equation (2.1) is called the **division identity for the division of** a **by** b; q is called the **quotient** and r the **remainder** in this division. The integer r is often called the **principal remainder** because it lies in the interval $0 \leqslant r < b$. We could use any suitable interval of length b for the remainder, e.g. $-\frac{1}{2}b \leqslant r < \frac{1}{2}b$.

2. Part of the importance of the theorem is the implication that, no matter what practical method we use for carrying out a division, e.g. long division, short division, synthetic division, etc., the quotient and principal remainder are independent of the method.

3. Extension of Theorem 2.1. *The statement of Theorem 2.1 holds with a any integer and no restriction on the integer q.*

Suppose that a is a negative integer so that $-a$ is a positive integer. By Theorem 2.1,

Thus
$$-a = q_1 b + r_1 \qquad \text{with} \quad 0 \leqslant r_1 \leqslant b - 1.$$
$$a = (-q_1)b + (-r_1) \quad \text{with} \quad -(b-1) \leqslant -r_1 \leqslant 0.$$

If $r_1 = 0$ we have the required result; if $r_1 > 0$, then

$$-(b-1) \leqslant -r_1 \leqslant -1 \quad \text{and so} \quad 1 \leqslant b - r_1 \leqslant b - 1.$$

Since
$$a = (-q_1 - 1)b + (b - r_1), \quad \text{we have}$$
$$a = qb + r \quad \text{with} \quad 0 \leqslant r < b,$$
where
$$q = -q_1 - 1 \quad \text{and} \quad r = b - r_1.$$

The uniqueness of q and r follows from that of q_1 and r_1.

4. When $r = 0$, i.e. when $a = qb$, we say that b is a *divisor* or *factor* of a or that b *divides* a and that a is a *multiple* of b.

If b divides a we write $b \mid a$; if b does not divide a we write $b \nmid a$. For example,
$$5 \mid 15 \quad \text{but} \quad 2 \nmid 5.$$

Explicit expressions for q, r in terms of integral part.

Suppose that x is a given real number and suppose that $n + 1$ denotes the smallest integer greater than x. Then this integer n is uniquely determined by x and $n \leqslant x < n + 1$, i.e. n is the largest integer less than or equal to x.

The integer n is called the **integral part** of the real number x and is denoted by $[x]$. The real number $t = x - [x]$ is called the **fractional part** of x; clearly

$$x = [x] + t \quad \text{and} \quad 0 \leqslant t < 1,$$

and $[x]$, t are uniquely determined by x.

For example, $\left[\dfrac{5}{2}\right] = 2$, $\left[-\dfrac{7}{4}\right] = -2$, $[\pi] = 3$, $[4 + \sqrt{2}] = 5$.

If we divide an integer a by a positive integer b, then

$$a = qb + r \quad \text{with} \quad 0 \leqslant r < b,$$

and so

$$\frac{a}{b} = q + \frac{r}{b} \quad \text{with} \quad 0 \leqslant \frac{r}{b} < 1.$$

Since q is an integer, it follows that $q = \left[\dfrac{a}{b}\right]$ and so $r = a - b\left[\dfrac{a}{b}\right]$.

For example, the quotient in the division of 367 by 13 is $\left[\dfrac{367}{13}\right] = 28$.

EXERCISE 6.1

1. Find the quotient and principal remainder for the divisions corresponding to the values of a and b in the following table, and verify their values in terms of integral parts.

	a	b
(i)	193	17
(ii)	897	101
(iii)	-581	23
(iv)	-682	31
(v)	2479	19

2. If $a \mid b$ and $b \mid c$, show that $a \mid c$; if a, b are positive integers such that $a \mid b$ and $b \mid a$, what can you say about a and b?

3. Show that, if a_1 and a_2 are multiples of an integer b, then $xa_1 + ya_2$ is also a multiple of b for any pair of integers x, y.

4. Show that, if $b_1 b_2 \mid a$, then $b_1 \mid a$ and $b_2 \mid a$. Is the converse of this result true? If so, prove it; if not, produce a counter-example.

5. Prove the following results on integral parts.

(i) If n is an integer, $[x + n] = [x] + n$.

(ii) $[x] + [-x] = 0$ or -1 according as x is an integer or not.

(iii) If n is a positive integer, then $\left[\dfrac{x}{n}\right] = \left[\dfrac{[x]}{n}\right]$.

(iv) For any real numbers x, y, $[x + y] \geqslant [x] + [y]$.

6. In the case of the division identity,

$$a = qb + r \quad \text{with} \quad -\tfrac{1}{2}b \leqslant r < \tfrac{1}{2}b,$$

show that

$$q = \left[\frac{a}{b} + \frac{1}{2}\right].$$

3. Number bases

The integer that is represented as 4703 in the standard decimal notation is the integer consisting of 4 thousands, 7 hundreds, zero tens and 3 units; in other words,

$$4703 = 4.10^3 + 7.10^2 + 0.10 + 3.$$

The expression on the right-hand side is of the form $4x^3 + 7x^2 + 0x + 3$ with $x = 10$, i.e. it is a polynomial in 10 whose coefficients are non-negative integers < 10.

The same integer can be expressed as

$$4703 = 1.8^4 + 1.8^3 + 1.8^2 + 3.8 + 7,$$

i.e. as a polynomial in 8 with non-negative coefficients < 8. This is called the representation of the integer **in the scale of** 8 or **to base** 8.

In fact any integer $g > 1$ can be used as a base for a unique representation of each positive integer as a polynomial in g with non-negative coefficients $< g$. We prove that this statement is true and indicate how to obtain the representation.

Theorem 3.1. *If g is an integer > 1, every positive integer a can be expressed uniquely in the form*

$$a = c_n g^n + c_{n-1} g^{n-1} + \ldots + c_1 g + c_0,$$

where the c_i are non-negative integers $\leqslant g - 1$.

Proof. On dividing a by g we have

$$a = a_1 g + c_0 \quad \text{with} \quad 0 \leqslant c_0 \leqslant g - 1 \quad \text{and} \quad a_1 \geqslant 0.$$

Similarly, dividing a_1 by g we have

$$a_1 = a_2 g + c_1 \quad \text{with} \quad 0 \leqslant c_1 \leqslant g - 1 \quad \text{and} \quad a_2 \geqslant 0,$$

and so

$$a = a_2 g^2 + c_1 g + c_0.$$

Now dividing a_2 by g we have

$$a_2 = a_3 g + c_2 \quad \text{with} \quad 0 \leqslant c_2 \leqslant g - 1 \quad \text{and} \quad a_3 \geqslant 0,$$

and so

$$a = a_3 g^3 + c_2 g^2 + c_1 g + c_0.$$

Since the sequence $\{1, g, g^2, g^3, g^4, \ldots\}$ is a sequence of positive integers in strictly increasing order, there is a power of g, g^n, say, with $n \geqslant 0$, for which

$$g^n \leqslant a < g^{n+1}. \tag{3.1}$$

If we proceed by successive divisions as above we reach the stage in which

$$a = a_n g^n + c_{n-1} g^{n-1} + \ldots + c_1 g + c_0, \tag{3.2}$$

with

$$0 \leqslant c_i \leqslant g-1 \quad \text{for each} \quad i = 0, 1, \ldots, n-1 \quad \text{and} \quad a_n \geqslant 0.$$

From (3.1) and (3.2), it follows that $0 < a_n \leqslant g-1$. If we denote a_n by c_n, we have an expression for a in the required form. The uniqueness of the expression follows from the fact that a quotient and principal remainder are uniquely determined in any division.

Note. The expression for a in terms of g is called the **representation of a in the scale of g** and g is called the **base** of the scale. We write

$$a = (c_n c_{n-1} \ldots c_1 c_0)_g, \quad \text{or simply} \quad a = c_n c_{n-1} \ldots c_1 c_0$$

if the base is clearly understood.

The g possible values $0, 1, 2, \ldots, g-1$ for the coefficients $c_0, \ldots, c_n$ are called the **digits** for the scale.

Example. Express the integer 71371 (in decimal notation) to base 7.

$$\begin{aligned}
71371 &= 10195.7 + 6 \\
&= (1456.7 + 3).7 + 6 \\
&= (208.7 + 0).7^2 + 3.7 + 6 \\
&= (29.7 + 5).7^3 + 0.7^2 + 3.7 + 6 \\
&= (4.7 + 1).7^4 + 5.7^3 + 0.7^2 + 3.7 + 6 \\
&= 4.7^5 + 1.7^4 + 5.7^3 + 0.7^2 + 3.7 + 6 \\
\therefore \quad 71371 &= (415036)_7.
\end{aligned}$$

For computing work, bases such as 2, 8 and 12 have considerable importance.

EXERCISE 6.2

1. Express each of the following integers (in decimal notation) in the scales of 2, 5, 7 and 9:
        (i) 81,   (ii) 579,   (iii) 1064,   (iv) 15287.

      Express the same integers in the scale of 12, denoting the eleventh and twelfth digits by t and e respectively.

2. Find the decimal form of each of the following integers:
        (i) $(10011101)_2$,   (ii) $(210102)_3$,   (iii) $(42031)_5$,   (iv) $(708010)_9$.

3. Perform the following additions in the scale of 7:
      (i)   305426 ,   (ii)   614103 ,   (iii)   55555
           + 24312   +203615   +22222

4. If $9322 = (4321)_g$, find the base g.

5. What is the decimal form of the largest integer representable by four digits in the scale of 5?

4. Least common multiple, greatest common divisor, euclidean algorithm

Suppose that we consider the set of integers $\{10, 24, 45\}$; any integer that is divisible by each member of the set is called a **common multiple** of the set. Clearly $10800 = 10.24.45$ is a positive common multiple. Consequently the set of positive common multiples is non-empty and so has a least member. This least member is called the **least common multiple** (l.c.m. for short) of 10, 24, 45 and will be denoted by [10, 24, 45]. The best way of finding the l.c.m. is to use factorisation in terms of prime numbers; since $10 = 2.5$, $24 = 2^3.3$ and $45 = 3^2.5$, the l.c.m. is $2^3.3^2.5$, i.e. 360. We shall return to this later when prime factorisation has been discussed [see Problem **9** of Exercise **6**]. At this stage we shall show, with the help of the division identity, that the set of all common multiples is completely determined when the l.c.m. is known. Since it is as easy to deal with this in the general case we shall consider a finite set $\{a_1, a_2, \ldots, a_n\}$ of non-zero integers and let $m = [a_1, a_2, \ldots, a_n]$ denote their l.c.m., i.e. their least positive common multiple. [Note that m exists since the set of their positive common multiples contains $|a_1 a_2 \ldots a_n|$ and so is non-empty.]

Theorem 4.1. *If m denotes the l.c.m. of the integers $a_1, a_2, \ldots, a_n$, then an integer l is a common multiple of $a_1, a_2, \ldots, a_n$ if and only if $m|l$.*

Proof. If $m|l$, then $a_i|m$ and $m|l$ so that $a_i|l$ for each $i = 1, 2, \ldots, n$. Thus l is a common multiple of $a_1, a_2, \ldots, a_n$.

Suppose now conversely that l is a common multiple of $a_1, a_2, \ldots, a_n$; we have to show that $m|l$. On dividing l by m we have

$$l = qm + r \quad \text{with} \quad 0 \leqslant r < m.$$

From this, $r = l - qm$. Now, for each $i = 1, 2, \ldots, n$,

$$a_i|m \quad \text{and} \quad a_i|l, \quad \text{so that} \quad a_i|(l - qm), \quad \text{i.e. } a_i|r.$$

Hence r is a non-negative common multiple of $a_1, a_2, \ldots, a_n$ that is $< m$. Since m is the least positive common multiple of $a_1, a_2, \ldots, a_n$, it follows that $r = 0$ and consequently that $m|l$.

Note. It follows that the set of all common multiples of $a_1, a_2, \ldots, a_n$ is $\{km : k \in \mathbf{Z}\}$, where m is their l.c.m. and $\mathbf{Z}$ is the set of all integers.

For example, the set of all common multiples of 10, 24, 45 is

$$\{360k : k \in \mathbf{Z}\} \quad \text{i.e.} \quad \{0, \pm 360, \pm 720, \pm 1080, \ldots\}.$$

As a sort of "dual" to common multiple we have the idea of common divisor. If $\{a_1, a_2, \ldots, a_n\}$ is a finite set of non-zero integers, any integer that is a divisor of each of $a_1, a_2, \ldots, a_n$ is called a **common divisor** of $a_1, a_2, \ldots, a_n$. Clearly 1 is a member of the set of *positive* common divisors and no common divisor can be greater than the smallest of the positive integers $|a_1|, |a_2|, \ldots, |a_n|$. Thus $a_1, a_2, \ldots, a_n$ have a **greatest common divisor** (g.c.d. for short); this is denoted by $(a_1, a_2, \ldots, a_n)$. For example, $(10, 24, 45) = 1$ and $(18, 24, 42, 66) = 6$. When the prime factors of the integers are known, the g.c.d. can be quickly obtained [see Problem 9 of Exercise 6], but there is no obvious method of obtaining the g.c.d. of even two given integers when the prime factors are not known. The simplest method involves successive uses of the division identity.

Consider $(5917, 1403)$ in which 5917 is the larger of the two integers involved.

$$\begin{array}{llll}
\text{Dividing} & 5917 \text{ by } 1403 \text{ we have:} & 5917 = 4.1403 + 305; & (4.1)_1 \\
\text{dividing} & 1403 \text{ by } 305 \text{ we have:} & 1403 = 4.305 + 183; & (4.1)_2 \\
\text{dividing} & 305 \text{ by } 183 \text{ we have:} & 305 = 1.183 + 122; & (4.1)_3 \\
\text{dividing} & 183 \text{ by } 122 \text{ we have:} & 183 = 1.122 + 61; & (4.1)_4 \\
\text{dividing} & 122 \text{ by } 61 \text{ we have:} & 122 = 2.61 + 0. & (4.1)_5
\end{array}$$

We now show that 61, the *last positive remainder* in this succession of division identities, is in fact $(5917, 1403)$, i.e. the g.c.d. of 5917 and 1403. Let d denote $(5917, 1403)$.

By working backwards from $(4.1)_5$ to $(4.1)_1$ in turn we see that 61 divides 122, 183, 305, 1403 and 5917, and so 61 is a common divisor of 5917 and 1403. Thus $61 \leqslant d$ since d is the largest common divisor.

From $(4.1)_1$, $305 = 5917 - 4.1403$, so that $d|305$ since $d|5917$ and $d|1403$.

By considering $(4.1)_1$ to $(4.1)_4$ in turn in this way, we see that d is a divisor of 305, 183, 122 and 61. It follows that $d \leqslant 61$.

From the inequations $61 \leqslant d \leqslant 61$ we deduce that $d = 61$, i.e.

$$(5917, 1403) = 61.$$

[In fact, $5917 = 61.97$ and $1403 = 61.23$.]

Note. Using $(4.1)_4$ to $(4.1)_1$ we have:

$$\begin{aligned}
61 &= 183 + (-1).122 \\
&= 183 + (-1).[305 + (-1).183] = (-1).305 + 2.183 \\
&= (-1).305 + 2.[1403 + (-4).305] = 2.1403 + (-9).305 \\
&= 2.1403 + (-9).[5917 + (-4).1403] = (-9).5917 + (38).1403,
\end{aligned}$$

so that

$$(5917, 1403) = x.5917 + y.1403,$$

where $x = -9$ and $y = 38$. In other words, we have been able, by using the succession of divisions, to express the g.c.d. of 5917 and 1403 as a sum of multiples of these two integers.

The step-by-step division process illustrated by the above example is called the **euclidean algorithm** (after the Greek mathematician Euclid) for the integers.

We can establish the process for any pair of positive integers a_1, a_2.

We can suppose that $a_1 \geqslant a_2$, by renaming the integers if necessary. On division,

$$a_1 = q_1 a_2 + a_3 \quad \text{with} \quad 0 \leqslant a_3 < a_2. \qquad (4.2)_1$$

If $a_3 = 0$, then $a_2 | a_1$ and clearly $(a_1, a_2) = a_2$.
If $a_3 > 0$, we proceed to divide a_2 by a_3, obtaining

$$a_2 = q_2 a_3 + a_4 \quad \text{with} \quad 0 \leqslant a_4 < a_3 < a_2 < a_1. \qquad (4.2)_2$$

If $a_4 > 0$, we proceed to divide a_3 by a_4, obtaining

$$a_3 = q_3 a_4 + a_5 \quad \text{with} \quad 0 \leqslant a_5 < a_4 < a_3 < a_2 < a_1, \qquad (4.2)_3$$

and so on.

In this way we obtain a strictly decreasing sequence of non-negative integers $a_1 > a_2 > a_3 > a_4 > a_5 > \ldots$. But the set of non-negative integers $< a_1$ is finite; thus the sequence must terminate after a finite number of steps. In other words there is a positive integer r such that the last two steps in the above division process are

$$a_{r-2} = q_{r-2} a_{r-1} + a_r \quad \text{with} \quad 0 < a_r < a_{r-1} < \ldots < a_2 < a_1, \qquad (4.2)_{r-2}$$
$$a_{r-1} = q_{r-1} a_r + 0. \qquad (4.2)_{r-1}$$

As in the earlier illustrative example, we show that the last non-negative remainder, namely a_r, is the g.c.d. of a_1 and a_2; i.e. if

$$d = (a_1, a_2), \quad \text{then} \quad d = a_r.$$

By working backwards from $(4.2)_{r-1}$ to $(4.2)_1$ in turn, we see that a_r divides $a_{r-1}, a_{r-2}, \ldots, a_2$ and a_1. Thus a_r is a common divisor of a_1, a_2, and so $a_r \leqslant d$ since d is the largest common divisor of a_1, a_2.

Similarly, by considering $(4.2)_1$ to $(4.2)_{r-2}$ in turn, we see that d divides $a_3, a_4, \ldots, a_{r-1}$ and a_r. It follows that $d \leqslant a_r$, and hence that $(a_1, a_2) = d = a_r$.

Also, from $(4.2)_{r-2}$, $a_r = a_{r-2} + (-q_{r-2}) \cdot a_{r-1}$; and, using $(4.2)_{r-3}$,

$$a_r = a_{r-2} + (-q_{r-2}) \cdot [a_{r-3} + (-q_{r-3} \cdot a_{r-2}]$$
$$= (-q_{r-2}) \cdot a_{r-3} + (1 + q_{r-2} q_{r-3}) \cdot a_{r-2}.$$

Proceeding backwards through the equations $(4.2)_{r-2}$ to $(4.2)_1$ in this way, as in the earlier worked example, we obtain an equation of the form

$$(a_1, a_2) = a_r = xa_1 + ya_2, \qquad (4.3)$$

where x, y are integers, i.e. the g.c.d. can be expressed as a sum of multiples of the integers involved.

Using (4.3) we obtain the following result which is an analogue for the case $n = 2$ of Theorem 4.1 for l.c.m.

Theorem 4.2. If $d = (a_1, a_2)$, then an integer k is a common divisor of the integers a_1, a_2 if and only if $k \mid d$.

Proof. If $k \mid d$, then $k \mid d$ and $d \mid a_i$ $(i = 1, 2)$. Thus $k \mid a_i$ $(i = 1, 2)$ and so k is a common divisor of a_1 and a_2.

Suppose now conversely that k is a common divisor of a_1 and a_2. Since $d = (a_1, a_2)$, there are integers x, y such that $d = xa_1 + ya_2$. But $k \mid a_1$ and $k \mid a_2$; so $k \mid (xa_1 + ya_2)$, i.e. $k \mid d$.

Example. The positive common divisors of 12 and 30 are the positive divisors of (12, 30), i.e. of 6, and so are 1, 2, 3, 6.

So far we have considered only two integers. The following result shows that in finding the g.c.d. of more than two integers we need only the process involving two integers.

Theorem 4.3. If $a_1, a_2, \ldots, a_n$ are given non-zero integers, and if $d_1 = a_1$, $d_2 = (d_1, a_2)$, $d_3 = (d_2, a_3), \ldots$, $d_n = (d_{n-1}, a_n)$ and $d = (a_1, a_2, \ldots, a_n)$, then $d = d_n$.

Proof. We first note that $d \mid a_i$ $(i = 1, 2, \ldots, n)$. Thus

$$d \mid d_1 \text{ and } d \mid a_2 \quad \text{and so, by Theorem 4.2,} \quad d \mid d_2 = (d_1, a_2);$$

$$d \mid d_2 \text{ and } d \mid a_3 \quad \text{and so, by Theorem 4.2,} \quad d \mid d_3 = (d_2, a_3);$$

$$\cdot \ \cdot \ \cdot \ \cdot \ \cdot \ \cdot \ \cdot \ \cdot \ \cdot \ \cdot \ \cdot \ \cdot \ \cdot \ \cdot \ \cdot \ \cdot \ \cdot \ \cdot \ \cdot$$

$$d \mid d_{n-1} \text{ and } d \mid a_n \quad \text{and so, by Theorem 4.2,} \quad d \mid d_n = (d_{n-1}, a_n).$$

Hence $d \leqslant d_n$.

On the other hand, since $d_n = (d_{n-1}, a_n)$, we have:

$$d_n \mid a_n \text{ and } d_n \mid d_{n-1} = (d_{n-2}, a_{n-1}).$$

But $d_{n-1} \mid a_{n-1}$ and $d_{n-1} \mid d_{n-2}$, and so

$$d_n \mid a_n, \ d_n \mid a_{n-1} \text{ and } d_n \mid d_{n-2} = (d_{n-3}, a_{n-2}).$$

But $d_{n-2} \mid a_{n-2}$ and $d_{n-2} \mid d_{n-3}$, and so

$$d_n \mid a_n, \ d_n \mid a_{n-1}, \ d_n \mid a_{n-2} \text{ and } d_n \mid d_{n-3} = (d_{n-4}, a_{n-3}).$$

Proceeding in this way, considering $d_n, d_{n-1}, \ldots, d_2$ in turn, we see that $d_n | a_i$ ($i = 1, 2, \ldots, n$), i.e. that d_n is a common divisor of $a_1, a_2, \ldots, a_n$. Thus $d_n \leqslant d$, the largest common divisor of $a_1, a_2, \ldots, a_n$.

From the pair of inequations $d \leqslant d_n \leqslant d$, it now follows that $d = d_n$.

Note. For each $r = 1, 2, \ldots, n$, $d_r = (a_1, a_2, \ldots, a_r)$. [Explain why.]

Example.   (i) $(12, 30, 21) = ((12, 30), 21) = (6, 21) = 3$;
           (ii) $(12, 30, 21, 7) = ((12, 30, 21), 7) = (3, 7) = 1$.

Corollary to Theorem 4.3. *If* $d = (a_1, a_2, \ldots, a_n)$, *then there are integers* $x_1, x_2, \ldots, x_n$ *such that*

$$d = x_1 a_1 + x_2 a_2 + \ldots + x_n a_n,$$

i.e. the g.c.d. is a sum of multiples of the integers $a_1, a_2, \ldots, a_n$.

Proof. In the notation of Theorem 4.3, there are, by (4.3), integers x_n and t_1 such that

$$d = d_n = x_n a_n + t_1 d_{n-1}. \tag{4.4}$$

Since $d_{n-1} = (d_{n-2}, a_{n-1})$, again using (4.3), there are integers y_1, y_2 such that

$$d_{n-1} = y_1 a_{n-1} + y_2 d_{n-2}. \tag{4.5}$$

Combining (4.4) and (4.5) we have

$$d = x_n a_n + x_{n-1} a_{n-1} + t_2 d_{n-2},$$

where x_{n-1} and t_2 are integers.

Proceeding similarly with

$$d_{n-2} = (d_{n-3}, a_{n-2}), \ldots, d_2 = (d_1, a_2) = (a_1, a_2),$$

we obtain the required result.

Using this result we can prove the following generalization of Theorem 4.2.

Theorem 4.4. *If* $d = (a_1, a_2, \ldots, a_n)$, *then an integer* k *is a common divisor of* $a_1, a_2, \ldots, a_n$ *if and only if* $k | d$.

Try to write out a proof which is a direct extension of that of Theorem 4.2.

Definition. Two integers a_1, a_2 for which $(a_1, a_2) = 1$ are said to be **relatively prime** or **prime to each other**. For such integers we can find integers x, y such that $xa_1 + ya_2 = 1$.

If a set of integers $\{a_1, a_2, \ldots, a_n\}$ is such that $(a_1, a_2, \ldots, a_n) = 1$, the set is said to be **relatively prime**; e.g. $\{4, 6, 7\}$ is such a set. Since $(4, 6) \neq 1$, the example shows that a relatively prime set of integers may contain pairs which are not prime to each other.

EXERCISE 6.3

1. For the pairs of integers a_1, a_2 given in the following table use the euclidean algorithm to determine $d = (a_1, a_2)$ and hence find integers x, y such that $d = xa_1 + ya_2$. Write down the positive common divisors of each pair of integers.

a_1	a_2
1147	851
1219	901
5213	2867
4277	2821

2. Use the results of problem **1** to determine
   (i) $(1147, 851, 407)$,   (ii) $(1219, 901, 1000)$,
   (iii) $(5213, 2867, a)$ for any positive integer a,
   (iv) $(4277, 2821, 845)$.
   Express each g.c.d. as a sum of multiples of the integers involved.
3. Show that, if a, b are positive integers with $a > b$, then
$$(a + b, a - b) = (a, b) \quad \text{or} \quad 2(a, b).$$
   [*Hint.* Let $d = (a, b), d_1 = (a + b, a - b)$ and use Theorem 4.2 to show that $d \mid d_1$ and $d_1 \mid 2d$.]
   Determine when the right-hand side is (a, b) and when $2(a, b)$.
4. Show that, if a, b, k are positive integers, then
$$(a + kb, b) = (a, b).$$
5. Show that $(a, b) = (a, b, a + b)$.
6. Given positive integers a, b, show that there are integers x, y such that $(x, y) = a$ and $[x, y] = b$ if and only if $a \mid b$.

5. Primes and the fundamental theorem of arithmetic

Any positive integer $p > 1$ whose only positive divisors are 1 and p itself is called a **prime number** or simply a **prime**. Any positive integer > 1 which is not a prime is said to be **composite**; such a number can be written as a product ab with $a > 1$ and $b > 1$. The first few primes are $2, 3, 5, 7, 11, 13, 17, \ldots$. Any prime which divides a given integer is called a **prime divisor** of that integer. To test whether an integer n is a prime or not we need only test for a possible prime divisor up to $\sqrt{n}$; for, if $ab = n$, then either $a \leqslant \sqrt{n}$ or $b \leqslant \sqrt{n}$. For example, to show that 67 is a prime we need only check, since $8 < \sqrt{67} < 9$, that none of the primes 2, 3, 5, 7 is a prime divisor of 67.

All the primes up to some given upper limit, e.g. 100, can be sifted out by the so-called **Sieve of Eratosthenes.** We list all the positive integers $\leqslant 100$:

1̸	2	3	4̸	5	6̸	7	8̸	9̸	1̸0̸
11	1̸2̸	13	1̸4̸	1̸5̸	1̸6̸	17	1̸8̸	19	2̸0̸
2̸1̸	2̸2̸	23	2̸4̸	2̸5̸	2̸6̸	2̸7̸	2̸8̸	29	3̸0̸
31	3̸2̸	3̸3̸	3̸4̸	3̸5̸	3̸6̸	37	3̸8̸	3̸9̸	4̸0̸
41	4̸2̸	43	4̸4̸	4̸5̸	4̸6̸	47	4̸8̸	4̸9̸	5̸0̸
5̸1̸	5̸2̸	53	5̸4̸	5̸5̸	5̸6̸	5̸7̸	5̸8̸	59	6̸0̸
61	6̸2̸	6̸3̸	6̸4̸	6̸5̸	6̸6̸	67	6̸8̸	6̸9̸	7̸0̸
71	7̸2̸	73	7̸4̸	7̸5̸	7̸6̸	7̸7̸	7̸8̸	79	8̸0̸
8̸1̸	8̸2̸	83	8̸4̸	8̸5̸	8̸6̸	8̸7̸	8̸8̸	89	9̸0̸
9̸1̸	9̸2̸	9̸3̸	9̸4̸	9̸5̸	9̸6̸	97	9̸8̸	9̸9̸	1̸0̸0̸

We cross out first 1; then the integers >2 and divisible by 2, i.e. all even integers >2; then (of those remaining) the integers >3 and divisible by 3; then (of those remaining) the integers >5 and divisible by 5; then (of those remaining) the integers >7 and divisible by 7; all the integers left are prime since 7 is the largest prime $\leqslant \sqrt{(100)}$.

The prime 2 is the only even prime, and all other even numbers are composite. Also any integer >5 whose last decimal digit is 5 is divisible by 5 and so is composite. It follows that the last digit of an odd prime ($\neq 5$) is 1, 3, 7 or 9.

Every integer $n>1$ has at least one prime divisor, e.g. its smallest divisor $n_1 > 1$; for, if n_1 were composite, say $n_1 = n_2 n_3$ with $n_2 > 1$ and $n_3 > 1$, then n_2 would be a divisor of n that is >1 but $<n_1$, contrary to the definition of n_1.

We use this result to prove that the set of primes is infinite. This result has been known for a long time, and in fact the proof given is an adaptation of one due to Euclid.

Theorem 5.1. *The set of primes is infinite.*

Proof. Suppose that the statement is false. Then the set of primes is finite and the primes can be labelled as $p_1, p_2, \ldots, p_n$, say.

Consider the integer

$$N = p_1 p_2 \ldots p_n + 1, \text{ which is } > 1.$$

Since N is of the form $qp_1 + 1$ and $1 < p_1$, it follows that N has principal remainder 1 on division by p_1 and so is not divisible by p_1. Similarly N is not divisible by $p_2, \ldots, p_n$, and hence N is not divisible by any prime.

But any integer >1 is divisible by at least one prime.

Hence we have a contradiction. It follows that the set of primes is infinite.

A glance at the list of primes <100 seems to suggest that the set of primes is scattered throughout the positive integers in a rather haphazard fashion. This is in many senses true, but there is a remarkable result, called the **Prime Number Theorem**, which shows that in one sense there is a certain smoothness about the regularity of distribution of the primes. If x is a positive real number, we denote by $\pi(x)$ the number of primes $\leqslant x$; e.g. $\pi(9{\cdot}3) = 4$, $\pi(30) = 10$, $\pi(100) = 25$. The Prime Number Theorem says that, as $x \to \infty$, $\dfrac{\pi(x)}{x/\log_e x} \to 1$; in other words, for large x, $\pi(x)$ is approximately $\dfrac{x}{\log_e x}$. The simplest proof of this result involves fairly advanced work in complex-variable theory. An elementary proof, not involving advanced mathematics but nevertheless extremely complicated, was obtained by the mathematicians Selberg and Erdös in 1948, and there have been other contributions since then.

The most important property of the set of primes is that, in a sense, the primes form the building bricks for the set of all positive integers; every integer >1 can be expressed in one and essentially only one way as a product of one or more primes.

$$\text{e.g.} \quad \begin{aligned} 540 &= 2.2.3.3.3.5 = 2^2.3^3.5, \\ 2072 &= 2.2.2.7.37 = 2^3.7.37, \\ 53 &= 53. \end{aligned}$$

This basic fact of mathematics is often called the **Fundamental Theorem of Arithmetic** or the **Unique Factorisation Theorem**.

We first prove that every positive integer $n>1$ can be expressed as a product of one or more primes. [We shall usually from now on omit "one or more".]

Theorem 5.2. *Every integer $n>1$ can be expressed as a product of primes.*

Proof. We use induction.

The statement is true for $n = 2$. Suppose that it is true for $n = 2, 3, \ldots, k$, and consider $k+1$.

If $k+1$ is a prime, the statement is true with $n = k+1$.

If $k+1$ is not a prime, then $k+1 = n_1 n_2$ where $2 \leqslant n_1 \leqslant k$ and $2 \leqslant n_2 \leqslant k$. By the inductive hypothesis, both n_1 and n_2 are products of primes and so $k+1 = n_1 n_2$ is a product of primes.

Hence the statement is true for $n = k+1$.

It is true for $n = 2$, and so for $n = 3$, and so for $n = 4$, and so on. Thus the result as stated is true.

The difficulty in proving the fundamental theorem is involved with the idea of uniqueness. We shall give a proof of this part of the main result based on the following theorem.

Theorem 5.3. *If p is a prime and if $p \mid ab$, where a, b are integers, then either $p \mid a$ or $p \mid b$.*

Proof. Either $p \mid a$ or $p \nmid a$. If $p \mid a$ the required result is true.

Suppose then that $p \nmid a$. In this case, a is prime to p, so that $(p, a) = 1$. We can find integers x, y such that $xp + ya = 1$. On multiplying each side by b, we have

$$xpb + yab = b. \tag{5.1}$$

Now, since $p \mid ab$, there is an integer k such that $ab = pk$. Using this in (5.1) we deduce that

$$p(xb + yk) = b, \text{ and so that } p \mid b.$$

This completes the proof.

Note. The corresponding statement with "a prime" replaced by "an integer" is not true; e.g. $4 \mid 2.6$, but $4 \nmid 2$ and $4 \nmid 6$.

The following theorem presents the main result. The proof is difficult and is included purely for completeness.

Theorem 5.4. (*Fundamental Theorem of Arithmetic*) *The expression for an integer $n > 1$ as a product of primes is unique apart from the order of the primes.*

Proof. Suppose that $n = p_1 p_2 \ldots p_r = q_1 q_2 \ldots q_s$, where $p_1, p_2, \ldots, p_r$ and $q_1, q_2, \ldots, q_s$ are primes ordered so that

$$p_1 \leqslant p_2 \leqslant \ldots \leqslant p_r \text{ and } q_1 \leqslant q_2 \leqslant \ldots \leqslant q_s.$$

We have to show that $r = s$ and that $p_i = q_i$ ($i = 1, 2, \ldots, r$).

We use induction on n. The result is true for $n = 2$. Suppose that it is true for $2, 3, \ldots, n-1$ and consider the integer n.

If n is a prime, the result is true; if not, n is composite and so in the above expressions for n, $r > 1$ and $s > 1$. Since

$$p_1 p_2 \ldots p_r = q_1 q_2 \ldots q_s,$$

it follows that $p_1 \mid q_1 q_2 \ldots q_s$. Using Theorem 5.3, we deduce that

$$\text{either } p_1 \mid q_1 \text{ or } p_1 \mid q_2 q_3 \ldots q_s.$$

If $p_1 \nmid q_1$, then, again using Theorem 5.3,

$$\text{either } p_1 | q_2 \text{ or } p_1 | q_3 \ldots q_s.$$

Proceeding in this way we see that $p_1 | q_i$ for some suffix i satisfying $1 \leqslant i \leqslant s$. But p_1, q_i are primes; thus $p_1 = q_i$.

Similarly, since $q_1 | p_1 p_2 \ldots p_r$, we deduce that $q_1 = p_j$ for some j such that $1 \leqslant j \leqslant r$.

Then

$$p_1 = q_i \geqslant q_1 = p_j \geqslant p_1, \text{ and so } p_1 = q_1.$$

It follows that

$$\frac{n}{p_1} = p_2 \ldots p_r = q_2 \ldots q_s.$$

But $2 \leqslant \dfrac{n}{p_1} \leqslant n-1$, so that the inductive hypothesis applies to the integer $\dfrac{n}{p_1}$. It follows that the products $p_2 \ldots p_r$ and $q_2 \ldots q_s$ are identical. This implies that $s = r$ and, by the ordering of the primes, that $p_i = q_i$ for each suffix i.

Hence the result is true for n, and the result as stated now follows by induction.

Notes. 1. The prime factorisation of a positive integer $n > 1$ can be written in the form

$$n = p_1^{\alpha_1} p_2^{\alpha_2} \ldots p_r^{\alpha_r},$$

where the p_i are primes satisfying $p_1 < p_2 < \ldots < p_r$ and the α_i are positive integers. The expression is called the **prime representation** or **decomposition** of n.

2. If n is any integer ($n \neq 0$ or ± 1), then we can write

$$n = \pm p_1^{\alpha_1} p_2^{\alpha_2} \ldots p_r^{\alpha_r}$$

in a unique way, the $+$ or $-$ applying according as n is positive or negative.

3. If n is any rational number ($n \neq 0$ or ± 1), then n can be uniquely expressed in the form

$$n = \pm p_1^{\alpha_1} p_2^{\alpha_2} \ldots p_r^{\alpha_r},$$

where the integers $\alpha_1, \alpha_2, \ldots, \alpha_r$ are positive or negative;

$$\text{e.g.} \quad \frac{2072}{6413} = \frac{2^3 . 7 . 37}{11^2 . 53} = 2^3 . 7 . 11^{-2} . 37 . 53^{-1}.$$

4. If the set of all primes is labelled $p_1, p_2, p_3, \ldots$, then the following product notation for a positive integer $n > 1$ is very useful,

$$n = \prod_{i=1}^{\infty} p_i^{\alpha_i} = p_1^{\alpha_1} p_2^{\alpha_2} p_3^{\alpha_3} \cdots,$$

where *only a finite number* of the integers α_i are non-zero.

Unique factorisation for the positive integers seems intuitively natural, but the following example shows that the idea is by no means natural, and may not hold in some sets closed under multiplication, i.e. sets in which the product of any two elements in the set is also in the set.

Example. Consider the set $S = \{4k + 1 : k \in \mathbf{Z}, k \geq 0\} = \{1, 5, 9, 13, \ldots\}$. Since $(4k_1 + 1)(4k_2 + 1) = 4(4k_1 k_2 + k_1 + k_2) + 1$, it follows that S is closed under multiplication.

An element $x \in S$ ($x > 1$) is called an *S-prime* if 1 and x are the only elements of S that divide x. Clearly $5, 9, 13, 17, 21, 29, \ldots$ are S-primes and $25 = 5.5$ is the smallest element of S greater than 1 that is not an S-prime. It is easy to show that every member of S can be expressed as a product of S-primes. It might seem natural for this factorisation to be unique, but in fact it is not;

$$\text{e.g.} \quad 441 = 21.21 = 9.49,$$

and 441, 21, 9, 49 are all elements of S, the integers 21, 9 and 49 being S-primes.

Show that 441 is the smallest member of S that can be expressed in two different ways as a product of S-primes, and find the next such member.

The study of primes is a difficult but fascinating branch of pure mathematical thought and has many unsolved problems. We list a few of these and at the same time some other unsolved problems in number theory.

1. Is every even integer > 2 the sum of two primes?
   [Goldbach conjectured in 1742 that the answer is *yes*.]

2. Is the set of primes of the form $x^4 + 1$ for some integer x infinite?
   [$17 = 2^4 + 1$ and $257 = 4^4 + 1$ are in the set.]

3. Are the only integer solutions of the equation $x^3 + y^3 + z^3 = 3$ those given by $(x, y, z) = (1, 1, 1), (4, 4, -5), (4, -5, 4), (-5, 4, 4)$?

4. Is the set of positive integers n for which each of the integers n and $n + 1$ has only one prime divisor infinite?

$[n = 2, 3, 4, 7, 8, 16, 31, 127, 256$ are the smallest of those at present known.$]$

5. Do there exist, for every integer $n > 1$, integers x, y, z such that

$$\frac{4}{n} = \frac{1}{x} + \frac{1}{y} + \frac{1}{z}?$$

6. W. H. Mills has proved that $\exists \theta \in \mathbf{R}$ such that the integral part of θ^{3^n} is a prime for all positive integers n. Can we determine any useful information about θ?

7. An integer of the form $\frac{1}{2}n(n+1)$ with n a positive integer is called a **triangular number** and one of the form $\frac{1}{6}n(n+1)(n+2)$ a **tetrahedral number**.

How many tetrahedral numbers are also triangular?
$[1, 10, 120, 1540$ and 7140 are examples.$]$

8. Does the digit 1 occur infinitely often in the decimal expansion of $\sqrt{2}$?

9. Does the sequence of nine digits 123456789 following one another in that order occur at least once in the decimal expansion of π?

10. Is the set of primes whose digits are all 1 infinite?

$$\left[\frac{10^2 - 1}{9} = 11 \text{ and } \frac{10^{23} - 1}{9} \text{ (with 23 digits) are examples.} \right]$$

EXERCISE 6.4

1. Express each of the following numbers as a product of powers of primes in increasing order
   (i) 490,   (ii) 1125,   (iii) 2728,   (iv) $\frac{330}{169}$,   (v) $\frac{833}{176}$,   (vi) $10^4 - 1$.

2. Use the sieve of Eratosthenes to find the primes < 200.

3. If the set of primes arranged in increasing order of magnitude is $\{p_1, p_2, p_3, \ldots\}$, find the smallest value of n for which $p_1 p_2 \ldots p_n + 1$ is composite.

4. Determine whether the following statements are true or false. If true, prove the result, and, if false, give a counterexample.
   (1) If p is a prime and $p | a^3$, then $p | a$.
   (2) If p is a prime and $p | a$ and $p | (a^2 + b^2)$, then $p | b$.
   (3) If $a^3 | b^2$, then $a | b$.
   (4) If p is a prime, $p | (a^2 + b^2)$ and $p | (b^2 + c^2)$, then $p | (a^2 + c^2)$.
   (5) If $m | (a^2 - 1)$, then $m | (a^4 - 1)$.
   (6) If $m | (a^4 - 1)$, then either $m | (a^2 - 1)$ or $m | (a^2 + 1)$.

5. Show that, if $a | mn$ and if $(a, m) = 1$, then $a | n$.
   [Model a proof on that of Theorem 5.3.]

6. Prove that, if $a|m$ and $b|m$ and if $(a, b) = 1$, then $ab|m$.
  Give an example to show that the corresponding statement with $(a, b) > 1$ is not true.

7. If $(a, b) = p$, a prime, what are the possible values of (a^2, b), (a^3, b) and (a^2, b^3)?

8. Let $E = \{2, 4, 6, 8, 10, \ldots\}$, i.e. the set of even positive integers. Show that E is closed under multiplication. An integer in E is called an *E-prime* if it cannot be expressed as the product of two or more members of E. What are the first six *E*-primes? Show that every member of E can be expressed as a product of *E*-primes, but that this factorisation is not unique. Find the first two members of E with more than one factorisation into *E*-primes.

6. Congruence notation, simple results on congruences

If a is any integer and m is a given positive integer, then, on dividing a by m, we obtain an equation of the form $a = q_1 m + r$, where $0 \leqslant r \leqslant m - 1$ and q_1 is an integer.

The integers a and r have many properties in common relative to m. For example, they have the same g.c.d. with m; for, $(a, m) = (r + q_1 m, m) = (r, m)$. [See problem 4 of Exercise 3.]

If the integer b also has remainder r on division by m, then $b = q_2 m + r$ where q_2 is an integer. Again $(b, m) = (r, m)$ and so $(a, m) = (b, m)$.

For such integers a, b, we have $a - b = (q_1 - q_2)m$, so that $m|(a - b)$.

It is very convenient to have a notation which recognizes the fact that two such integers a, b behave in many respects similarly with respect to m, i.e. are in a certain sense equivalent to each other with respect to m and each is equivalent to their common remainder r on division by m. The notation used (introduced by the famous mathematician Gauss) is called the congruence notation for the integers.

We write

$$a \equiv b \ (\mathrm{mod} \ m) \tag{6.1}$$

when $m|(a - b)$ and say that a is **congruent to** b **modulo** m. The integer m is called the **modulus** of the **congruence** (6.1). It follows that $a \equiv b \ (\mathrm{mod} \ m)$ if and only if $a - b = km$ for some integer k;

e.g. $18 \equiv 5 \ (\mathrm{mod} \ 13)$ since $18 - 5 = 1.13,$
        $-3 \equiv 18 \ (\mathrm{mod} \ 7)$ since $-3 - 18 = (-3).7.$

If $m \nmid (a - b)$, we write $a \not\equiv b \ (\mathrm{mod} \ m)$ and say that a and b are **incongruent** modulo m; e.g. $4 \not\equiv -2 \ (\mathrm{mod} \ 7)$ since $7 \nmid (4 + 2 = 6)$.

The congruence notation defines a relation on $\mathbf{Z}$, the set of all integers, called the **congruence relation** modulo m. We first prove that this is an equivalence relation on $\mathbf{Z}$.

Theorem 6.1. *The congruence relation modulo m is an equivalence relation on* $\mathbf{Z}$.

Proof.   (i) *For any $a \in \mathbf{Z}$, $a \equiv a$ (mod m)*; for, $a - a = 0$ and $m \mid 0$.

(ii) $a \equiv b$ *(mod m)* $\Rightarrow b \equiv a$ *(mod m)*; for,

$$a \equiv b \text{ (mod } m) \Rightarrow a - b = km \quad \text{for some integer } k$$
$$\Rightarrow b - a = (-k)m$$
$$\Rightarrow b \equiv a \text{ (mod } m).$$

(iii) $a \equiv b$ *(mod m) and* $b \equiv c$ *(mod m)* $\Rightarrow a \equiv c$ *(mod m)*; for,

$$\left. \begin{array}{l} a \equiv b \text{ (mod } m) \\ \text{and} \quad b \equiv c \text{ (mod } m) \end{array} \right\} \Rightarrow \left. \begin{array}{l} a - b = km \text{ for some } k \in \mathbf{Z} \\ \text{and } b - c = lm \text{ for some } l \in \mathbf{Z} \end{array} \right\}$$
$$\Rightarrow (a - b) + (b - c) = (k + l)m$$
$$\Rightarrow a - c = (k + l)m$$
$$\Rightarrow a \equiv c \text{ (mod } m).$$

By (i), (ii), (iii), the congruence relation modulo m is reflexive, symmetric and transitive and so is an equivalence relation on $\mathbf{Z}$.

It follows from Theorem 6.1 that the congruence relation modulo m partitions $\mathbf{Z}$ into disjoint equivalence classes called the **congruence classes modulo** m. Can we identify these? A clue to the answer lies in the fact that congruence modulo m is closely associated with division by the integer m and so with the possible remainders on division by m.

If a is any integer, then, on dividing a by m, we have

$$a = qm + r, \quad \text{where } 0 \leqslant r \leqslant m - 1.$$

The remainder r is uniquely determined by a and $a - r = qm$, so that

$$a \equiv r \text{ (mod } m).$$

No pair of the possible remainders $0, 1, 2, \ldots, m-1$ are congruent to each other modulo m, and every integer is congruent to one and only one of these m integers. There are thus m congruence classes modulo m and each class contains one and only one of the possible remainders $0, 1, 2, \ldots, m-1$ on division by m. Any set of m integers, one from each of these m congruence classes, is called a **complete set of residues modulo** m or simply a C.S.R. (mod m). The simplest such set is $\{0, 1, 2, \ldots, m-1\}$, e.g. $\{0, 1, 2, 3, 4\}$ is a C.S.R. (mod 5); every integer is congruent modulo 5 to one and only one of the integers 0, 1, 2, 3, 4.

$\{1, 2, 3, 4, 5\}$ is also a C.S.R. (mod 5).

$\{0, \pm 1, \pm 2\}$ is another C.S.R. (mod 5). This set has the advantage for arithmetical work that the largest integer in the set, namely 2, is smaller numerically than the integer of largest numerical value in any other set; the set is often called the **least absolute set of residues** modulo 5.

Similarly, $\{0, 1, 2, 3, 4, 5, 6, 7\}$, $\{1, 2, 3, 4, 5, 6, 7, 8\}$ and $\{0, \pm 1, \pm 2, \pm 3, 4\}$ are all complete sets of residues modulo 8.

Note. In testing whether a set of integers is a C.S.R. (mod m) we have to check two things, (i) that the set has m members and (ii) that no pair are congruent to each other (mod m).

Addition, subtraction and multiplication of congruences.

Theorem 6.2. *If $a \equiv b$ (mod m) and $c \equiv d$ (mod m), then*

$$\text{(i)} \quad a+c \equiv b+d \pmod{m},$$
$$\text{(ii)} \quad a-c \equiv b-d \pmod{m},$$
$$\text{(iii)} \quad ac \equiv bd \pmod{m}.$$

Proof. We have $a-b = km$ and $c-d = lm$ for some integers k, l.

Thus, $(a+c)-(b+d) = (k+l)m$, and so $a+c \equiv b+d \pmod{m}$,
$(a-c)-(b-d) = (k-l)m$, and so $a-c \equiv b-d \pmod{m}$,
$ac - bd = a(c-d)+d(a-b) = (al+dk)m$, and so
$ac \equiv bd \pmod{m}$.

It follows that we can add, subtract and multiply congruences.

For ordinary equality, if $na = nb$ and $n \neq 0$, then $a = b$. The corresponding result for congruences, given in the following theorem, is not so simple.

Theorem 6.3. *If*

$$na \equiv nb \pmod{m},$$

where n is an integer, then

$$a \equiv b \left(\bmod \frac{m}{(n, m)} \right),$$

where (n, m) is the g.c.d. of n and m.

Proof. If we denote (n, m) by d and write $m = dm_1$, $n = dn_1$, then $(m_1, n_1) = 1$.

$$na \equiv nb \pmod{m} \Rightarrow n(a-b) = km \text{ for some } k \in \mathbf{Z}$$
$$\Rightarrow dn_1(a-b) = kdm_1$$
$$\Rightarrow n_1(a-b) = km_1$$
$$\Rightarrow m_1 \,|\, n_1(a-b)$$
$$\Rightarrow m_1 \,|\, (a-b), \text{ since } (m_1, n_1) = 1 \qquad (6.2)$$
$$\Rightarrow a \equiv b \left(\bmod m_1 = \frac{m}{(n, m)} \right).$$

[See Problem **5** of Exercise 4 for the step numbered (6.2).]

Note. As a special case of this result we have:

If $\qquad (n, m) = 1$, then $\dfrac{m}{(n, m)} = m$, and so

$$na \equiv nb \ (\mathrm{mod}\ m) \Rightarrow a \equiv b \ (\mathrm{mod}\ m).$$

e.g. $\qquad 2a \equiv 2b \ (\mathrm{mod}\ 5) \Rightarrow a \equiv b \ (\mathrm{mod}\ 5)$, since $(2, 5) = 1$.

We note that, if $(n, m) \neq 1$, then

$$na \equiv nb \ (\mathrm{mod}\ m) \not\Rightarrow a \equiv b \ (\mathrm{mod}\ m).$$

e.g. $\qquad 10 \equiv 6 \ (\mathrm{mod}\ 4)$, but $5 \not\equiv 3 \ (\mathrm{mod}\ 4)$.

Congruence equations.

(1) An integer x is such that

$$
\begin{aligned}
& x+5 \equiv 3 && (\mathrm{mod}\ 7) && (6.3) \\
\Leftrightarrow\ & x+5-5 \equiv 3-5 && (\mathrm{mod}\ 7) \\
\Leftrightarrow\ & \qquad x \equiv -2 && (\mathrm{mod}\ 7) \\
\Leftrightarrow\ & \qquad x \equiv 5 && (\mathrm{mod}\ 7)
\end{aligned}
$$

We say that $x \equiv 5 \ (\mathrm{mod}\ 7)$ is the **solution** of the congruence equation (6.3) in the sense that every integer in the congruence class of integers $\equiv 5$ (mod 7) satisfies (6.3) and these are the only integers satisfying (6.3).

(2) Consider now the congruence equation

$$x^2 \equiv 1 \ (\mathrm{mod}\ 8). \qquad (6.4)$$

We first note that, if an integer x_0 satisfies (6.4) and if $x \equiv x_0 \ (\mathrm{mod}\ 8)$, then x also satisfies (6.4). For, by multiplication of congruences, $x^2 \equiv x_0^2$ (mod 8); but $x_0^2 \equiv 1 \ (\mathrm{mod}\ 8)$, and so $x^2 \equiv 1 \ (\mathrm{mod}\ 8)$. Thus every integer in the congruence class (mod 8) to which x_0 belongs satisfies (6.4). We say that (6.4) has a *root* $x \equiv x_0 \ (\mathrm{mod}\ 8)$.

It follows that, in finding the roots of (6.4), we need consider only a C.S.R. (mod 8) and find the solutions in this set.

Consider the set $\{0, 1, 2, 3, 4, 5, 6, 7\}$ and check the following table:

$x \equiv$	0	1	2	3	4	5	6	7	(mod 8)
$x^2 \equiv$	0	1	4	1	0	1	4	1	(mod 8)

In the second line each square is formed and reduced (mod 8); e.g.

$$5^2 = 25 \equiv 1 \ (\mathrm{mod}\ 8).$$

It follows that (6.4) has four roots, namely $x \equiv 1, 3, 5$ and $7 \ (\mathrm{mod}\ 8)$.

(3) Consider now the linear congruence equation

$$4x \equiv 8 \ (\text{mod } 12). \tag{6.5}$$

Again, in looking for the roots of this congruence, we need consider only a C.S.R. (mod 12), e.g. $\{0, 1, 2, 3, 4, 5, 6, 7, 8, 9, 10, 11\}$ and draw up a table as follows:

$x \equiv$	0	1	2	3	4	5	6	7	8	9	10	11	(mod 12)
$4x \equiv$	0	4	8	0	4	8	0	4	8	0	4	8	(mod 12)

It follows that (6.5) has four roots, namely $x \equiv 2, 5, 8$ and 11 (mod 12).

(4) We prove the following result for a general linear congruence.
The linear congruence

$$ax \equiv b \ (\text{mod } m) \tag{6.6}$$

has a root if and only if $(a, m) \,|\, b$.

Suppose first that (6.6) has a root x_0 so that $ax_0 \equiv b$ (mod m). Then $ax_0 - b = km$ for some integer k, so that $b = ax_0 - km$. Now

$$(a, m) \,|\, (ax_0 - km)$$

and consequently $\qquad (a, m) \,|\, b.$

To prove the converse, suppose that $(a, m) \,|\, b$. Then $b = (a, m)b_1$ for some integer b_1. Using the euclidean algorithm we can find integers x_1, y_1 such that

$$x_1 a + y_1 m = (a, m).$$

Multiplying each side by b_1, we deduce that

$$a(x_1 b_1) + y_1 b_1 m = b,$$

and so that $\qquad a(x_1 b_1) \equiv b$ (mod m).

Hence (6.6) has the root $\qquad x \equiv x_1 b_1$ (mod m).

This completes the proof of the stated result.

Note. In the special case in which $(a, m) = 1$, the condition $(a, m) \,|\, b$ is satisfied for every integer b and consequently (6.6) always has a solution. This solution is unique modulo m since

$$ax_1 \equiv b \ (\text{mod } m) \quad \text{and} \quad ax_2 \equiv b \ (\text{mod } m)$$
$$\Rightarrow ax_1 \equiv ax_2 \ (\text{mod } m)$$
$$\Rightarrow x_1 \equiv x_2 \ (\text{mod } m), \text{ since } (a, m) = 1.$$

Example. The linear congruence $2x \equiv 5 \pmod 4$ has no root since $(2, 4) = 2$ and $2 \nmid 5$.

On the other hand the linear congruence $2x \equiv 5 \pmod 7$ has a solution since $(2, 7) = 1$; it is $x \equiv 6 \pmod 7$.

The nature of the set of roots can easily be checked by considering a C.S.R. for the modulus involved.

Addition and multiplication tables $\pmod m$. By taking a C.S.R. $\pmod m$ or a suitable subset of a C.S.R. $\pmod m$ we can draw up addition and multiplication tables which illustrate how the sets behave under addition and multiplication $\pmod m$.

Example 1. Addition table for the set $\{0, 1, 2, 3, 4\}$ $\pmod 5$.

+ (mod 5)	0	1	2	3	4
0					
1					
2					
3					
4					

Figure 1

+ (mod 5)	0	1	2	3	4
0	0	1	2	3	4
1	1	2	3	4	0
2	2	3	4	0	1
3	3	4	0	1	2
4	4	0	1	2	3

Figure 2

The elements 0, 1, 2, 3, 4 of the set are listed in the display column as shown in Figure 1 and also, in the same order, in the display row. The box of squares is then filled in, as indicated in Figure 2, in the following way:

If a denotes an element in the display column and b an element in the display row, then $a+b$ reduced modulo 5 is placed in the square at the intersection of the row in which a lies and the column in which b lies; e.g. $2+4 \equiv 1 \pmod 5$ and 1 is placed in the square in which the row determined by 2 and the column determined by 4 intersect.

In discussing columns and rows in the table (Figure 2) we ignore the display column and display row.

Properties of addition $\pmod 5$ can be checked from the table.
A glance at the table checks that the entries in the table all belong to the given set $\{0, 1, 2, 3, 4\}$; we say that the set is **closed** under addition modulo 5. This simply corresponds to the fact that if a, b are any two integers then $a+b$ is congruent modulo 5 to a unique integer in the set $\{0, 1, 2, 3, 4\}$.

Each row and column contains a permutation of the set $\{0, 1, 2, 3, 4\}$.

For any two integers a, b, we have $a+b = b+a$, and so

$$a+b \equiv b+a \text{ (mod 5)},$$

i.e. addition (mod 5) is **commutative**. This could easily have been deduced from the table, since it clearly holds if and only if the table is symmetrical about the main diagonal, running from the first position in the first row down to the last position in the last row.

If a, b, c are any integers, then $a+(b+c) = (a+b)+c$, and so

$$a+(b+c) \equiv (a+b)+c \text{ (mod 5)},$$

i.e. addition (mod 5) is **associative**. This can be checked from the table, but is a long and tedious process.

If a is any integer, then $0+a = a+0$, and so

$$0+a \equiv a \equiv a+0 \text{ (mod 5)}.$$

Thus 0 is an **identity** for addition (mod 5). This is clear from the table since the row and column determined by 0 preserve the order of the elements in the set $\{0, 1, 2, 3, 4\}$.

If a is any integer in the set $\{0, 1, 2, 3, 4\}$, the table shows that there is an integer b *in the set* such that $a+b \equiv 0 \equiv b+a$ (mod 5); e.g. $a = 0$ has $b = 0$, $a = 1$ has $b = 4$, $a = 2$ has $b = 3$, $a = 3$ has $b = 2$, and $a = 4$ has $b = 1$. This b is called the **additive inverse** (mod 5) of element a in the given set.

Example 2. <u>Multiplication table for the set $\{1, 2, 3, 4, 5, 6\}$ (mod 7).</u>

(mod 7)	1	2	3	4	5	6
1	1	2	3	4	5	6
2	2	4	6	1	3	5
3	3	6	2	5	1	4
4	4	1	5	2	6	3
5	5	3	1	6	4	2
6	6	5	4	3	2	1

Each row and column is a permutation of the given set and so the set is closed under multiplication (mod 7).

Again properties of multiplication (mod 7) can be checked from the table.

For any two integers a, b, we have $ab = ba$, and so

$$ab \equiv ba \ (\text{mod } 7),$$

i.e. multiplication (mod 7) is commutative. This is indicated in the table by the fact that the table is symmetrical about the main diagonal.

Similarly, for any three integers a, b, c, we have $a(bc) = (ab)c$, and so

$$a(bc) \equiv (ab)c \ (\text{mod } 7),$$

i.e. multiplication (mod 7) is associative. This could be checked from the table, but the process would be much too tedious.

If a is any integer, then $1.a = a = a.1$, and so

$$1.a \equiv a \equiv a.1 \ (\text{mod } 7),$$

i.e. 1 is an **identity for multiplication** (mod 7). This is clear from the table since the row and column determined by 1 preserve the order of the elements in the set $\{1, 2, 3, 4, 5, 6\}$.

If a is any integer in the set $\{1, 2, 3, 4, 5, 6\}$, the table shows that there is an integer b *in the set* such that $ab \equiv 1 \equiv ba$ (mod 7); e.g. $a = 1$ has $b = 1$, $a = 2$ has $b = 4$, $a = 3$ has $b = 5$, $a = 4$ has $b = 2$, $a = 5$ has $b = 3$, and $a = 6$ has $b = 6$. This b is called the **multiplicative inverse** (mod 7) of element a in the given set.

A glance at the table shows that the linear congruence

$$4x \equiv 2 \ (\text{mod } 7)$$

has root $x \equiv 4$ (mod 7).

Example 3. <u>Multiplication table for the set $\{2, 4, 6, 8\}$ (mod 10).</u>

(mod 10)	2	4	6	8
2	4	8	2	6
4	8	6	4	2
6	2	4	6	8
8	6	2	8	4

Verify the following facts from the table:

(1) The set $\{2, 4, 6, 8\}$ is closed under multiplication (mod 10).

(2) 6 is an identity for the set under multiplication (mod 10).

(3) Each element in the set has a multiplicative inverse (mod 10) in the set.

(4) The linear congruence $4x \equiv 2$ (mod 10) has one root in the set.

Note. In connection with such tables we note that, if m is any modulus and a, b, c any integers, then, since $a+b = b+a$ and $a+(b+c) = (a+b)+c$, we have

$$a+b \equiv b+a \ (\text{mod } m) \quad \text{and} \quad a+(b+c) \equiv (a+b)+c \ (\text{mod } m),$$

so that addition (mod m) is always commutative and associative.

Since $ab = ba$ and $a(bc) = (ab)c$, we also have

$$ab \equiv ba \ (\text{mod } m) \quad \text{and} \quad a(bc) \equiv (ab)c \ (\text{mod } m),$$

so that multiplication (mod m) is always commutative and associative.

Fermat's Theorem. Suppose that p is a prime and that a is an integer prime to p. Consider the set of positive powers of a, namely

$$A = \{a, a^2, a^3, a^4, \ldots\}.$$

Since every integer is congruent modulo p to one of the integers $0, 1, \ldots, p-1$, the set A has only a finite number of integers that are distinct modulo p. As a result, there are distinct positive integers r, s with $r > s$ for which $a^r \equiv a^s$ (mod p), and so $a^{r-s} \equiv 1$ (mod p). A natural question to ask is whether there is some positive power of a, the same for all such a, that is congruent to 1 (mod p). In this connection Fermat proved the following result.

Theorem 6.4. *If* $(a, p) = 1$, *then* $a^{p-1} \equiv 1$ (mod p).

Proof. The set

$$\{0, 1, 2, \ldots, p-1\} \tag{6.7}$$

is a C.S.R. (mod p).

The set

$$\{0, a, 2a, \ldots, (p-1)a\} \tag{6.8}$$

is also a C.S.R. (mod p). For, it contains p members; also, if ra and sa are members of the set, then

$$ra \equiv sa \ (\text{mod } p) \Rightarrow r \equiv s \ (\text{mod } p) \quad [\text{using Theorem 6.3}]$$
$$\Rightarrow r = s,$$

since r, s are members of the C.S.R. (6.7). Thus the integers in (6.8) are incongruent (mod p) in pairs. Hence (6.8) is a C.S.R. (mod p).

It follows that the integers in set (6.8) are congruent (mod p) to the integers in set (6.7) but not necessarily in the same order. In particular, the integers in the set $\{a, 2a, \ldots, (p-1)a\}$ are congruent (mod p) to the integers in the set $\{1, 2, \ldots, p-1\}$ in some order. By multiplication of congruences it follows that

$$a.2a.\ldots.(p-1)a \equiv 1.2.\ldots.(p-1) \ (\text{mod } p),$$

i.e.

$$a^{p-1}.(p-1)! \equiv (p-1)! \ (\text{mod } p).$$

Since $((p-1)!, p) = 1$, we deduce, by using Theorem 6.3, that

$$a^{p-1} \equiv 1 \ (\text{mod } p).$$

Example. If $p = 13$ and $a = 2$, then $2^{12} \equiv 1 \ (\text{mod } 13)$.

Corollary to Theorem 6.4. *If p is a prime and a is any integer, then*

$$a^p \equiv a \ (\text{mod } p).$$

Proof. If $(a, p) = 1$, then $a^{p-1} \equiv 1 \ (\text{mod } p)$, and so, on multiplication by a,

$$a^p \equiv a \ (\text{mod } p).$$

If $(a, p) \neq 1$, then $p \mid a$; in this case $a^p \equiv 0 \equiv a \ (\text{mod } p)$, and so

$$a^p \equiv a \ (\text{mod } p).$$

There is an extension of Theorem 6.4 (due to Euler) in a form involving a general modulus m. For completeness this will be presented later in the general exercise [see Problem **23** of Exercise 6].

EXERCISE 6.5

1. Find the integers in the set $\{0, 1, 2, 3, 4, 5, 6\}$ that are congruent (mod 7) to the integers
   (i) 33,   (ii) -17,   (iii) 35,   (iv) 23,   (v) -145.
2. List all the integers x in the set $\{x \in \mathbf{Z} : 1 \leqslant x \leqslant 100, x \equiv 6 \ (\text{mod } 11)\}$.
3. Find a complete set of residues modulo 17 that contains only multiples of 3.
4. Every integer in the set $\{x : x = 6k+1$ for some $k \in \mathbf{Z}\}$ satisfies the condition $x \equiv 1$ (mod 6). Express in set notation the integers which satisfy
   (i) $x \equiv 5 \ (\text{mod } 12)$,   (ii) $x \equiv 2 \ (\text{mod } 13)$.
5. Show that, if $a \equiv b \ (\text{mod } m)$ and $d \mid m, d > 0$, then $a \equiv b \ (\text{mod } d)$.
6. Show that, if $a \equiv b \ (\text{mod } m)$ and $c \equiv d \ (\text{mod } m)$, then
   $$ax + cy \equiv bx + dy \ (\text{mod } m) \quad \text{and} \quad a^2 + c^2 \equiv b^2 + d^2 \ (\text{mod } m).$$
7. Prove that, if $\{ax_1, ax_2, \ldots, ax_m\}$ is a C.S.R. (mod m) and if $(a, m) = 1$, then $\{x_1, x_2, \ldots, x_m\}$ is also a C.S.R. (mod m).
8. Solve the congruence equations
   (i) $x + 7 \equiv 2 \ (\text{mod } 13)$,   (ii) $x + 31 \equiv 17 \ (\text{mod } 37)$,   (iii) $2x \equiv 3 \ (\text{mod } 9)$,
   (iv) $4x \equiv 6 \ (\text{mod } 10)$,   (v) $3x \equiv 4 \ (\text{mod } 6)$,   (vi) $x^2 \equiv -1 \ (\text{mod } 13)$,
   (vii) $x^2 \equiv -1 \ (\text{mod } 7)$,   (viii) $x^3 \equiv 1 \ (\text{mod } 8)$,   (ix) $x^4 \equiv 1 \ (\text{mod } 6)$.

9. Show that any integer which is a square is congruent (mod 10) to 0, 1, 4, 5, 6 or 9.

10. Show that, for every integer n, $n^7 - n \equiv 0 \pmod{42}$.

11. Draw up the tables of addition (mod 6) and multiplication (mod 6) for the set $\{0, 1, 2, 3, 4, 5\}$.

Discuss the question of closure and the existence of identity elements and inverse elements for these operations.

12. Draw up the table of multiplication (mod 12) for the set $\{1, 5, 7, 11\}$ and examine the question of closure and the existence of an identity element and inverse elements.

EXERCISE 6

1. Show that, if a and b are integers, then $a + b$ is odd if and only if exactly one of a, b is odd.

2. Show that, if x and y are odd integers, then $x^2 + y^2$ is even but not divisible by 4. Deduce that, if at least one of x, y is > 1, then $x^2 + y^2$ is not a prime.

3. A positive integer n equals $a_r a_{r-1} \ldots a_1 a_0$ in decimal notation, i.e.
$$n = a_r 10^r + a_{r-1} 10^{r-1} + \ldots + a_1 10 + a_0.$$
Show that

(i) $3 \mid n \Leftrightarrow 3 \mid (a_0 + a_1 + \ldots + a_r)$,
(ii) $9 \mid n \Leftrightarrow 9 \mid (a_0 + a_1 + \ldots + a_r)$,
(iii) $2 \mid n \Leftrightarrow 2 \mid a_0$,
(iv) $4 \mid n \Leftrightarrow 4 \mid (a_1 a_0)$, i.e. $4 \mid (a_1 10 + a_0)$,
(v) $8 \mid n \Leftrightarrow 8 \mid (a_2 a_1 a_0)$.

4. Find integers x, y, z such that $8x + 12y + 15z = 1$.

5. The **Fibonacci Numbers** are the integers $1, 2, 3, 5, 8, 13, \ldots$, in which each integer after the second is the sum of the two preceding integers. Show, by induction, that two consecutive Fibonacci numbers are relatively prime.

6. If a, b are positive integers such that $(a, b) = 1$ and ab is a perfect square, show that a and b are perfect squares.

7. Prove that, if a, m, n are integers, then
$$(a, mn) = 1 \Leftrightarrow (a, m) = 1 \quad \text{and} \quad (a, n) = 1.$$

8. Show that, if n is a positive integer, then
$$\left[\frac{n+1}{2}\right] + \left[\frac{n+2}{2^2}\right] + \left[\frac{n+2^2}{2^3}\right] + \left[\frac{n+2^3}{2^4}\right] + \ldots = n,$$
where $[\ \ ]$ means "integral part of".
[*Hint.* Express n in the scale of 2. This problem is difficult.]

9. If
$$m = \prod_{i=1}^{\infty} p_i^{\alpha_i} \quad \text{and} \quad n = \prod_{i=1}^{\infty} p_i^{\beta_i}, \quad \text{show that}$$
$$(m, n) = \prod_{i=1}^{\infty} p_i^{\min(\alpha_i, \beta_i)} \quad \text{and} \quad [m, n] = \prod_{i=1}^{\infty} p_i^{\max(\alpha_i, \beta_i)},$$
where $\min(\alpha_i, \beta_i)$ means the smaller of α_i, β_i and $\max(\alpha_i, \beta_i)$ the larger of α_i, β_i.

10. Prove the following results for integers:

(i) If $a > 0$ and $b > 0$, then $(a, b)[a, b] = ab$.
(ii) If $m > 0$, then $(ma, mb) = m(a, b)$ and $[ma, mb] = m[a, b]$.
(iii) $(a, [b, c]) = [(a, b), (a, c)]$.
(iv) $[a, (b, c)] = ([a, b], [a, c])$.
(v) If $(a, b) = c$, then $(a^2, b^2) = c^2$.

11. Prove that an odd prime is of the form $4k+1$ or $4k-1$ for some integer k. Prove that the product of a finite number of integers of the form $4k+1$ is also of this form. Deduce that the set of primes of the form $4k-1$ is infinite.

12. If n_1, n_2 are positive integers, verify that
$$2^{n_1 n_2} - 1 = (2^{n_1} - 1)(2^{(n_2-1)n_1} + 2^{(n_2-2)n_1} + \ldots + 2^{n_1} + 1),$$
and deduce that, if $2^n - 1$ is a prime, then n is a prime.

[A prime of the form $2^p - 1$, where p is a prime, is called a **Mersenne prime**. The number of known primes of this form keeps increasing with the development of computers; 23 of them are at present known, the largest being $2^{11213} - 1$.]

13. If m is odd, verify that
$$2^{2^r m} + 1 = (2^{2^r} + 1)(2^{2^r(m-1)} - 2^{2^r(m-2)} + \ldots - 2^{2^r} + 1),$$
and deduce that, if $2^n + 1$ is a prime, then n is a power of 2.

[A prime of the form $2^{2^r} + 1$ is called a **Fermat prime**. At present the only primes of this form known are the five given by $r = 0, 1, 2, 3, 4$.]

14. If $n = p_1^{\alpha_1} p_2^{\alpha_2} \ldots p_r^{\alpha_r} (n > 1)$, show that

(i) $d(n)$, the number of positive divisors of n, is given by the formula
$$d(n) = \prod_{i=1}^{r} (1 + \alpha_i),$$

(ii) $\sigma(n)$, the sum of the positive divisors of n, is given by the formula
$$\sigma(n) = \prod_{i=1}^{r} \frac{p_i^{\alpha_i + 1} - 1}{p_i - 1}.$$

15. An integer n is called a **perfect number** if $\sigma(n) = 2n$. Verify that 6, 28, 496 are perfect numbers.

Show that, if $2^p - 1$ is a prime, then $2^{p-1}(2^p - 1)$ is a perfect number.

From Problem 12, at present 23 even perfect numbers are known. It is not known whether there is an odd perfect number.

16. Show that, if $a \equiv b \pmod{m}$, then $(a, m) = (b, m)$.

17. Let $f(x)$ be a polynomial in x with integral coefficients. Show that, if $a \equiv b \pmod{m}$, then $f(a) \equiv f(b) \pmod{m}$.

18. Solve the simultaneous congruence equations
$$x \equiv 1 \pmod{2},$$
$$x \equiv 2 \pmod{3}.$$

19. If a is an odd integer, show that $a^2 \equiv 1 \pmod{8}$, and deduce, by induction, that $a^{2^n} \equiv 1 \pmod{2^{n+2}}$ for all positive integers n.

20. The binomial coefficient
$$\binom{p}{r} = \frac{p(p-1)(p-2)\ldots(p-r+1)}{r!}$$
is an integer. Show that, if $1 \leqslant r \leqslant p-1$ and if p is a prime, then
$$\binom{p}{r} \equiv 0 \pmod{p}.$$

Hence show that
$$(a+b)^p \equiv a^p + b^p \pmod{p}.$$

Give an example to show that this result is not true when p is composite.

21. (i) Show that, if $m > 2$, the set $\{1^2, 2^2, \ldots, m^2\}$ is not a C.S.R. $\pmod{m}$.

(ii) Show that the sets (a) $\{2, 4, 6, \ldots, 2m\}$ and (b) $\{1, 3, 5, \ldots, 2m-1\}$ are complete sets of residues $\pmod{m}$ if and only if m is odd.

22. Show that (i) $a^{13} - a \equiv 0 \pmod{2730}$ for every integer a,

(ii) $a^{17} - a \equiv 0 \pmod{8160}$ for every odd integer a.

23. If n is a positive integer, the number of positive integers $\leqslant n$ that are prime to n is denoted by $\phi(n)$. The function ϕ defined on the set of positive integers in this way is called **Euler's function**.

Verify that $\phi(1) = 1$, $\phi(2) = 1$, $\phi(3) = 2$, $\phi(4) = 2$, $\phi(5) = 4$, $\phi(6) = 2, \ldots$.

Prove that, if p is a prime, then $\phi(p) = p - 1$ and $\phi(p^\alpha) = p^{\alpha-1}(p-1)$.

Any set of $\phi(n)$ integers such that each is congruent (mod n) to one of the positive integers $\leqslant n$ that are prime to n is called a **reduced set of residues** (mod n) [an R.S.R. (mod n) for short]. Each integer in such a set is prime to n. Each complete set of residues (mod n) contains one and only one reduced set of residues (mod n).

e.g. for $n = 12$, $\{0, 1, 2, 3, 4, 5, 6, 7, 8, 9, 10, 11\}$ is a C.S.R. (mod 12) and $\{1, 5, 7, 11\}$ is a R.S.R. (mod 12);

Similarly, for $n = 7$, $\{0, 1, 2, 3, 4, 5, 6\}$ is a C.S.R. (mod 7) and $\{1, 2, 3, 4, 5, 6\}$ is a R.S.R. (mod 7).

Euler proved the following extension of Theorem 6.4:

$$\text{If } (a, m) = 1, \quad \text{then} \quad a^{\phi(m)} \equiv 1 \text{ (mod } m).$$

A proof can be modelled on that of Theorem 6.4 as follows:

(i) Let $n = \phi(m)$ and let $\{x_1, x_2, \ldots, x_n\}$ be a R.S.R. (mod m).

(ii) Show that $\{ax_1, ax_2, \ldots, ax_n\}$ is also a R.S.R. (mod m).

(iii) Deduce the result from $\displaystyle\prod_{i=1}^{n} (ax_i) \equiv \prod_{i=1}^{n} x_i \text{ (mod } m).$

Note. It can be proved that ϕ is multiplicative in the following sense. If $(m, n) = 1$, then

$$\phi(mn) = \phi(m).\phi(n).$$

Deduce that, if $n = p_1^{\alpha_1} p_2^{\alpha_2} \ldots p_r^{\alpha_r}$ is the prime decomposition of n, then

$$\phi(n) = p_1^{\alpha_1 - 1} p_2^{\alpha_2 - 1} \ldots p_r^{\alpha_r - 1}(p_1 - 1)(p_2 - 1) \ldots (p_r - 1).$$

Complex Numbers

1. Introduction. An informal approach.

Chapter 4 outlined the construction of the basic system $\mathbf{N}$ of natural numbers and its extension through $\mathbf{Z}$ and $\mathbf{Q}$ to the system $\mathbf{R}$ of real numbers. Each extension enables us to solve equations which were previously unsolvable. Thus in $\mathbf{Q}$ we can solve any equation $ax+b = 0$ with $a \neq 0$ and $a, b \in \mathbf{Z}$ (or indeed $a, b \in \mathbf{Q}$). Moreover, in $\mathbf{R}$ (but not in $\mathbf{Q}$), for each non-negative number k, the equation

$$x^2 = k \tag{1.1}$$

is solvable. It follows that we can solve every equation

$$ax^2 + bx + c = 0 \tag{1.2}$$

with $a, b, c \in \mathbf{R}$, $a \neq 0$ and $b^2 \geqslant 4ac$, for (1.2) is equivalent to

$$(2ax+b)^2 = b^2 - 4ac.$$

However there is no element x of $\mathbf{R}$ satisfying (1.1) when $k < 0$; in particular

$$x^2 = -1 \tag{1.3}$$

has no solution in $\mathbf{R}$. Consequently (1.2) is not satisfied by any real number if $b^2 < 4ac$.

Assume now, tentatively, that the extension process can be continued to yield a number system containing $\mathbf{R}$ and also a solution of (1.3). If the elements of the new system are to be easily handled, and indeed worthy of the name "number", they should, we feel, satisfy as many as possible of the properties listed for $\mathbf{Q}$ in Chapter 4, Section 3, p. 63. Among these are (omitting the quantifier $\forall$ for brevity)

$$\begin{array}{cc} (x+y)+z = x+(y+z) & (xy)z = x(yz) \\ x+y = y+x & xy = yx \end{array} \right\} \tag{1.4}$$
$$x(y+z) = xy+xz.$$

117

Let a, b be any two real numbers. By assumption these are in the new system, which also contains an element i such that $i^2 = -1$. (Electrical engineers often use the symbol j instead of i.) We must be able to form the product bi, then the sum $a+bi$. If now $a'+b'i$, with a', $b' \in \mathbf{R}$, is a second "number" formed in the same way, then assuming the properties (1.4) we find that

$$(a+bi)+(a'+b'i) = (a+a')+(b+b')i, \tag{1.5}$$

$$(a+bi)(a'+b'i) = (aa'-bb')+(ab'+a'b)i. \tag{1.6}$$

Write out the details for yourself.

Example 1. (i) $(1+2i)+(2+3i) = 3+5i$.
(ii) $(1+2i)(2+3i) = -4+7i$.
(iii) $(1+2i)^2 = (1+2i)(1+2i) = -3+4i$.

(1.5) and (1.6) indicate that the sum and product of numbers $a+bi$ will be expressible in the same form. Also the sum and product are clearly commutative, as we wished them to be. You may like to verify the associative property too; for multiplication this involves several lines of working. Since we want to keep the new system as small as possible, we shall not look for any more elements—our exploration will suggest strongly that those already considered do everything we can reasonably expect. A number system with the properties we have assumed does indeed exist, a typical member being of the form $a+bi$ (a, $b \in \mathbf{R}$) and called a **complex number**. When we have gained practice in working with these numbers according to the rules which we are gradually building up we shall start afresh in Section **2**. With experience to guide us we shall then set up the new system on a more rigorous basis. Suffice it for the moment to say that not only can the system be constructed, but its use opens up wide horizons in pure mathematics; it also has far-reaching applications to physical problems.

We include the real number a in the form $a+bi$ by regarding it as $a+0.i$. Also i is included as $0+1.i$. We have

$$0+(a+bi) = (0+a)+bi = a+bi,$$
$$1(a+bi) = 1a+(1b)i = a+bi$$

so 0 and 1 are respectively additive and multiplicative identity elements.

To sum up, the working rules established so far amount to saying that, as far as addition and multiplication are concerned the complex numbers $a+bi$, $c+di$, ... are manipulated according to the familiar rules of algebra, and i^2 is replaced by -1 whenever it occurs.

Positive powers of a complex number are defined in the usual way. In particular

$$i^3 = i.i^2 = i(-1) = -i, i^4 = -i.i = -(-1) = 1,$$
$$i^5 = i, i^6 = -1, i^7 = -i, i^8 = 1, \ldots$$

the sequence repeating itself cyclically.

Notation and terminology. We denote the set of complex numbers by **C**. A typical element of **C** is usually denoted by a single letter such as z. If $z = x + yi$ with $x, y \in \mathbf{R}$ then x is called the **real part** of z and denoted by $\mathcal{R}z$; y is called the **imaginary part** of z, denoted by $\mathcal{I}z$. The latter term is a relic of the name "imaginary numbers" by which complex numbers used to be known. Note that it is y and not yi (as you might perhaps expect) which is the "imaginary part".

Example 2. (i) $\mathcal{R}i = 0$, $\mathcal{I}i = 1$, $\mathcal{R}(1+2i) = 1$, $\mathcal{I}(1+2i) = 2$. (ii) For all $z \in \mathbf{C}$, $z = \mathcal{R}z + (\mathcal{I}z)i$.

Remember that $\mathbf{R} \subset \mathbf{C}$, i.e. every real number is also a complex number. A number of the form yi ($y \in \mathbf{R}$, $y \neq 0$) is called a **pure imaginary** number. Sometimes in speaking of a complex number we want to exclude the special case of a real number (e.g. in Section **8**)—we may then speak of a **non-real** number.

By working in **C** we can solve all those quadratic equations with real coefficients which were previously unsolvable.

Example 3. The equation

$$z^2 - 8z + 25 = 0$$

is equivalent to $(z-4)^2 = -9$. Now $(3i)^2 = 3^2 i^2 = -9$ and also $(-3i)^2 = -9$. This suggests putting

$$z - 4 = 3i \text{ or } -3i$$

whence
$$z = 4 + 3i \text{ or } 4 - 3i$$

(we naturally write $4 - 3i$ rather than $4 + (-3i)$). That these values of z *do* satisfy the equation can be verified by direct calculation. There are no more roots—this will appear in Section **8**.

Subtraction in **C** is easy. Find for yourself a complex number z satisfying

$$z + (2 + 3i) = 1 + 4i.$$

We write this number z as $(1+4i)-(2+3i)$. The general rule should be apparent:

$$(a+bi)-(a'+b'i) = (a-a')+(b-b')i.$$

In particular $a+bi$ has $-a-bi$ as its **additive inverse**.

To investigate the possibility of **division** in $\mathbf{C}$ we seek a multiplicative inverse for a complex number. It is easy to see that i has an inverse, because

$$i(-i) = -i^2 = -(-1) = 1$$

so that $i^{-1} = -i$. What is the inverse of $-i$? As a more general example we look for an inverse of $2+i$, that is, a number $a+bi$ such that $(a+bi)(2+i) = 1$. This leads to the equations $2a-b = 1$, $a+2b = 0$ which have the unique solution $a = \frac{2}{5}$, $b = -\frac{1}{5}$. Thus

$$(2+i)^{-1} = \tfrac{2}{5}-\tfrac{1}{5}i = \tfrac{1}{5}(2-i).$$

It is now easy to discover the general rule, that, if x and y are not both 0, the complex number $z = x+yi$ has inverse

$$z^{-1} = \frac{x}{x^2+y^2} - \frac{y}{x^2+y^2}i = \frac{x-yi}{x^2+y^2}. \tag{1.7}$$

The condition that x and y are not both 0, i.e. that $z \neq 0$, ensures that the denominator x^2+y^2 is non-zero. Thus every complex number except 0 has a multiplicative inverse. In fact $\mathbf{C}$, like $\mathbf{Q}$ and $\mathbf{R}$, is a **field**.

Exercises. (a) Evaluate $(1+2i)(2+i)(3-i)$, $(3-2i)^3+(3+2i)^3$, $(2+i)^5$. Use the Binomial Theorem for the last of these.

(b) Show that $(1+i)^4 = -4$ and find three other complex numbers z satisfying $z^4 = -4$.

(c) Solve the equations $z^2+4z+53 = 0$, $2z^2+3z+2 = 0$.

(d) If $z = x+yi$ with x, $y \in \mathbf{R}$ find $\mathscr{R}(z^2)$, $\mathscr{I}(z^3)$, $\mathscr{R}[(1+z)/(1-z)]$ ($z \neq 1$).

(e) Find $[(1-i)(2+3i)]^{-1}$ in two ways.

2. Construction of the system C

So far we have merely *assumed* the existence of a number system containing $\mathbf{R}$ and a number i such that $i^2 = -1$. We have however found strong evidence for the existence of a **field** with these properties. Now we indicate how to construct such a system using as "raw material" just the real numbers and the algebraic operations upon them.

Each of the "informal" complex numbers $x+yi$ is specified by two real numbers x and y. So we make our fresh start by considering the set $\mathbf{R}^2 = \mathbf{R} \times \mathbf{R}$ of all ordered pairs (x, y) of real numbers. Initially $\mathbf{R}^2$ is just a set without any "algebraic structure". We have to define operations of addition and multiplication on $\mathbf{R}^2$. How are we to do this? A similar problem arose in Sections 2 and 3 of Chapter 4 when we were constructing $\mathbf{Z}$ and $\mathbf{Q}$. There we were guided by intuitively familiar properties of integers and fractions (rational numbers). Here too we have experience to guide us, namely the exploratory work of Section 1. As far as addition is concerned this suggests that we define the sum of two elements of $\mathbf{R}^2$ by

$$(x_1, y_1) + (x_2, y_2) = (x_1 + x_2, y_1 + y_2). \tag{2.1}$$

(Remember too the way in which vectors are added.) The $+$ on the right-hand side of (2.1) represents the known operation of addition in $\mathbf{R}$. It is easy to check that addition, so defined, is associative and commutative, that it has $(0, 0)$ as additive identity (zero) element, and that (x, y) has additive inverse $-(x, y) = (-x, -y)$.

The definition of multiplication analogous to (2.1) would be

$$(x_1, y_1)(x_2, y_2) = (x_1 x_2, y_1 y_2). \tag{2.2}$$

This however is unsatisfactory—it does not have all the properties we require. Moreover it is not in accord with the investigations of Section 1. These suggest [see (1.6)] the definition

$$(x_1, y_1)(x_2, y_2) = (x_1 x_2 - y_1 y_2, x_1 y_2 + x_2 y_1). \tag{2.3}$$

It is straightforward to check the associative and commutative properties; the former is a little long but you may already have done it, in effect, while studying Section 1. The element $(1, 0)$ has the (multiplicative) identity property: $\forall (x, y) \in \mathbf{R}^2$,

$$(1, 0)(x, y) = (1.x - 0.y, 1.y + 0.x) = (x, y).$$

The inverse of an element (x, y) other than $(0, 0)$ can be sought just as in Section 1: we find

$$(x, y)^{-1} = \left(\frac{x}{x^2 + y^2}, \frac{-y}{x^2 + y^2} \right) \tag{2.4}$$

this being meaningful since $x^2 + y^2 \neq 0$.

We must not forget the distributive property, linking addition and multiplication:

$$(x_1, y_1)[(x_2, y_2) + (x_3, y_3)] = (x_1, y_1)(x_2, y_2) + (x_1, y_1)(x_3, y_3),$$

but once again verification is straightforward.

Exercise. Investigate the operation of multiplication defined on $\mathbf{R}^2$ by (2.2). Why is it less satisfactory than (2.3)?

We have now proved that, with addition and multiplication defined by (2.1) and (2.3), the set $\mathbf{R}^2$ becomes a **field** (cf. Chapter 10). Next we concentrate our attention on the subset of $\mathbf{R}^2$ consisting of all elements of the form $(x, 0)$, $x \in \mathbf{R}$. Denote this subset temporarily by $\mathbf{S}$. The sum and product of two elements of $\mathbf{S}$ are given by

$$(x_1, 0)+(x_2, 0) = (x_1+x_2, 0), \tag{2.5}$$

$$(x_1, 0)(x_2, 0) = (x_1 x_2, 0) \tag{2.6}$$

and are themselves in $\mathbf{S}$. Moreover (2.5) and (2.6) show that as far as the operations of addition and multiplication are concerned (and, along with these, subtraction and division) the symbols $(x, 0)$ *behave exactly like the real numbers x*. Since these operations are our primary concern, and it is clearly more convenient to write x than $(x, 0)$, we shall do this henceforth.

Note. This situation may be described more technically by using a term which will be mentioned briefly in Chapters 9 and 10. By associating with $x \in \mathbf{R}$ the element $(x, 0)$ of $\mathbf{S}$ we obtain a mapping from $\mathbf{R}$ to $\mathbf{S}$. This mapping is a bijection (why?). $\mathbf{R}$ and $\mathbf{S}$ are algebraic systems of the same type—both fields. ($\mathbf{R}^2$ is also a field and $\mathbf{S}$ being a subset of $\mathbf{R}^2$ which is itself a field, may be called a "subfield" of $\mathbf{R}^2$.) (2.5) and (2.6) may be expressed by saying that the bijection "preserves" the operations of addition and multiplication—it is an "isomorphism". What we are doing above is "identifying" the fields $\mathbf{S}$ and $\mathbf{R}$ which are isomorphic (essentially the same). Another example of isomorphism involving $\mathbf{C}$ is given at the end of Chapter 10.

Consider now the element $(0, 1)$ of $\mathbf{R}^2$. By (2.3), $(0, 1)^2 = (0, 1)(0, 1) = (-1, 0)$ which we are now writing simply as -1. So -1 has a square root $(0, 1)$ in $\mathbf{R}^2$. Following the lead of Section 1 we denote $(0, 1)$ by i. We then have $(0, y) = (y, 0)(0, 1)$ or briefly yi, and finally

$$(x, y) = (x, 0)+(0, y) = x+yi.$$

We see that we have validated all the provisional work of Section 1. We can therefore return to the notation of that section, writing $\mathbf{C}$ rather than $\mathbf{R}^2$ and $x+yi$ instead of (x, y). (Occasionally the form $x+iy$ is more convenient; the two are equal by the commutative property of multiplication). In particular the zero and (multiplicative) identity elements $(0, 0)$ and $(1, 0)$ become just 0 and 1.

An important property of C. The fact that every non-zero complex number has an inverse leads at once to

Theorem 2.1. *If z, $w \in C$ and $zw = 0$ then $z = 0$ or $w = 0$.*

Proof. Either $z = 0$, in which case there is nothing more to do, or $z \neq 0$. In the latter case z has an inverse z^{-1} and

$$w = 1w = (z^{-1}z)w = z^{-1}(zw) = z^{-1}.0 = 0.$$

As an immediate consequence we have the **cancellation property**:

Theorem 2.2. *If z, w_1, $w_2 \in C$, $z \neq 0$ and $zw_1 = zw_2$, then $w_1 = w_2$.*

Proof. $zw_1 = zw_2 \Rightarrow z(w_1 - w_2) = 0$. Since $z \neq 0$, Theorem 2.1 shows that $w_1 - w_2 = 0$ and so $w_1 = w_2$.

Theorem 2.1 can be restated thus:

If z, $w \in C$ and $z \neq 0$, $w \neq 0$, then $zw \neq 0$.

Exercise. Prove by induction that if $n \in N$ and $z_1, \ldots, z_n$ are n non-zero complex numbers, then $z_1 z_2 \ldots z_n \neq 0$.

It follows that if a product of complex numbers is 0, then at least one of the factors is 0.

Equating real and imaginary parts. It is implicit in the definition of $R^2 = R \times R$ that

$$(x_1, y_1) = (x_2, y_2) \Leftrightarrow x_1 = x_2 \text{ and } y_1 = y_2.$$

It often happens that we are told, or know as a result of some calculations, that two complex numbers are equal. We can then be certain that their real parts are equal and likewise their imaginary parts.

Example 4. Given that x, $y \in R$ and $(x+yi)^2 = 5 - 12i$, find x and y.

We have $x^2 - y^2 + 2xyi = 5 - 12i$, which is equivalent to

$$x^2 - y^2 = 5, \qquad 2xy = -12, \qquad xy = -6.$$

It follows that $x^4 - 5x^2 - 36 = 0$, $x^2 = 9$ or $x^2 = -4$. The second possibility must be rejected since $x \in R$. The first gives $x = 3$, when $y = -2$; or $x = -3$, $y = 2$. Check by substituting in the original equation.

This process is called **equating real and imaginary parts**. It will be exploited in Section 7.

In this respect it is interesting to compare the construction of C by means of ordered pairs with the corresponding constructions for Z and

Q sketched in Chapter 4. The latter were more elaborate in that they involved equivalence relations. In **Q**, for instance, we cannot "equate numerators and denominators"—from $x_1/y_1 = x_2/y_2$ with $x_1, x_2 \in \mathbf{Z}$, $y_1, y_2 \in \mathbf{N}$ we cannot deduce without further information that $x_1 = x_2$ and $y_1 = y_2$. Of all the extensions of number system listed in

$$\mathbf{N} \subset \mathbf{Z} \subset \mathbf{Q} \subset \mathbf{R} \subset \mathbf{C}$$

the last is the easiest!

A warning. Although the passage from **R** to **C** brings many advantages, particularly in the solution of equations, one thing is lost! **R**, like **Q** [see Chapter 4, Section 3] is an **ordered field**. **C** is *not* an ordered field. To see this, suppose there existed in **C** an order relation expressed by < and > satisfying the properties listed on p. 63. It is an immediate consequence of these properties that if $x \neq 0$ then $x^2 > 0$. In particular $1 = 1^2 > 0$. Also $-1 = i^2 > 0$, so $1 + (-1) > 1 + 0$. Thus $1 > 0$ and $0 > 1$ both hold, which is impossible.

Remembering that $\mathbf{R} \subset \mathbf{C}$ we can sum this up by saying: The inequality signs $<, \leqslant, >, \geqslant$ must not be used in conjunction with complex numbers unless we know that the particular numbers involved are in fact real.

Exercise. Find the complex numbers whose square is $-15 + 8i$.

3. Geometrical representation of complex numbers

Let E^2 be the familiar Euclidean plane of elementary geometry and set up a rectangular coordinate system in E^2. By associating with the complex number $x + yi (x, y \in \mathbf{R})$ the point with coordinates (x, y) in E^2 we obtain a mapping from **C** to E^2. This mapping is a bijection; it provides an interesting geometrical representation of complex numbers. Under this representation the algebraic operations of addition and multiplication in **C** correspond to simple geometrical operations in E^2. We shall study these here and in Section **6**. Complex numbers can, indeed, be used to prove geometrical theorems. We shall not say anything about this aspect, but we shall mention some simple geometrical transformations in Section **6**. [See **Topics to Explore**, p. 132.]

In this context E^2 is called the **Argand diagram** (named after J. R. Argand, 1768–1822), the **Gauss plane** (after the famous German mathematician C. F. Gauss, 1777–1855) or simply the **complex plane**. The x-axis is now called the **real axis**; its points correspond to the real numbers. The y-axis is renamed the **imaginary axis**; its points other than the origin O represent the pure imaginary numbers.

Another way of looking at the representation suggests itself when we consider addition. Let $z_1 = x_1 + y_1 i$ and $z_2 = x_2 + y_2 i$ be two complex numbers with representative points $P_1(x_1, y_1)$ and $P_2(x_2, y_2)$. Then $z_1 + z_2 = (x_1 + x_2) + (y_1 + y_2)i$ is represented by the point $P_3(x_1 + x_2, y_1 + y_2)$. Using vector notation we see that

$$\vec{OP_1} + \vec{OP_2} = \vec{OP_3}. \tag{3.1}$$

This suggests that for some purposes it may be convenient to associate with $x + yi$ the vector $\vec{OP}$ rather than the point $P(x, y)$. If we do so, addition of complex numbers corresponds exactly to vector addition. Observe that (2.1) is just (3.1) expressed in terms of components.

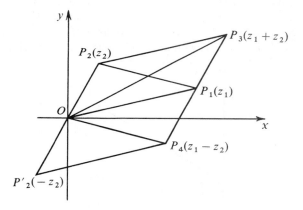

The point P'_2 representing $-z_2$ is the image of P_2 under the **half-turn** about O. Consequently the number $z_1 - z_2 = z_1 + (-z_2)$ corresponds to the point P_4 where $OP_1 P_4 P'_2$, like $OP_1 P_3 P_2$, is a parallelogram, or alternatively to the vector

$$\vec{OP_4} = \vec{P_2 P_1} = \vec{OP_1} - \vec{OP_2}.$$

The geometrical description of multiplication will be dealt with in Section **6**, but some simple cases are explored in Exercises (b), (c), (d) below.

Exercises. (a) What figure is formed by the points of E^2 representing the complex numbers $1, i, -1, -i$?

(b) Let k be a non-zero real number. Prove that for each $z \in C$ the point representing kz is obtained from the point representing z by applying the **dilatation** $[O, k]$ with centre O and constant k. What is the corresponding operation in vector algebra?

(*c*) Prove that for each $z \in \mathbf{C}$ the point representing iz is obtained from that representing z by applying the **positive quarter-turn** (rotation through $90°$ anticlockwise) about O. Denote this rotation by R and let H be the half-turn about O. Then $R^2 = R \circ R = H$. This is the geometrical interpretation of the fundamental relation $i^2 = -1$.

(*d*) The points P, Q distinct from O represent the numbers z, w. Show that $\angle POQ = 90°$ if and only if z/w is pure imaginary.

4. The modulus and argument of a complex number

We have seen that addition of complex numbers corresponds to vector addition in the complex plane. On the other hand the expression (1.6) for the product of two complex numbers is not suggestive of any simple geometrical process. We introduce now an alternative way of writing complex numbers which is well adapted to deal with multiplication.

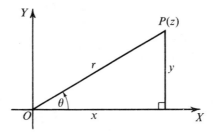

As in Section **3** we represent the complex number $z = x + yi$ by the point $P(x, y)$. Let r, θ be the **polar coordinates** of P, using O as **pole** and OX as **initial line**. This means that $r = OP$ and $\theta = \angle XOP$, the angle being measured from OX to OP with the usual convention "anticlockwise positive, clockwise negative." Thus

$$x = \mathscr{R}z = r \cos \theta, \qquad y = \mathscr{I}z = r \sin \theta.$$

Examples (i) If $z = i$ then $r = 1$ and $\theta = \frac{1}{2}\pi$.

(ii) If $z = 1 - i$ then $r = \sqrt{2}$, $\theta = -\frac{1}{4}\pi$. (We use *radian* measure for θ.) Alternatively, $\theta = 2\pi - \frac{1}{4}\pi = \frac{7}{4}\pi$.

The polar coordinate $r = OP$ is a *non-negative* real number called the **modulus** of z and denoted by $|z|$. Read this symbol as "mod z". For example, $|1 - i| = \sqrt{2}$. Generally, for $z = x + yi$, $|z| = \sqrt{(x^2 + y^2)}$. [**Note.** The sign $\sqrt{}$ always denotes the non-negative square root.] If $z = x$, a real number, then

$$|z| = \sqrt{(x^2)} = \begin{cases} x & \text{if } x \geqslant 0; \\ -x & \text{if } x < 0. \end{cases}$$

This is the absolute value of x as defined in Chapter 3, Section 1, Example 2 and denoted by $|x|$. The two uses of the sign $|\ |$ are therefore consistent.

Exercise. Prove that $\mathscr{R}z \leqslant |\mathscr{R}z| \leqslant |z|$, $\mathscr{I}z \leqslant |\mathscr{I}z| \leqslant |z|$.

Distance between two points of the complex plane. Let P_1, P_2 be two points corresponding to the numbers z_1, z_2. Then we saw in Section 3 that $z_1 - z_2$ is represented by the point P_4 where $\overrightarrow{OP_4} = \overrightarrow{P_2P_1}$. Hence

$$\text{distance } P_2P_1 = \text{distance } OP_4 = |z_1 - z_2|.$$

The argument of a complex number.

The polar angle θ is called the **argument** (or **amplitude**, or **phase**) of z. The argument is *not defined* when $z = 0$. Moreover the angle θ is not *uniquely* determined by z (see Example (ii) above). For given values of x and y, not both 0, if the equations

$$\cos\theta = \frac{x}{\sqrt{(x^2 + y^2)}}, \qquad \sin\theta = \frac{y}{\sqrt{(x^2 + y^2)}}$$

are satisfied by a particular value θ_0 of θ, then they hold also for $\theta = \theta_0 + 2\pi$, $\theta_0 + 4\pi, \ldots, \theta_0 - 2\pi, \theta_0 - 4\pi, \ldots$ and generally for $\theta = \theta_0 + 2n\pi$ ($n \in \mathbf{Z}$), but for no other values of θ. The notation arg z may be used to denote any of these values. Sometimes however it is convenient to impose a restriction such as

$$-\pi < \theta \leqslant \pi, \tag{4.1}$$

or $\theta \in (-\pi, \pi]$ in the notation of Chapter 2, Section 4 (iv), whereupon θ is completely determined by z for $z \neq 0$. With this restriction 'arg' becomes a mapping from $\mathbf{C} - \{0\}$ to $\mathbf{R}$ (*Note*. Here $-$ is the sign for *relative difference* introduced in Chapter 2, Section 5, IV, and $\{0\}$ is the set consisting of the number 0 only. So $\mathbf{C} - \{0\}$ denotes the set of all complex numbers excluding 0, and $z \in \mathbf{C} - \{0\}$ would be a convenient way of abbreviating "z is a non-zero complex number.") The unique value of arg z satisfying (4.1) is called the **principal value**.

Exercise. Find the modulus and the principal value of the argument for each of the complex numbers $1 + i$, $-3 + 3i$, $-5i$, -10, $i\sqrt{3} - 1$, $-\sqrt{2}(1 + i)$, $2 - 3i$, using tables for the last.

The complex number z can now be written in **polar** or **modulus-argument form**, as

$$z = r\cos\theta + ir\sin\theta = r(\cos\theta + i\sin\theta)$$
$$= |z|\{\cos(\arg z) + i\sin(\arg z)\}.$$

In particular a number of modulus 1 is of the form $\cos\theta + i\sin\theta$. The product of two such numbers is especially noteworthy:

$$(\cos\theta + i\sin\theta)(\cos\phi + i\sin\phi)$$
$$= (\cos\theta\cos\phi - \sin\theta\sin\phi) + i(\cos\theta\sin\phi + \sin\theta\cos\phi)$$
$$= \cos(\theta + \phi) + i\sin(\theta + \phi).$$

What is the modulus of the product?

For two general complex numbers

$$z = r(\cos\theta + i\sin\theta),$$
$$w = s(\cos\phi + i\sin\phi)$$

we have
$$zw = rs(\cos\theta + i\sin\theta)(\cos\phi + i\sin\phi)$$
$$= rs[\cos(\theta + \phi) + i\sin(\theta + \phi)].$$

So *any two complex numbers may be multiplied by*
  (i) *multiplying their moduli*
and (ii) *adding their arguments.*

Example 5. Let $z = 1 + i\sqrt{3}$, $w = \sqrt{3} - i$. We find that

$$|z| = |w| = 2, \qquad \arg z = \tfrac{1}{3}\pi, \qquad \arg w = -\tfrac{1}{6}\pi$$

(the arguments here being principal values). Consequently $|zw| = 4$ and $\arg(zw) = \tfrac{1}{6}\pi$, so

$$zw = 4(\tfrac{1}{2}\sqrt{3} + \tfrac{1}{2}i) = 2(\sqrt{3} + i).$$

Check this by direct multiplication.

Step (i) in the multiplication process is expressed by the simple relation

$$|zw| = |z||w|. \tag{4.2}$$

Corresponding to step (ii) we might write

$$\arg(zw) = \arg z + \arg w \tag{4.3}$$

but this needs a little care in interpretation, in view of the fact that a complex number does not determine its argument completely. (4.3) is true in the sense that if any values of $\arg z$ and $\arg w$ are substituted in the right-hand side the resulting sum is one of the values of the left-hand side, and all such values can be so obtained. If, however, we mean by 'arg z,' etc., the *principal values* of the arguments then (4.3) does not necessarily hold. See Exercise (*c*) below. This is one reason why the restriction to principal values is sometimes inconvenient.

The polar form of complex numbers is ideally suited for multiplication, but awkward for addition. There is no way of simplifying the expression

$$r(\cos \theta + i \sin \theta) + s(\cos \phi + i \sin \phi)$$

in general. It is very important to remember that, except in special cases, $|z+w| \neq |z|+|w|$. However the three moduli are related by the important *inequality*

$$|z+w| \leqslant |z|+|w| \qquad (4.4)$$

which has a simple geometrical interpretation. Let P, Q, R be the points representing $z, w, z+w$ respectively. If, as in the figure, O, P, Q are not collinear, we have from the triangle OPR

$$OR < OP + PR = OP + OQ,$$

i.e.
$$|z+w| < |z| + |w|.$$

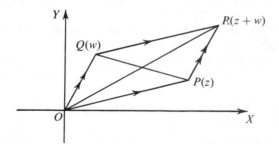

The same conclusion is valid when O, P, Q are collinear, with O between P and Q (draw a figure) so that

$$w = kz \qquad (4.5)$$

with k real and negative. If however (4.5) holds with $k>0$ then P, Q are on the same side of O and we have

$$|z+w| = |z|+|w|$$

(draw another figure). For other proofs of (4.4), which is called the **triangle inequality** in virtue of the geometrical interpretation just given, see Exercise (*d*) below and Exercise 9, No. 10, p. 152.

Exercise. (*a*) (i) Prove that $\frac{1}{2}\pi - \theta$ is a value of $\arg(\sin\theta + i\cos\theta)$. Determine the principal value when $-\pi < \theta \leqslant \pi$.

(ii) Prove that if $z = 1 + \cos\theta + i\sin\theta$ then $|z| = 2|\cos\frac{1}{2}|$. (Remember that $|z| \geqslant 0$ always!) Show that, if $0 \leqslant \theta < \pi$ then $\frac{1}{2}\theta$ is the principal value of $\arg z$. What happens if $\pi < \theta < 2\pi$? Draw a figure for each case.

(*b*) Describe geometrically the complex numbers z for which $|z| = |z-1|$. Show that there is just one value of z satisfying

$$|z| = |z-1| = |z-i|.$$

(*c*) (i) Prove that if $z = w = -1$ and 'arg' means principal value then $\arg(zw) \neq \arg z + \arg w$.

(ii) Prove that, for principal values,

$$\arg(zw) = \arg z + \arg w + \varepsilon,$$

where $\varepsilon = 0$, 2π or -2π. Illustrate by examples.

(*d*) Let $z = r(\cos\theta + i\sin\theta)$, $w = s(\cos\phi + i\sin\phi)$. Calculate $|z+w|^2$ in terms of r, s, θ, ϕ. Use the inequality $\cos(\theta-\phi) \leqslant 1$ to prove (4.4) and investigate when equality occurs.

(*e*) [This may be omitted at a first reading].

(i) Without some limitation such as (4.1) 'arg' would not be a mapping. Why?

(ii) When the restriction (4.1) is imposed, what is the image of the mapping 'arg'?

(iii) Why would it not be appropriate to use $(-\pi, \pi)$ or $[-\pi, \pi]$ instead of $(-\pi, \pi]$ in restricting 'arg'? Could we have used $[-\pi, \pi)$? $[0, 2\pi)$?

(iv) With each $z \in \mathbf{C}$ is associated its modulus $|z| \in \mathbf{R}$. Does this define a mapping from $\mathbf{C}$ to $\mathbf{R}$? If so:

What is its image?   Is it injective?   Is it surjective?

5. Conjugate complex numbers

Recall from Sections **1** and **2** that the multiplicative inverse z^{-1} of the non-zero complex number $z = x + yi$ is given by $z^{-1} = (x - yi)/(x^2 + y^2)$. We call $x - yi$ the complex number **conjugate** to z, or briefly "the conjugate of z", and denote it by $\bar{z}$. Thus

$$\bar{2} = 2, \qquad \bar{i} = -i, \qquad \overline{3 - 2i} = 3 + 2i.$$

Since the modulus of z is $|z| = \sqrt{(x^2 + y^2)}$ we have $z^{-1} = \bar{z}/|z|^2$. This can be written

$$z\bar{z} = |z|^2 \tag{5.1}$$

and is one of several useful relations involving $\bar{z}$. (5.1) holds for $z = 0$, too. We also have

$$z + \bar{z} = x + yi + x - yi = 2x$$

and similarly $z - \bar{z} = 2yi$, so

$$x = \mathscr{R}z = \frac{1}{2}(z + \bar{z}), \qquad y = \mathscr{I}z = \frac{1}{2i}(z - \bar{z}) \qquad (5.2)$$

Moreover

$$z \in \mathbf{R} \Leftrightarrow \bar{z} = z,$$

z is pure imaginary or $0 \Leftrightarrow \bar{z} = -z$.

Write, temporarily, $w = \bar{z}$. Then $w = x - yi$ and we have $\bar{w} = x - (-y)i = x + yi = z$. Hence

$$\bar{\bar{z}} = z. \qquad (5.3)$$

Putting $z_1 = x_1 + y_1 i$, $z_2 = x_2 + y_2 i$ with $x_1, x_2, y_1, y_2 \in \mathbf{R}$ it is easy to prove

$$\text{(i) } \overline{z_1 + z_2} = \bar{z}_1 + \bar{z}_2, \qquad \text{(ii) } \overline{z_1 z_2} = \bar{z}_1 \bar{z}_2. \qquad (5.4)$$

Example 6. Prove that for all z, w in $\mathbf{C}$,

$$\mathscr{R}(z\bar{w}) = \tfrac{1}{2}(z\bar{w} + \bar{z}w).$$

We have by (5.2), $\quad 2\mathscr{R}(z\bar{w}) = z\bar{w} + \overline{z\bar{w}}$

$$= z\bar{w} + \bar{z}\bar{\bar{w}} \quad \text{by (5.4)(ii)}$$
$$= z\bar{w} + \bar{z}w \quad \text{by (5.3)}.$$

It follows incidentally that $z\bar{w} + \bar{z}w$ is always real.

Exercises. (a) (i) Prove that $|\bar{z}| = |z|$ for all $z \in \mathbf{C}$.
   (ii) Prove that $\overline{1/z} = 1/\bar{z}$ for all $z \in \mathbf{C} - \{0\}$. [See *Note* on p. 127.]
   (iii) How are the principal values of arg z and arg $\bar{z}$ related ($z \neq 0$)?
   (b) Prove that for all z, $w \in \mathbf{C}$, $z\bar{w} - \bar{z}w$ is pure imaginary or zero.
   (c) Define a mapping $\phi: \mathbf{C} \to \mathbf{C}$ by $\phi(z) = \bar{z}$. Use (5.3) to prove that ϕ is a bijection. What can you say about $\phi \circ \phi$?

6. Geometrical interpretation of complex numbers (continued)

We can now complete our study of the geometrical description of multiplication. Let P, Q, P' be the points in the complex plane representing z, w, z' respectively, where

$$z' = wz \qquad (6.1)$$

Suppose first that $|w| = 1$ so that $w = \cos\alpha + i\sin\alpha$ for some angle α. Then

$$OP' = |z'| = |w||z| = 1.|z| = |z| = OP$$
and $$\angle XOP' = \arg z' = \arg w + \arg z = \alpha + \angle XOP$$

(we do not insist on principal values here). Hence P' is obtained from P by *rotation* about the origin O through angle α.

If, finally, w is a general non-zero complex number,

$$w = k(\cos\alpha + i\sin\alpha)$$

where $k \neq 0$, this rotation must be combined with the dilatation $[O, k]$. Note that the rotation and dilatation commute—draw a figure to illustrate this. The product is a transformation of a type known as a **spiral similarity**. It transforms any triangle into a **directly similar** triangle.

The above geometrical description of multiplication can be expressed briefly as follows. Let A, P, Q, P' be the points representing 1, z, w, wz respectively. Then triangles OAQ, OPP' are directly similar. Draw a figure.

Exercise. Since $zw = wz$ the roles of $P(z)$ and $Q(w)$ in this description can be interchanged. So there is another pair of similar triangles in the figure. What are they?

The relation (6.1), with w a non-zero constant, can be regarded as an equation of a **transformation** under which $P'(z')$ is the image of $P(z)$. Dilatations and rotations with centre O appear as special cases. Other kinds of transformation have equations which can be expressed simply in terms of complex numbers.

Topics to explore. I. Remember that addition of complex numbers is just vector addition. So the equation $z' = z + b$, with b constant, corresponds to

$$\overrightarrow{OP'} = \overrightarrow{OP} + \overrightarrow{OB}$$

where $\overrightarrow{OB}$ is a constant vector. This is a **translation**.

II. Let $z' = \bar{z}$. Then if P is (x, y), P' is $(x, -y)$. This transformation is **reflection** in the real axis. What property of reflection is expressed by equation (5.3)? Show that $z' = -\bar{z}$ is also a reflection. Use these equations, together with (5.3), to work out the compositions (in both orders) of these reflections. Do they commute? What is the product transformation? Consider the reflection in the line $y = x$ and try to express its equation in terms of complex numbers.

III. What kind of transformation is represented by $z' = 2a - z$ where a is constant?

IV. Revise, or find out about, the type of transformation called a **glide-reflection**. Prove that $z' = \bar{z} + 1$ represents a glide-reflection.

[*Note.* All the transformations mentioned in I–IV are **isometries**. See Section **5** of Chapter 9.]

7. De Moivre's Theorem

In Section **4** we have already proved and used the very important relation

$$(\cos\theta + i\sin\theta)(\cos\phi + i\sin\phi) = \cos(\theta + \phi) + i\sin(\theta + \phi). \quad (7.1)$$

Putting $\phi = \theta$ we deduce that

$$(\cos\theta + i\sin\theta)^2 = \cos 2\theta + i\sin 2\theta. \quad (7.2)$$

On working out the left-hand side of this equation we find

$$(\cos^2\theta - \sin^2\theta) + i(2\sin\theta\cos\theta)$$

which incorporates the familiar expressions for $\cos 2\theta$ and $\sin 2\theta$. We now prove a generalization of (7.2) which is named after the French mathematician Abraham de Moivre (1667–1754).

Theorem 7.1 (De Moivre's theorem). *For all $n \in \mathbf{N}$ and $\theta \in \mathbf{R}$*

$$(\cos\theta + i\sin\theta)^n = \cos n\theta + i\sin n\theta. \quad (7.3)$$

Proof. We use induction on n.

Consider $n = 1$. Each side of (7.3) then has the value $\cos\theta + i\sin\theta$, so (7.3) is true for $n = 1$.

Take any $k \geqslant 1$ and suppose that (7.3) is true for $n = k$, so that

$$(\cos\theta + i\sin\theta)^k = \cos k\theta + i\sin k\theta.$$

Consider $n = k + 1$:

$$\begin{aligned}
(\cos\theta + i\sin\theta)^{k+1} &= (\cos\theta + i\sin\theta)(\cos\theta + i\sin\theta)^k \\
&= (\cos\theta + i\sin\theta)(\cos k\theta + i\sin k\theta) \\
&= \cos(\theta + k\theta) + i\sin(\theta + k\theta),
\end{aligned}$$

by (7.1) with $\phi = k\theta$,

$$= \cos(k+1)\theta + i\sin(k+1)\theta.$$

Thus, if the statement (7.3) is true for $n = $ any $k \geqslant 1$, it is also true for $n = k + 1$. Hence, by induction, the theorem is established.

Example 7. Evaluate $(\cos \frac{3}{8}\pi + i \sin \frac{3}{8}\pi)^4$.

We have, by de Moivre's theorem,

$$(\cos \tfrac{3}{8}\pi + i \sin \tfrac{3}{8}\pi)^4 = \cos 4\,(\tfrac{3}{8}\pi) + i \sin 4\,(\tfrac{3}{8}\pi)$$
$$= \cos \tfrac{3}{2}\pi + i \sin \tfrac{3}{2}\pi = 0 + i(-1) = -i.$$

Example 8. Show that for all $n \in \mathbf{N}$ and $\theta \in \mathbf{R}$,

$$(\cos \theta - i \sin \theta)^n = \cos n\theta - i \sin n\theta. \tag{7.4}$$

We have $\cos \theta - i \sin \theta = \cos(-\theta) + i \sin(-\theta)$, so by de Moivre's theorem

$$(\cos \theta - i \sin \theta)^n = \cos n\,(-\theta) + i \sin n\,(-\theta)$$
$$= \cos n\theta - i \sin n\theta.$$

It follows from de Moivre's theorem that if z is a complex number expressed in polar form as $z = r(\cos \theta + i \sin \theta)$, and if $n \in \mathbf{N}$, then

$$z^n = r^n(\cos n\theta + i \sin n\theta). \tag{7.5}$$

Example 9. Evaluate $(1+i)^{11}$.

We have

$$1 + i = \sqrt{2}\left(\frac{1}{\sqrt{2}} + i\frac{1}{\sqrt{2}}\right) = \sqrt{2}\left(\cos \frac{1}{4}\pi + i \sin \frac{1}{4}\pi\right)$$

so using (7.5),

$$(1+i)^{11} = 2^{11/2}\left(\cos \frac{11}{4}\pi + i \sin \frac{11}{4}\pi\right) = 2^{11/2}\left(\cos \frac{3}{4}\pi + i \sin \frac{3}{4}\pi\right)$$
$$= 2^{11/2}\left(-\frac{1}{\sqrt{2}} + i\frac{1}{\sqrt{2}}\right) = 2^5(i-1) = 32(i-1).$$

Note. We have

$$(\cos \theta + i \sin \theta)^{-1} = \cos \theta - i \sin \theta$$

by (1.7), so the result (7.4) of Example **8** may be written

$$(\cos \theta + i \sin \theta)^{-n} = \cos(-n)\theta + i \sin(-n)\theta$$

Here the negative index $-n$ is interpreted for complex numbers in the same way as for real numbers. This may be regarded as a generalization of de Moivre's theorem to negative integer indices. With the convention that $z^0 = 1$ for all $z \in \mathbf{C} - \{0\}$ we can say that (7.3) is true for all $n \in \mathbf{Z}$ and $\theta \in \mathbf{R}$.

Exercises. (*a*) Evaluate

$$\left(\cos\frac{5}{24}\pi+i\sin\frac{5}{24}\pi\right)^4, \quad \left(\cos\frac{7}{12}\pi+i\sin\frac{7}{12}\pi\right)^{-3}, \quad (1-i)^7, \quad (\sqrt{3}+i)^9.$$

(*b*) Simplify $(\cos 2\theta+i\sin 2\theta)^5/(\cos 3\theta+i\sin 3\theta)^3$ and $(\cos\theta+i\sin\theta)^8(\cos\theta-i\sin\theta)^4$.

(*c*) Prove that for all $n\in\mathbf{Z}$ and $\theta\in\mathbf{R}$

$$(1+\cos\theta+i\sin\theta)^n = 2^n\cos^n\tfrac{1}{2}\theta(\cos\tfrac{1}{2}n\theta+i\sin\tfrac{1}{2}n\theta).$$

The next two examples illustrate important applications of de Moivre's theorem.

Example 10. Express $\cos 4\theta$ and $\sin 4\theta$ in terms of $\cos\theta$ and $\sin\theta$.

Using de Moivre's theorem and the binomial theorem we have
$\cos 4\theta+i\sin 4\theta$

$$\begin{aligned}
&= (\cos\theta+i\sin\theta)^4\\
&= \cos^4\theta+4\cos^3\theta.i\sin\theta+6\cos^2\theta.i^2\sin^2\theta\\
&\qquad\qquad\qquad\qquad +4\cos\theta.i^3\sin^3\theta+i^4\sin^4\theta\\
&= \cos^4\theta+4i\cos^3\theta\sin\theta-6\cos^2\theta\sin^2\theta-4i\cos\theta\sin^3\theta+\sin^4\theta.
\end{aligned}$$

Equating real parts (see Section **2**, p. 123) we deduce that

$$\cos 4\theta = \cos^4\theta-6\cos^2\theta\sin^2\theta+\sin^4\theta,$$

while the imaginary parts give

$$\sin 4\theta = 4\cos^3\theta\sin\theta-4\cos\theta\sin^3\theta.$$

Exercise. Use the relation $\cos^2\theta+\sin^2\theta = 1$ to deduce from these results that

$$\cos 4\theta = 8\cos^4\theta-8\cos^2\theta+1,$$
$$\sin 4\theta = 4\sin\theta(2\cos^3\theta-\cos\theta).$$

Show also that $\qquad\cos^4\theta = \tfrac{1}{8}(\cos 4\theta+4\cos 2\theta+3).\qquad\qquad(7.6)$

Results of the type (7.6) can also be obtained by applying de Moivre's theorem in a slightly different way:

Example 11. Express $\cos^6\theta$ in terms of cosines of integer multiples of θ.

Put $z = \cos\theta+i\sin\theta$. Then we have

$$2\cos\theta = z+z^{-1}.$$

(Compare (5.2); note that $z^{-1} = \bar{z}$ here.) Hence

$$2^6 \cos^6 \theta = (z + z^{-1})^6$$

$$= z^6 + \binom{6}{1} z^5.z^{-1} + \binom{6}{2} z^4.z^{-2} + \binom{6}{3} z^3.z^{-3} + \binom{6}{4} z^2.z^{-4}$$

$$+ \binom{6}{5} z.z^{-5} + z^{-6}$$

by the binomial theorem,

$$= (z^6 + z^{-6}) + 6(z^4 + z^{-4}) + 15(z^2 + z^{-2}) + 20.$$

Now for all $n \in \mathbf{N}$, $\qquad z^n = \cos n\theta + i \sin n\theta$

and $\qquad\qquad\qquad z^{-n} = \cos n\theta - i \sin n\theta,$

so $\qquad\qquad\qquad z^n + z^{-n} = 2 \cos n\theta.$ $\hfill (7.7)$

Using (7.7) with $n = 2, 4, 6$ we have

$$2^6 \cos^6 \theta = 2 \cos 6\theta + 12 \cos 4\theta + 30 \cos 2\theta + 20,$$

so $\qquad\qquad \cos^6 \theta = \tfrac{1}{32}(\cos 6\theta + 6 \cos 4\theta + 15 \cos 2\theta + 10).$

Check. Put $\theta = 0$. Both sides reduce to 1.

$\sin^6 \theta$ can be dealt with similarly, starting with $2i \sin \theta = z - z^{-1}$. For odd powers of $\sin \theta$ we must use

$$z^n - z^{-n} = 2i \sin n\theta \hfill (7.8)$$

instead of (7.7).

Exercises. (a) Express $\cos 5\theta$ and $\sin 5\theta$ in terms of $\cos \theta$ and $\sin \theta$ by means of de Moivre's theorem. Hence express $\cos 5\theta$ as a polynomial in $\cos \theta$. Show that, for $\sin \theta \neq 0$, $\sin 5\theta/\sin \theta$ can also be expressed as a polynomial in $\cos \theta$. What is the value of this polynomial when $\theta = 0$? Is it correct to say that $\sin 5\theta/\sin \theta = 5$ when $\theta = 0$?

(b) (i) Express $\cos^5 \theta$ in terms of $\cos \theta$, $\cos 3\theta$ and $\cos 5\theta$.

(ii) Express $\sin^7 \theta$ in terms of sines of multiples of θ.

Topic to explore.   Let $c = \cos \theta$.   The relations $\cos 2\theta = 2c^2 - 1$, $\cos 3\theta = 4c^3 - 3c$, $\cos 4\theta = 8c^4 - 8c^2 + 1$ suggest that, for $n \in \mathbf{N}$, $\cos n\theta$ is a polynomial in c of degree n. Prove this by induction on n. [*Hint:* $\cos (n+1)\theta = 2 \cos n\theta \cos \theta - \cos (n-1)\theta$.] Guess a formula for the coefficient of c^n. Prove your conjecture by induction.

For what natural numbers n can $\sin n\theta$ be expressed as a polynomial in $\sin \theta$ alone?

8. Algebraic equations

We have seen that every quadratic equation

$$az^2 + bz + c = 0 \qquad (8.1)$$

where a, b, $c \in \mathbf{R}$ ($a \neq 0$) can be solved in $\mathbf{C}$. It is natural to ask, "What happens if a, b, c are allowed to be complex numbers?" Is it necessary to expand the number system still further, thus opening up the prospect of an endless sequence of extensions, becoming steadily more elaborate? The exciting (and fortunate!) answer is "No". Even if a, b, c are complex the equation (8.1) still has a solution in $\mathbf{C}$. Such equations with complex coefficients are not particularly important in applications, and it is not worth while practising their solution extensively. However the result is so striking, and its theoretical consequences so far-reaching, that it is a good idea to consider just one simple illustration.

Example 12. Solve the equation

$$z^2 + (4 - 2i)z + (8i - 2) = 0. \qquad (8.2)$$

The procedure is the same as for an equation with real coefficients—we "complete the square", writing (8.2) in the form

$$(z + 2 - i)^2 = 2 - 8i + (2 - i)^2 = 5 - 12i.$$

In Example 4 (Section 2) we found that $5 - 12i$ has exactly two square roots in $\mathbf{C}$, viz. $\pm(3 - 2i)$. Hence (8.2) has two solutions given by

$$z + 2 - i = 3 - 2i \quad \text{or} \quad -3 + 2i,$$
$$z = 1 - i, \qquad -5 + 3i.$$

It is evident from Example 12 that the crucial step in the solution of this type of equation involves finding the square roots of a given complex number. Example 4 illustrated one way of doing this. That every complex number $w = r(\cos \theta + i \sin \theta)[r \in \mathbf{R}, r \geqslant 0]$ has a square root follows from de Moivre's theorem because

$$[\sqrt{r}(\cos \tfrac{1}{2}\theta + i \sin \tfrac{1}{2}\theta)]^2 = r(\cos \theta + i \sin \theta) = w.$$

Here $\sqrt{r}$ has its usual meaning—the non-negative square root of r. Hence $v = \sqrt{r}(\cos \tfrac{1}{2}\theta + i \sin \tfrac{1}{2}\theta)$ is a square root of w; so too is $-v$, since $(-v)^2 = v^2$. There are no more, for the equation $z^2 = w$ can be written $z^2 - v^2 = 0$ or

$$(z - v)(z + v) = 0. \qquad (8.3)$$

By Theorem 2.1, (8.3) holds if and only if $z - v = 0$ or $z + v = 0$, $z = v$ or $z = -v$.

However the story does not finish with quadratic equations. We can consider cubic equations such as $z^3+z^2-z+2=0$, and so on. The general problem concerns the **algebraic equation** or **polynomial equation** of degree n

$$a_0 z^n + a_1 z^{n-1} + a_2 z^{n-2} + \ldots + a_{n-1} z + a_n = 0. \tag{8.4}$$

Here $n \in \mathbf{N}$, $a_j \in \mathbf{C}$ $(0 \leqslant j \leqslant n)$ and $a_0 \neq 0$; the last condition ensures that the term in z^n actually appears. By multiplying (8.4) throughout by a_0^{-1} we obtain an **equivalent** equation in which the **leading term** (term of "highest degree") is simply z^n. Thus there is no loss of generality in assuming at the outset that $a_0 = 1$, and henceforward we work with the equation

$$z^n + a_1 z^{n-1} + \ldots + a_{n-1} z + a_n = 0. \tag{8.5}$$

For $n = 3$, 4 or 5 the equation is called **cubic, quartic** or **quintic**.

The next theorem is one of the most important in mathematics:

Theorem 8.1 (The Fundamental Theorem of Algebra). *The equation* (8.5), *where* $a_1, \ldots, a_n$ *are complex numbers, has a root in* **C**.

Remember that $\mathbf{R} \subset \mathbf{C}$, so Theorem 8.1 holds in particular when $a_1, \ldots a_n$ are real. However, as $z^2 + 1 = 0$ shows, there may not be a root in **R**. Unfortunately the proof of Theorem 8.1 lies beyond the scope of this book. We shall therefore take it for granted and deduce an important consequence.

Denote the left-hand side of (8.5) by $f(z)$. It is a polynomial of degree n, with leading term z^n.

Theorem 8.2. *There exist complex numbers* $\alpha_1, \alpha_2, \ldots, \alpha_n$ *(not necessarily all distinct) such that*

$$f(z) = (z - \alpha_1)(z - \alpha_2) \ldots (z - \alpha_n). \tag{8.6}$$

Illustration. Let $f(z) = z^3 + 5z^2 + 3z - 9$. It is easy to check that $z = 1$ is a root of $f(z) = 0$. We find that

$$z^3 + 5z^2 + 3z - 9 = (z - 1)(z^2 + 6z + 9) = (z - 1)(z + 3)(z + 3).$$

In this simple example $\alpha_1 = 1$, $\alpha_2 = 3$, $\alpha_3 = 3$ are integers. Of course this will not happen in general.

Proof of Theorem 8.2. We use induction. Consider $n = 1$. Then $f(z) = z + a_1$ and the theorem is seen to be true by taking $\alpha_1 = -a_1$.

Take any $k \geqslant 1$ and suppose that the theorem is true for all polynomials with leading term z^k. We call this supposition the "induction hypothesis".

Consider $n = k+1$. Let $f(z)$ be the polynomial

$$f(z) = z^{k+1} + a_1 z^k + \ldots + a_k z + a_{k+1}.$$

By the Fundamental Theorem, the equation $f(z) = 0$ has a root, $z = \alpha_1$ say. Then

$$f(\alpha_1) = \alpha_1^{k+1} + a_1 \alpha_1^k + \ldots + a_k \alpha_1 + a_{k+1} = 0,$$

so $f(z) = f(z) - f(\alpha_1) =$

$$(z^{k+1} - \alpha_1^{k+1}) + a_1 (z^k - \alpha_1^k) + \ldots + a_k (z - \alpha_1).$$

Now for $r \in \mathbf{N}$, $r > 1$ we have

$$z^r - \alpha_1^r = (z - \alpha_1)(z^{r-1} + \alpha_1 z^{r-2} + \ldots + \alpha_1^{r-1}) \qquad (8.7)$$

Hence
$$f(z) = (z - \alpha_1) g(z),$$

where $g(z)$ is a polynomial with leading term z^k. By the induction hypothesis there exist complex numbers, which we may call $\alpha_2, \ldots, \alpha_{k+1}$ (not necessarily all distinct), such that

$$g(z) = (z - \alpha_2) \ldots (z - \alpha_{k+1}).$$

Then
$$f(z) = (z - \alpha_1)(z - \alpha_2) \ldots (z - \alpha_{k+1}).$$

Thus if the result is true for $n = $ any $k \geqslant 1$ it is also true for $n = k+1$. Hence the theorem is established by induction. We describe (8.6) by saying that $f(z)$ has been factorized into *linear* factors.

Exercises. (a) Write out the cases $r = 2, 3, 4$ of (8.7) in full. Show that the first two terms of $g(z)$ are $z^k + (a_1 + \alpha_1) z^{k-1} + \ldots$

(b) Let $f(z) = z^4 + 3z^3 - 3z^2 - 11z - 6$. Verify that $z = -1$ and $z = 2$ are roots of $f(z) = 0$. Factorize $f(z)$ completely as in (8.6).

From (8.6) we see that

$$f(\alpha_1) = f(\alpha_2) = \ldots = f(\alpha_n) = 0.$$

Hence $\alpha_1, \alpha_2, \ldots, \alpha_n$ are roots of $f(z) = 0$. Moreover no number α not in this list is a root; for, $f(\alpha) = (\alpha - \alpha_1)(\alpha - \alpha_2) \ldots (\alpha - \alpha_n)$, and, if none of the factors is zero, then $f(\alpha) \neq 0$ (see the Exercise following Theorem 2.2). Consequently:

Theorem 8.3. *The equation $f(z) = 0$ of degree n cannot have more than n distinct roots.*

A number α which appears once only among $\alpha_1, \alpha_2, \ldots, \alpha_n$ is called a **simple root** of $f(z) = 0$. If α appears exactly twice in the list it is a **double root**, and so on. Double, triple, ... roots are referred to generally as

multiple or **repeated roots**. They have **multiplicity** $2, 3, \ldots$. A simple root has multiplicity 1.

Example 13. The equation
$$z^6 - 3z^5 + 6z^3 - 3z^2 - 3z + 2 = 0$$
may be written $\quad (z-1)^3(z+1)^2(z-2) = 0.$

It has a triple root $z = 1$, a double root $z = -1$ and a simple root $z = 2$. The multiplicities 3, 2, 1 of the roots sum to 6, the degree of the equation.

This is a general property, sometimes expressed by saying: A polynomial equation of degree n with coefficients in $\mathbf{C}$ has n roots in $\mathbf{C}$, *counted according to multiplicity*.

Theorem 8.1 states a property of the field $\mathbf{C}$ of complex numbers: *Every algebraic equation with coefficients in the field has a root in the field.* We express this by saying that the field $\mathbf{C}$ is *algebraically closed*. On the other hand the field $\mathbf{R}$ of real numbers is not algebraically closed, because the algebraic equation $z^2 + 1 = 0$ has its coefficients in $\mathbf{R}$ but has no root in $\mathbf{R}$.

Equations with real coefficients have an important property which in its simplest form is illustrated by the equation $z^2 - 8z + 25 = 0$ solved in Example 3, Section 1. The roots of this equation are $z = 4 + 3i$ and $z = 4 - 3i$. Observe that these are *conjugate complex numbers*.

Theorem 8.4. *If $z = \alpha$ is a root of the equation*
$$z^n + a_1 z^{n-1} + \ldots + a_{n-1} z + a_n = 0 \tag{8.8}$$
where $a_1, \ldots, a_n$ are **real** *then $z = \bar{\alpha}$ is a root of* (8.8).

Proof. Since $z = \alpha$ is a root of (8.8) we have
$$\alpha^n + a_1 \alpha^{n-1} + \ldots + a_{n-1} \alpha + a_n = 0. \tag{8.9}$$

Applying the operation of "complex conjugation" to both sides of (8.9) and using (5.4) (i) and (ii) we deduce that
$$\bar{\alpha}^n + \bar{a}_1 \bar{\alpha}^{n-1} + \ldots + \bar{a}_{n-1} \bar{\alpha} + \bar{a}_n = 0.$$

Now each a_j is real, so $\bar{a}_j = a_j$ and
$$\bar{\alpha}^n + a_1 \bar{\alpha}^{n-1} + \ldots + a_{n-1} \bar{\alpha} + a_n = 0,$$
showing that $z = \bar{\alpha}$ satisfies (8.8).

Note. If α is real then $\bar{\alpha} = \alpha$ and the result of Theorem 8.4 is "trivial"—it does not tell us anything we did not already know. We certainly can *not* infer from it that $z = \alpha$ is a *double* root of (8.8)!

Before investigating further the case of *non-real* α we give an illustration.

Example 14. Verify that $z = 1 + i$ is a root of the equation

$$z^4 + 4z^3 - 8z + 20 = 0,$$

express the left-hand side as a product of two quadratic factors with real coefficients and solve the equation completely. Denote the left-hand side by $f(z)$.

We have $(1+i)^2 = 2i$, $(1+i)^3 = 2(i-1)$, $(1+i)^4 = -4$ so $f(1+i) = -4 + 8(i-1) - 8(1+i) + 20 = 0$. Since the coefficients of $f(z)$ are real $z = \overline{1+i} = 1-i$ is also a root. Corresponding to these two roots we have the factors $z - 1 - i$ and $z - 1 + i$ whose product is $z^2 - 2z + 2$. We quickly find that

$$f(z) = (z^2 - 2z + 2)(z^2 + 6z + 10)$$

and the remaining roots are those of $z^2 + 6z + 10 = 0$, namely the conjugate pair $z = -3 \pm i$.

Returning to the general equation (8.8) with real coefficients and a root $z = \alpha$, observe that if α is *non-real* then $\bar{\alpha} \neq \alpha$ so $(z - \alpha)$ and $(z - \bar{\alpha})$ are *distinct* factors of the left-hand side $f(z)$ of (8.8). They combine to give the quadratic factor

$$(z - \alpha)(z - \bar{\alpha}) = z^2 - (\alpha + \bar{\alpha})z + \alpha\bar{\alpha}$$
$$= z^2 - 2(\mathcal{R}\alpha)z + |\alpha|^2 \qquad (8.10)$$

[by (5.1) and (5.2)]. (8.10) has *real* coefficients. If $n > 2$ the factor (8.10) can now be removed from $f(z)$, the quotient $g(z)$ being a polynomial of degree $n - 2$ which also has real coefficients. [In Example **14**, with $\alpha = 1 + i$, $g(z) = z^2 + 6z + 10$.] This argument may be repeated as long as non-real roots remain to be considered. We reach the conclusion

Theorem 8.5. *If* $f(z) = z^n + a_1 z^{n-1} + a_2 z^{n-2} + \ldots + a_n$, *where* $a_1, a_2, \ldots, a_n$ *are* **real** *then* $f(z)$ *can be expressed as a product of real factors each of which is either linear or a quadratic of the form* (8.10).

Consequently non-real roots of $f(z) = 0$ occur in conjugate pairs. We imply by this rather more than might be inferred directly from the earlier Theorem 8.4, namely that *conjugate roots have the same multiplicity*.

Example 15. If $f(z) = z^6 - z^5 - 2z^3 - 3z^2 - z - 2$
then $\qquad\qquad f(z) = (z^2 + 1)^2(z + 1)(z - 2)$.

The conjugate roots $z = \pm i$ arising from $(z^2+1)^2 = 0$ both have multiplicity 2.

Exercises. (*a*) Verify that $z = 2-i$ is a root of the equation

$$z^4 - 2z^3 - z^2 + 2z + 10 = 0$$

and solve the equation completely.

(*b*) Find an equation of degree 6 with real coefficients which has $1+i$ as a double root and 1 and -1 as simple roots.

An important consequence of Theorem 8.5 is

Theorem 8.6. *If $f(z)$ is a polynomial of **odd** degree with **real** coefficients then $f(z) = 0$ has at least one real root.*

For instance every cubic equation with real coefficients has at least one real root.

Proof. The contrapositive form of the statement is that if $f(z)$ has real coefficients and all the roots of $f(z) = 0$ are non-real then the degree of $f(z)$ is even. This follows at once from Theorem 8.5 because all the real factors are then quadratic.

Relations between the roots and coefficients of an algebraic equation. It is familiar that if $a, b \in \mathbf{R}$ and $z = \alpha$ and $z = \beta$ are the roots of the quadratic equation

$$z^2 + az + b = 0$$

then $\qquad\qquad\qquad \alpha + \beta = -a, \qquad \alpha\beta = b. \qquad\qquad\qquad (8.11)$

Remind yourself of the proof of the relations (8.11) and observe that it also applies when $a, b \in \mathbf{C}$. These relations have analogues for equations of higher degree which we shall illustrate by considering the cubic

$$z^3 + az^2 + bz + c = 0. \qquad\qquad (8.12)$$

Theorem 8.7. *If $z = \alpha, \beta, \gamma$ are the roots of (8.12) then*

$$\alpha + \beta + \gamma = -a, \qquad \beta\gamma + \gamma\alpha + \alpha\beta = b, \qquad \alpha\beta\gamma = -c.$$

(Note the alternation of signs on the right-hand sides.)

Proof. We have

$$\begin{aligned} z^3 + az^2 + bz + c &= (z-\alpha)(z-\beta)(z-\gamma) \\ &= z^3 - (\alpha+\beta+\gamma)z^2 + (\beta\gamma+\gamma\alpha+\alpha\beta)z - \alpha\beta\gamma. \end{aligned}$$

(Check this in detail.) Equating the coefficients of z^2, z and the constant terms gives the results at once.

Example 16. For the equation $z^3 - 5z + 2 = 0$ we have

$$\alpha + \beta + \gamma = 0, \qquad \beta\gamma + \gamma\alpha + \alpha\beta = -5, \qquad \alpha\beta\gamma = -2.$$

Check these by using the values $\alpha = 2, \beta = -1 + \sqrt{2}, \gamma = -1 - \sqrt{2}$.

Example 17. If α, β, γ are the roots of the equation

$$z^3 - 2z^2 + 3z - 1 = 0$$

find $\alpha^2 + \beta^2 + \gamma^2$ and $\alpha^3 + \beta^3 + \gamma^3$.

We have $\alpha + \beta + \gamma = 2, \beta\gamma + \gamma\alpha + \alpha\beta = 3, \alpha\beta\gamma = 1$.

Hence $\alpha^2 + \beta^2 + \gamma^2 = (\alpha + \beta + \gamma)^2 - 2(\beta\gamma + \gamma\alpha + \alpha\beta) = 4 - 6 = -2$.

To calculate $\alpha^3 + \beta^3 + \gamma^3$ we may use the relation

$$\alpha^3 + \beta^3 + \gamma^3 - 3\alpha\beta\gamma = (\alpha + \beta + \gamma)(\alpha^2 + \beta^2 + \gamma^2 - \beta\gamma - \gamma\alpha - \alpha\beta)$$

which you should verify. This gives

$$\alpha^3 + \beta^3 + \gamma^3 = 3 + 2(-2 - 3) = -7.$$

For another method see Exercise (c) below.

Exercises. (a) For the roots α, β, γ of the equation

$$z^3 + 2z^2 - 5z - 6 = 0$$

find $\alpha + \beta + \gamma, \alpha^2 + \beta^2 + \gamma^2$ and $\alpha^3 + \beta^3 + \gamma^3$. Find one root by trying simple values, solve the equation completely and check your previous results.

(b) If α, β, γ are the roots of $z^3 + az^2 + bz + c = 0$ express $\beta^2\gamma^2 + \gamma^2\alpha^2 + \alpha^2\beta^2$ in terms of a, b, c.

(c) Calculate $\alpha^3 + \beta^3 + \gamma^3$ in Example **17** by writing down the three relations expressing the fact that α, β, γ are the roots of the equation, and adding.

(d) Observe that in Example **17**, $\alpha^2 + \beta^2 + \gamma^2 < 0$. What can you infer about the roots?

Topics to explore. I. Consider the results corresponding to Theorems 8.2 and 8.7 for the equation $az^3 + bz^2 + cz + d = 0$.

II. Find results analogous to those of Theorem 8.7 for a quartic equation. As an exercise prove that if $\alpha, \beta, \gamma, \delta$ are the roots of $z^4 + z - 1 = 0$ then $\alpha^4 + \beta^4 + \gamma^4 + \delta^4 = 4$.

9. Roots of unity

We know from the work of Section **8** that the equation

$$z^n - 1 = 0 \tag{9.1}$$

has n roots, counted according to multiplicity. One of them is, of course, $z = 1$. These roots, which we shall find to be all simple, are the nth **roots of unity**.

Example 18. (i) For $n = 2$ the roots of (9.1) are $z = 1$ and $z = -1$. (ii) For $n = 4$ the equation factorizes as

$$(z-1)(z+1)(z^2+1) = 0$$

and its roots are $z = 1, -1, i, -i$.

De Moivre's theorem shows that the number

$$z = \cos(2\pi/n) + i \sin(2\pi/n)$$

is an nth root of unity. For

$$z^n = \cos 2\pi + i \sin 2\pi = 1 + i.0 = 1.$$

Similarly $\cos(4\pi/n) + i \sin(4\pi/n)$ is a root and so, generally, is

$$z = \cos(2k\pi/n) + i \sin(2k\pi/n) \tag{9.2}$$

for all $k \in \mathbf{Z}$.

Example 19. Take $n = 3$. Putting $k = 0$ in (9.2) gives $z = 1$.

For $k = 1$, $z = \cos(2\pi/3) + i \sin(2\pi/3) = \frac{1}{2}(-1 + i\sqrt{3})$;
for $k = 2$, $z = \cos(4\pi/3) + i \sin(4\pi/3) = \frac{1}{2}(-1 - i\sqrt{3})$.

Observe that these last two roots are complex conjugates; they are the roots of $z^2 + z + 1 = 0$. However $k = 3$ gives $z = 1$ again, $k = 4$ gives the same root as $k = 1$, and as k increases the roots already found recur "cyclically". Moreover nothing new is obtained by taking $k < 0$. That no more roots appear is only to be expected, by Theorem 8.3.

Exercise. Prove that each of the numbers $\frac{1}{2}(-1 \pm i\sqrt{3})$ is the square of the other. They are often denoted by ω and ω^2 (it is of little consequence which is chosen to be ω).

  (i) What figure in the complex plane is formed by the points representing 1, ω and ω^2?

  (ii) Prove that $-\omega$ is a 6th root of unity.

  (iii) Form the product $(\alpha + \omega\beta + \omega^2\gamma)(\alpha + \omega^2\beta + \omega\gamma)$, remembering that $\omega^3 = 1$ and $\omega^2 + \omega + 1 = 0$. Referring to the relation used in Example **17** express $\alpha^3 + \beta^3 + \gamma^3 - 3\alpha\beta\gamma$ as a product of three factors, each linear in α, β, γ.

The general situation is that n distinct nth roots of unity are obtained by giving to k in (9.2) the n values

$$k = 0, 1, \ldots, n-1 \tag{9.3}$$

The resulting values of z are distinct because the angles $2k\pi/n$ are distinct and lie in the interval $[0, 2\pi)$. By Theorem 8.3 there are no more roots of (9.1). If the roots we have found were not all simple, their multiplicities would have sum $> n$ and this is impossible.

Instead of (9.3) any n consecutive integers might be used, such as $k = 1, 2, \ldots, n$. When n is odd, so that $n = 2m+1$ with $m \in \mathbf{W}$, a particularly convenient choice is

$$k = -m, \ldots, -1, 0, 1, \ldots, m.$$

This displays at once the pairing of the non-real roots as complex conjugates:

$$z = \cos(2k\pi/n) \pm i \sin(2k\pi/n) \tag{9.4}$$

for $k = 1, 2, \ldots, m = \tfrac{1}{2}(n-1)$.

Exercise. How do the conjugate pairs arise when (9.3) is used? (Distinguish the cases n even and odd, remembering that there are *two* real roots in the first case).

The geometrical picture is simple. For example the points in the complex plane representing the 4th roots $1, i, -1, -i$ are the vertices of a *square*. What did you discover about the points corresponding to $1, \omega, \omega^2$? Generally we see that the nth roots of unity all satisfy the relation $|z| = 1$. (Prove this in two ways: using (9.2) and directly from (9.1).) Hence the corresponding points all lie on the circle $x^2 + y^2 = 1$ which has centre O and radius 1 and is often called the *unit circle* (with centre O). Moreover they are equally spaced round the circle (why?) and are therefore the vertices of a *regular polygon* with n vertices (briefly, a "regular n-gon"). For $n = 5$ we have a regular pentagon, for $n = 6$ a regular hexagon, and so on.

Exercise. Find all the 6th roots of unity and mark the corresponding points in a diagram.

Roots of a general complex number. De Moivre's theorem may be used to find the nth roots of any given complex number w, that is, the roots of the equation

$$z^n = w. \tag{9.5}$$

If $w = 0$ then (9.5) is satisfied only by $z = 0$, which is a root of multiplicity n. Leaving this case aside, put

$$w = r(\cos\theta + i\sin\theta)$$

where $r = |w| > 0$ and θ is a value of arg w, say the principal value to fix things. Then de Moivre's theorem shows that each of the numbers

$$z = \sqrt[n]{r}\left[\cos\frac{\theta + 2k\pi}{n} + i\sin\frac{\theta + 2k\pi}{n}\right], \qquad (9.6)$$

where $k = 0, 1, \ldots, n-1$, is a root of (9.5). Here $\sqrt[n]{r}$ denotes the unique positive real nth root of r. The existence of this root is intuitively clear but depends ultimately on deep basic properties of the real field **R**. The n numbers given by (9.6) are all different (why?) so they comprise all the nth roots of w.

Example 20. Find the 4th roots of -16.

We have $-16 = 16(\cos\pi + i\sin\pi)$ and $16 = 2^4$, so that $\sqrt[4]{16} = 2$ and the roots are

$$2\left[\cos\frac{\pi + 2k\pi}{4} + i\sin\frac{\pi + 2k\pi}{4}\right], \qquad (k = 0, 1, 2, 3).$$

There are easily found to be $\sqrt{2}(1+i)$, $\sqrt{2}(-1+i)$, $\sqrt{2}(-1-i)$ and $\sqrt{2}(1-i)$.

Exercises. (a) Find the cube roots of $-8i$.

(b) What happens if a value of arg w other than the principal value is taken for θ in (9.6)?

(c) What type of figure is formed by the points in the complex plane corresponding to the roots obtained in Example **20**?

Topic to explore. Primitive roots. Let $\omega = \frac{1}{2}(-1+i\sqrt{3})$ [see Example **19**]. Then ω, $-\omega$, -1 are all 6th roots of unity. For each of these numbers ε find all the different values of ε^k where $k \in \mathbf{Z}$. You should find that one of the numbers yields *six* values of ε^k, and that these are all the 6th roots of unity, but the others give fewer values. An nth root of unity, ε, such that every nth root can be expressed in the form ε^k for some $k \in \mathbf{Z}$ is called a *primitive* nth root of unity. Can you find another primitive 6th root? Find all the primitive nth roots for $n = 4, 5$. A primitive nth root of unity can also be described as an nth root which does not satisfy $z^m = 1$ for any natural number $m < n$. Try to find a criterion for deciding when

$\cos (2k\pi/n) + i \sin (2k\pi/n)$ is a primitive nth root. Find a quartic equation satisfied by the primitive 8th roots (of which there are four).

Some simple factorizations. The polynomial $z^n - 1$ whose zeros are the nth roots of unity, and certain related and similar polynomials, can easily be resolved into real factors as described in Section **8**.

Exercise. Show that $z^2 + z + 1$ is a factor of $z^6 - 1$ and factorize the latter polynomial completely.

If n is *odd* there is only one real nth root of unity, namely $z = 1$. The remaining roots satisfy the equation

$$z^{n-1} + z^{n-2} + \ldots + z + 1 = 0,$$

the left-hand side of which resolves into $\frac{1}{2}(n-1)$ real quadratic factors.

Example 21. Find the real quadratic factors of

$$f(z) = z^6 + z^5 + z^4 + z^3 + z^2 + z + 1.$$

The non-real 7th roots of unity are the numbers

$$\cos (2k\pi/7) \pm i \sin (2k\pi/7), \qquad k = 1, 2, 3.$$

[Take $n = 7$, $m = 3$ in (9.4).] These give rise to real quadratic factors

$$z^2 - 2z \cos (2k\pi/7) + 1, \qquad k = 1, 2, 3$$

of $f(z)$. Hence

$$f(z) = (z^2 - 2z \cos \tfrac{2}{7}\pi + 1)(z^2 - 2z \cos \tfrac{4}{7}\pi + 1)(z^2 - 2z \cos \tfrac{6}{7}\pi + 1).$$

Exercise. By putting $z = 1$ in this result and using the relation $1 - \cos 2\theta = 2 \sin^2 \theta$, prove that

$$\sin \tfrac{1}{7}\pi \sin \tfrac{2}{7}\pi \sin \tfrac{3}{7}\pi = \tfrac{1}{8}\sqrt{7}.$$

(*Note.* You will have to take a square root. Observe that the product on the left is *positive*.)

When n is *even* both 1 and -1 are nth roots of unity and $z^n - 1$ has $(z-1)$ and $(z+1)$ as factors. The polynomial of degree $n-2$ which remains after removal of these has $\frac{1}{2}(n-2)$ real quadratic factors. For $n = 6$ see the Exercise at the beginning of this Section.

Exercise. Prove that $z^2 - \sqrt{2}\,z + 1$ is a factor of $z^8 - 1$ and factorize $z^8 - 1$ completely into real linear and quadratic factors.

Example 22. Express $z^4 + 16$ as a product of two real quadratic factors.

The roots of $z^4 + 16 = 0$ were found in Example 20 to be $\sqrt{2}(1 \pm i)$, $\sqrt{2}(-1 \pm i)$. The first pair are conjugate and give rise to the factor $z^2 - 2\sqrt{2}z + 4$. Dealing similarly with the second pair we obtain

$$z^4 + 16 = (z^2 - 2\sqrt{2}z + 4)(z^2 + 2\sqrt{2}z + 4).$$

Check by multiplication.

10. Some series involving complex numbers

A familiar and important example of a "series" is the **geometric series**
$$a + ar + ar^2 + \ldots + ar^{n-1} + \ldots,$$

where a is the **first term** and r the **common ratio**. The **sum to n terms**, or nth **partial sum** of this series is

$$S_n = a + ar + ar^2 + \ldots + ar^{n-1} = \sum_{j=1}^{n} ar^{j-1}.$$

We have

$$rS_n = ar + ar^2 + \ldots + ar^{n-1} + ar^n,$$

so $\qquad (1-r)S_n = a - ar^n = a(1 - r^n).$

Hence, if $r \neq 1$,

$$S_n = \frac{a(1 - r^n)}{1 - r}. \tag{10.1}$$

Exercise. What is the value of S_n when $r = 1$?

In previous work with this series a and r have been real numbers. We now see, however, that the evaluation of S_n applies equally well if a, r are complex. In particular, putting $a = 1$ and $r = z$, we find that for all $z \in \mathbf{C} - \{1\}$

$$1 + z + z^2 + \ldots + z^{n-1} = \frac{1 - z^n}{1 - z}. \tag{10.2}$$

Also $\qquad z + z^2 + z^3 + \ldots + z^n = \frac{z(1 - z^n)}{1 - z}. \tag{10.3}$

Interesting deductions can be made from these results by taking z in polar form.

Example 23. Prove that, if θ is real and $\theta \neq 2k\pi$ ($k \in \mathbf{Z}$), then

$$\cos \theta + \cos 2\theta + \ldots + \cos n\theta = \frac{\cos \frac{1}{2}(n+1)\theta \sin \frac{1}{2}n\theta}{\sin \frac{1}{2}\theta},$$

$$\sin \theta + \sin 2\theta + \ldots + \sin n\theta = \frac{\sin \frac{1}{2}(n+1)\theta \sin \frac{1}{2}n\theta}{\sin \frac{1}{2}\theta}.$$

Note. The condition $\theta \neq 2k\pi$ ($k \in \mathbf{Z}$) ensures that $\sin \frac{1}{2}\theta \neq 0$.

In (10.3) put $z = \cos \theta + i \sin \theta$ with $\theta \neq 2k\pi$ ($k \in \mathbf{Z}$), so that $z \neq 1$. By de Moivre's theorem the left-hand side of (10.3) is then

$$(\cos \theta + \cos 2\theta + \ldots + \cos n\theta) + i(\sin \theta + \sin 2\theta + \ldots + \sin n\theta) = C + iS \text{ say.}$$

To evaluate the right-hand side we first calculate

$$\begin{aligned} 1 - z^n &= 1 - \cos n\theta - i \sin n\theta \\ &= 2 \sin^2 \tfrac{1}{2}n\theta - 2i \sin \tfrac{1}{2}n\theta \cos \tfrac{1}{2}n\theta \\ &= -2i \sin \tfrac{1}{2}n\theta (\cos \tfrac{1}{2}n\theta + i \sin \tfrac{1}{2}n\theta). \end{aligned}$$

Using this, and also the special case $n = 1$, we find that

$$\begin{aligned} C + iS &= \frac{2i \sin \frac{1}{2}n\theta (\cos \theta + i \sin \theta)(\cos \frac{1}{2}n\theta + i \sin \frac{1}{2}n\theta)}{2i \sin \frac{1}{2}\theta (\cos \frac{1}{2}\theta + i \sin \frac{1}{2}\theta)} \\ &= \frac{\sin \frac{1}{2}n\theta}{\sin \frac{1}{2}\theta} \left[\cos \tfrac{1}{2}(n+1)\theta + i \sin \tfrac{1}{2}(n+1)\theta \right]. \end{aligned}$$

Equating real and imaginary parts now gives the results required:

$$\begin{aligned} C &= \cos \tfrac{1}{2}(n+1)\theta \sin \tfrac{1}{2}n\theta / \sin \tfrac{1}{2}\theta, \\ S &= \sin \tfrac{1}{2}(n+1)\theta \sin \tfrac{1}{2}n\theta / \sin \tfrac{1}{2}\theta. \end{aligned}$$

Exercise (a) Prove one of these results by induction.

(b) Evaluate the right-hand side of (10.3) in another way by putting $z = w^2$, $w = \cos \frac{1}{2}\theta + i \sin \frac{1}{2}\theta$, and observing that

$$\frac{z(1 - z^n)}{1 - z} = \frac{w^{n+1}(w^n - w^{-n})}{w - w^{-1}}$$

[cf. (7.8)].

(c) By putting $a = \alpha^{n-1}$ and $r = z/\alpha$ in (10.1) show that

$$z^n - \alpha^n = (z - \alpha)(z^{n-1} + \alpha z^{n-2} + \ldots + \alpha^{n-1})$$

[cf. (8.7)].

The same technique can be applied to other identities besides (10.3).

Example 24. Prove that, for all real θ,

$$1 + \binom{n}{1}\cos\theta + \binom{n}{2}\cos 2\theta + .. + \binom{n}{r}\cos r\theta + .. + \cos n\theta = 2^n \cos^n \tfrac{1}{2}\theta \cos \tfrac{1}{2}n\theta.$$

By de Moivre's theorem $\cos r\theta$ is the real part of $(\cos\theta + i\sin\theta)^r$, so we recognize the sum in question as the real part of

$$1 + \binom{n}{1}z + \binom{n}{2}z^2 + \ldots + \binom{n}{r}z^r + \ldots + z^n$$

where $z = \cos\theta + i\sin\theta$. By the binomial theorem this is

$$\begin{aligned}
(1+z)^n &= (1 + \cos\theta + i\sin\theta)^n = [2\cos\tfrac{1}{2}\theta(\cos\tfrac{1}{2}\theta + i\sin\tfrac{1}{2}\theta)]^n \\
&= 2^n \cos^n \tfrac{1}{2}\theta(\cos\tfrac{1}{2}n\theta + i\sin\tfrac{1}{2}n\theta),
\end{aligned}$$

using de Moivre's theorem again. The result follows on picking out the real part of this expression.

Exercise (*a*) Write the result of Example 24 using $\sum$ notation.

Evaluate

$$\sum_{r=0}^{n} \binom{n}{r}\sin r\theta.$$

(*b*) Is $2\cos\tfrac{1}{2}\theta$ always the modulus of $1 + \cos\theta + i\sin\theta$? Is this of any significance in the above work?

(*c*) Show that, if $z = \cos\theta + i\sin\theta$, then

$$(1 - z\cos\theta)^n = (-i\sin\theta)^n(\cos n\theta + i\sin n\theta).$$

Deduce that, if n is even, then

$$\sum_{r=0}^{n} (-1)^r \binom{n}{r}\cos^r\theta\cos r\theta = (-1)^{\frac{1}{2}n}\cos n\theta \sin^n\theta.$$

What is the corresponding result for n odd?

Infinite geometric series. Let us return to (10.1) with $a, r \in \mathbf{R}$. If $-1 < r < 1$, which we can also write as $|r| < 1$, then r^n becomes numerically smaller and smaller as n increases; by taking n large enough we can make r^n as small as we please. We say that "r^n tends to 0 as n tends to infinity", or briefly, $r^n \to 0$ as $n \to \infty$. The **sum** (to infinity) of the series $a + ar + ar^2 + \ldots + ar^{n-1} + \ldots$

is

$$\frac{a(1-0)}{1-r} = \frac{a}{1-r},$$

and we write $\displaystyle\sum_{n=1}^{\infty} ar^{n-1} = a/(1-r)$ for $|r| < 1$.

Similar conditions apply in the complex case, e.g. to (10.2). As a measure of the "magnitude" of a complex number we may use its modulus. Now $|z^n| = |z|^n$, so if $|z| < 1$ the "magnitude" of z^n can be made as small as we please by taking n sufficiently large. Again we say "$z^n \to 0$ as $n \to \infty$" and the sum to infinity of $1 + z + z^2 + \ldots$ is $(1 - z)^{-1}$. In other words

$$\sum_{n=0}^{\infty} z^n = (1 - z)^{-1}, \qquad |z| < 1. \tag{10.4}$$

In these circumstances the series is said to be **convergent**. For $|z| \geqslant 1$ the sum (to infinity) does not exist—the series is **divergent**.

Example 25. Find the sum of the series

$$\sum_{n=0}^{\infty} r^n \cos n\theta = 1 + r \cos \theta + \ldots + r^n \cos n\theta + \ldots$$

for $|r| < 1$.

We recognize this as the real part of (10.4) when $z = r(\cos \theta + i \sin \theta)$. So we require $\mathscr{R}(1 - z)^{-1}$. Now

$$(1 - z)^{-1} = \frac{1 - \bar{z}}{|1 - z|^2} = \frac{1 - r(\cos \theta - i \sin \theta)}{(1 - r \cos \theta)^2 + (r \sin \theta)^2}$$

$$= \frac{1 - r \cos \theta + ir \sin \theta}{1 - 2r \cos \theta + r^2}.$$

Hence $\qquad \displaystyle\sum_{n=0}^{\infty} r^n \cos n\theta = \frac{1 - r \cos \theta}{1 - 2r \cos \theta + r^2}, \qquad |r| < 1.$

Exercise (a) Find the sum $\displaystyle\sum_{n=0}^{\infty} r^n \sin n\theta$ for $|r| < 1$.

(b) Prove that if $\theta \neq k\pi$ $(k \in \mathbf{Z})$ then

$$\sum_{n=0}^{\infty} \cos^n \theta \cos n\theta = 1.$$

Be sure to explain where you use the condition $\theta \neq k\pi$.

EXERCISE 7

1. Evaluate (i) $(2 - i)^3(2 + i)^3$, (ii) $(2 - 3i)/(1 + 2i)$, (iii) $(1 + i)^{-3}$.
2. For what values of the real number x is $(x + i)(1 - xi)$ (i) real, (ii) pure imaginary?
3. Find the smallest $n \in \mathbf{N}$ such that $(\sqrt{3} + i)^n$ is (i) pure imaginary, (ii) real. Describe the sets of natural numbers with these properties.

4. Prove that if P, Q are the points of the complex plane representing the numbers z, w respectively, then the midpoint of PQ corresponds to $\frac{1}{2}(z+w)$. Find the number represented by that point of trisection of PQ that is nearer to P. Prove that if S is the point of PQ defined by $\vec{PS} = k\,\vec{PQ}$ $(k \in \mathbf{R})$ then S corresponds to $(1-k)z + kw$.

5. Points P_1, P_2, P_3, P_4 of the complex plane represent the numbers z_1, z_2, z_3, z_4 respectively. (i) Assuming that P_1, P_2, P_3 are not collinear find the number corresponding to the centroid of triangle $P_1P_2P_3$. (ii) What kind of figure is formed by the four points when $z_1 + z_3 = z_2 + z_4$?

6. Mark in the plane the points representing the numbers $z_1 = 2 - i$, $z_2 = 1 + 3i$ and $z_3 = (1-i)z_1 + iz_2$. What can you say about the triangle which they form?

7. Find the modulus and the principal value of the argument of $(1-i)(\sqrt{3}+i)$.

8. The angles θ, ϕ lie in the interval $(-\frac{1}{2}\pi, \frac{1}{2}\pi)$ and
$$z = (\cos\theta + \cos\phi) + i(\sin\theta + \sin\phi).$$
Prove that $|z| = 2\cos\frac{1}{2}(\theta - \phi)$ and find the principal value of $\arg z$.

9. Find the number k such that $2k = |1 - ki|$.

10. (i) Prove that for all $z, w \in \mathbf{C}$, $|z+w|^2 = |z|^2 + |w|^2 + 2\mathcal{R}(z\bar{w})$. (ii) Use the fact that $\mathcal{R}u \leqslant |u|$ for all $u \in \mathbf{C}$ to deduce that $|z+w|^2 \leqslant (|z|+|w|)^2$. (iii) Deduce the triangle inequality. (Compare this with the proof for the real case given in (iii) of Section 5, Chapter 4).

11. (i) Prove that for all $z, w \in \mathbf{C}$, $|z-w| \geqslant |z|-|w|$ and $|z-w| \geqslant |w|-|z|$. Deduce that
$$|z-w| \geqslant ||z| - |w||.$$
(ii) Prove that if $z_1, z_2, z_3 \in \mathbf{C}$ then
$$|z_1 + z_2 + z_3| \leqslant |z_1| + |z_2| + |z_3|.$$

12. Prove that if $z \in \mathbf{C} - \{0\}$ then $z\bar{z}^{-1} + \bar{z}z^{-1}$ is a real number lying in the interval $[-2, 2]$.

13. Prove that for all $z \in \mathbf{C}$,
$$|z + i\bar{z}|^2 = 2|z|^2 + 2\mathcal{I}(z^2).$$

14. Prove that if $z \in \mathbf{C} - \{0\}$ then
$$z + z^{-1} \in \mathbf{R} \Leftrightarrow z \in \mathbf{R} \text{ or } |z| = 1.$$

15. (i) By writing the equation $z' = 1 + i - iz$ in the form $z' - 1 = -i(z-1)$ show that it represents the quarter-turn clockwise about the point of the complex plane corresponding to the number 1. (ii) Prove that the quarter-turn anticlockwise about the point corresponding to i has equation $z' = iz + 1 + i$.

16. (i) Simplify $(\cos 5\theta + i\sin 5\theta)^2 (\cos 3\theta - i\sin 3\theta)^3$.
(ii) Prove that $[\cos(\theta + \frac{1}{4}\pi) + i\sin(\theta + \frac{1}{4}\pi)]^8 = \cos 8\theta + i\sin 8\theta$.
(iii) Show that $(\sin\theta + i\cos\theta)^n = \cos n(\frac{1}{2}\pi - \theta) + i\sin n(\frac{1}{2}\pi - \theta)$.
(iv) For what numbers $n \in \mathbf{N}$ is $(\sin\theta + i\cos\theta)^n = \sin n\theta + i\cos n\theta$?

17. Using tables find the principal value of $\arg(8+i)$. Hence find the smallest $n \in \mathbf{N}$ for which the real part of $(8+i)^n$ is negative.

18. (i) Express $\cos 8\theta$ in terms of $\cos\theta$ and $\sin\theta$ by means of de Moivre's theorem.
(ii) Express $\cos^5\theta \sin^3\theta$ in terms of sines of integer multiples of θ.

19. (i) Evaluate $(1 - 4i)^2$.   (ii) Solve the equation
$$z^2 + 2(1-i)z + (15 + 6i) = 0.$$

20. Use de Moivre's theorem to find the square roots of $8 + 8\sqrt{3}i$.

21. Solve the equations
(i) $z^3 + 3z^2 + 4z - 8 = 0$,   (ii) $z^4 + 5z^2 + 4 = 0$,   (iii) $(z+1)^4 = z^4$.

22. The equation $z^3 + (i-2)z^2 - 4i = 0$ has a real root. By putting $z = x \in \mathbf{R}$ and equating real and imaginary parts find this root.

23. Find the equation of degree 5 with leading term z^5 which has $z = 1$ as a double root and $z = -1$ as a triple root.

24. Prove that the quartic equation $z^4 + z - 1 = 0$ has at least two real roots.

25. Let α, β, γ be the roots of the equation $z^3 + z^2 + z - 2 = 0$. Find $\alpha^2 + \beta^2 + \gamma^2$, $\beta^2\gamma^2 + \gamma^2\alpha^2 + \alpha^2\beta^2$ and $\alpha^2\beta^2\gamma^2$. Find a cubic equation whose roots are $\alpha^2, \beta^2, \gamma^2$.

26. The equation $z^3 + az + b = 0$ has $z = \alpha$ as a double root. Prove that $a = -3\alpha^2$, $b = 2\alpha^3$ and hence that $4a^3 + 27b^2 = 0$.

27. Express in the form $\cos\theta + i\sin\theta$: (i) the cube roots of $(1 + i)/\sqrt{2}$, (ii) the 4th roots of $\frac{1}{2}(1 + i\sqrt{3})$.

28. Find the 6th roots of 729.

29. The non-zero complex numbers z and t satisfy the equation $z^4 = t^3$. Which (if any) of the following statements is true?
   (i) Exactly one cube root of z is a 4th root of t.
   (ii) Every cube root of z is a 4th root of t.

30. Express $z^4 + 1$ as a product of two real quadratic factors.

31. Prove by equating coefficients that the polynomial $z^4 + z^3 + z^2 + z + 1$ has a factorization of the form
$$(z^2 + az + 1)(z^2 + bz + 1)$$
where $a, b \in \mathbf{R}$. Deduce that $\cos\frac{2}{5}\pi = \frac{1}{4}(\sqrt{5} - 1)$.

32. Express $z^8 + z^7 + z^6 + z^5 + z^4 + z^3 + z^2 + z + 1$ as a product of four real quadratic factors. By putting $z = -1$ evaluate
$$\cos\frac{1}{9}\pi \cos\frac{2}{9}\pi \cos\frac{4}{9}\pi.$$

33. Prove that if $\alpha = \cos\frac{2}{7}\pi + i\sin\frac{2}{7}\pi$ then $\alpha + \alpha^2 + \alpha^4$ and $\alpha^3 + \alpha^5 + \alpha^6$ are the roots of $z^2 + z + 2 = 0$. Hence show that
$$\alpha + \alpha^2 + \alpha^4 = \frac{1}{2}(-1 + i\sqrt{7}).$$
Prove also that $\alpha + \alpha^6$, $\alpha^3 + \alpha^4$, $\alpha^2 + \alpha^5$ are the roots of
$$z^3 + z^2 - 2z - 1 = 0.$$

34. Prove that if a, b, c are complex numbers such that
$$a + b + c = a^2 + b^2 + c^2 = 0$$
then $a^4 + b^4 + c^4 = 0$.

35. $ABCDEFG$ is a regular polygon of seven sides inscribed in a circle of unit radius. H is the midpoint of the small arc AB. Prove that $HA.HB.HC.HD.HE.HF.HG = 2$.

36. Expand $(z - \cos\theta)^n$ by the binomial theorem. Prove that if n is even then
$$\cos n\theta - \binom{n}{1}\cos(n-1)\theta\cos\theta + \binom{n}{2}\cos(n-2)\theta\cos^2\theta - \ldots + (-1)^n\cos^n\theta$$
$$= (-1)^{n/2}\sin^n\theta.$$

37. Let $T_n = 1 + 2z + 3z^2 + \ldots + nz^{n-1}$.
   Prove that $(1 - z)T_n = (1 + z + z^2 + \ldots + z^{n-1}) - nz^n$ and hence that
$$T_n = \frac{1 - (n+1)z^n + nz^{n+1}}{(1 - z)^2}.$$
It can be proved that if $|z| < 1$ then $(n+1)z^n \to 0$ and $nz^{n+1} \to 0$ as $n \to \infty$. Assuming this,

find $\qquad\qquad \sum_{n=1}^{\infty} nz^{n-1} \quad$ for $\quad |z| < 1.$

Matrices

1. Introduction

In what follows the small letters used will represent elements of a given algebraic structure such as a ring, integral domain or field [see Chapter 10]. In fact the structure will almost always be the field of rational numbers $\mathbf{Q}$, or the real field $\mathbf{R}$, or the complex field $\mathbf{C}$, or the integral domain of integers $\mathbf{Z}$, and all that we require are the well-known properties of addition and multiplication for the particular system chosen. Let F denote some such structure; matrices can be defined in terms of elements of F. These matrices are said to be **matrices over** F and F is called the set of **scalars** or **numbers** for the matrices. From now on, F will be rarely mentioned, but will be assumed to have been chosen beforehand.

Consider the pair of linear equations

$$\begin{aligned}
y_1 &= a_{11}x_1 + a_{12}x_2, \\
y_2 &= a_{21}x_1 + a_{22}x_2,
\end{aligned} \tag{1.1}$$

where $a_{11}, a_{12}, a_{21}, a_{22}$ are given elements of F. The pair of equations (1.1) may be regarded as a transformation or mapping which associates with each vector $\begin{bmatrix} x_1 \\ x_2 \end{bmatrix}$ a unique vector $\begin{bmatrix} y_1 \\ y_2 \end{bmatrix}$. The mapping is completely determined by the rectangular array of numbers

$$\begin{bmatrix} a_{11} & a_{12} \\ a_{21} & a_{22} \end{bmatrix},$$

which remains on the right-hand side of (1.1) when x_1, x_2 are removed. The notation used has been chosen in order to emphasize the importance of the position of each number in such an array. Thus

a_{11} is the number in the 1st row and 1st column of the array,
a_{12} is the number in the 1st row and 2nd column of the array,

a_{21} is the number in the 2nd row and 1st column of the array,

a_{22} is the number in the 2nd row and 2nd column of the array.

In a general notation,

a_{ij} is the number in the ith row and jth column of the array.

Similarly, the set of linear equations

$$y_1 = a_{11}x_1 + a_{12}x_2,$$
$$y_2 = a_{21}x_1 + a_{22}x_2,$$
$$y_3 = a_{31}x_1 + a_{32}x_2$$

has, associated with it, the rectangular array of coefficients

$$\begin{bmatrix} a_{11} & a_{12} \\ a_{21} & a_{22} \\ a_{31} & a_{32} \end{bmatrix}.$$

This array has 3 rows and 2 columns.

More generally, the following set of m linear equations in n variables $x_1, x_2, \ldots, x_n$,

$$y_1 = a_{11}x_1 + a_{12}x_2 + \ldots + a_{1n}x_n,$$
$$y_2 = a_{21}x_1 + a_{22}x_2 + \ldots + a_{2n}x_n,$$
$$\cdot \quad \cdot \quad \cdot \quad \cdot \quad \cdot \quad \cdot \quad \cdot \quad \cdot \quad \cdot$$
$$y_m = a_{m1}x_1 + a_{m2}x_2 + \ldots + a_{mn}x_n$$

has, associated with it, the array

$$\begin{bmatrix} a_{11} & a_{12} \ldots a_{1n} \\ a_{21} & a_{22} \ldots a_{2n} \\ \cdot & \cdot \quad \cdot \quad \cdot \\ a_{m1} & a_{m2} \ldots a_{mn} \end{bmatrix}. \qquad (1.2)$$

This array has m rows and n columns.

A rectangular array of elements of F of the form (1.2) is called a **matrix of order** or **type** $m \times n$ **over** F. We often denote the matrix by a capital letter, e.g. A, and write

$$A = \begin{bmatrix} a_{11} & a_{12} \ldots a_{1n} \\ a_{21} & a_{22} \ldots a_{2n} \\ \cdot & \cdot \quad \cdot \quad \cdot \\ a_{m1} & a_{m2} \ldots a_{mn} \end{bmatrix},$$

or $A = [a_{ij}]_{m \times n}$, or simply $A = [a_{ij}]$, when the order of A is clearly understood or is irrelevant. The elements a_{ij} are called the **entries** or **elements** of A, a_{ij} being the entry in the ith row and jth column.

If $m = n$, i.e. if A is of order $n \times n$ (i.e. if A has n rows and n columns), then A is called a **square matrix of order** n.

If the entries of A are real numbers, we call A a **real matrix**; similarly we define in a corresponding way, a **rational matrix**, a **complex matrix**, an **integral matrix** (i.e. a matrix with integer entries), etc.

An $m \times 1$ matrix is of the form $\begin{bmatrix} a_{11} \\ a_{21} \\ \vdots \\ a_{m1} \end{bmatrix}$, and is called a

column matrix or vector of dimension m.

A $1 \times n$ matrix is of the form $[a_{11} a_{12} \ldots a_{1n}]$, and is called a **row matrix or vector of dimension** n.

A 1×1 matrix is of the form $[a_{11}]$. We shall identify this matrix with the scalar a_{11}.

2. Equality, addition and multiplication by a scalar for matrices

Equality. Two matrices $A = [a_{ij}]$ and $B = [b_{ij}]$ are said to be **equal** if and only if *they have the same order* and $a_{ij} = b_{ij}$ *for each pair of suffixes i, j* (i.e. the entries of one are equal respectively to the corresponding entries of the other); in other words A and B are equal if and only if they are identical. When A and B are equal we write $A = B$.

Example 1. Find the values of x and y in $\mathbf{Q}$ for which

$$\begin{bmatrix} 3 & x+1 \\ y & 5 \end{bmatrix} = \begin{bmatrix} x-1 & 5 \\ x-3y & 5 \end{bmatrix}.$$

The matrices are equal $\Leftrightarrow \begin{cases} 3 = x-1, \\ x+1 = 5, \\ y = x-3y, \\ 5 = 5 \end{cases} \Leftrightarrow$ and $y = 1$.

Exercise. Why are there no values of x, y, z for which

$$\begin{bmatrix} x-1 & y+2 \\ z & 1 \end{bmatrix} = \begin{bmatrix} 2 & 3 \\ -1 & 0 \end{bmatrix}?$$

Note that

$$\begin{bmatrix} 1 \\ 2 \end{bmatrix} \neq \begin{bmatrix} 1 & 0 \\ 2 & 0 \end{bmatrix}.$$

Addition of matrices. If $A = [a_{ij}]$ and $B = [b_{ij}]$, we define $A+B$, *when and only when A and B are of the same order,* $m \times n$ say, by writing

$$A+B = [a_{ij}+b_{ij}].$$

Thus $A+B$ is the $m \times n$ matrix
$$\begin{bmatrix} a_{11}+b_{11} & a_{12}+b_{12} \ldots a_{1n}+b_{1n} \\ a_{21}+b_{21} & a_{22}+b_{22} \ldots a_{2n}+b_{2n} \\ \cdot \quad \cdot \quad \cdot \quad \cdot \\ a_{m1}+b_{m1} & a_{m2}+b_{m2} \ldots a_{mn}+b_{mn} \end{bmatrix}.$$

$A+B$ is *not defined* if A, B are of different orders.

Example 2. $\begin{bmatrix} 1 & 0 & -1 \\ 2 & -1 & 3 \end{bmatrix} + \begin{bmatrix} -2 & \frac{1}{2} & 5 \\ -1 & 2 & 0 \end{bmatrix} = \begin{bmatrix} -1 & \frac{1}{2} & 4 \\ 1 & 1 & 3 \end{bmatrix}$;

but $\begin{bmatrix} 2 & 1 \\ 3 & -5 \end{bmatrix} + \begin{bmatrix} 6 \\ 4 \end{bmatrix}$ has no meaning.

The zero $m \times n$ matrix is the $m \times n$ matrix with all its entries 0; it is denoted by $O_{m \times n}$ or simply by O. It plays a special role in matrix addition.

If $A = [a_{ij}]$ is any $m \times n$ matrix, then

$$A+O = [a_{ij}+0] = [a_{ij}] = A, \quad \text{and, similarly,} \quad O+A = A.$$

The negative of an $m \times n$ matrix $A = [a_{ij}]$. We look for an $m \times n$ matrix $X = [x_{ij}]$ such that $A+X = O = X+A$.

Now $\qquad A+X = O \Leftrightarrow [a_{ij}]+[x_{ij}] = [0]$
$$\Leftrightarrow a_{ij}+x_{ij} = 0 \quad \forall i, j$$
$$\Leftrightarrow x_{ij} = -a_{ij} \quad \forall i, j$$
$$\Leftrightarrow X = [-a_{ij}].$$
Similarly $\qquad X+A = O \Leftrightarrow X = [-a_{ij}].$

The matrix $[-a_{ij}]$ is denoted by $-A$ and called the **negative of** A. We have shown in introducing this matrix that

$$A+(-A) = O = (-A)+A.$$

From the definition of $-A$, it is clear that $-(-A) = A$.

Subtraction for matrices is defined, as in the case of number systems, by writing $A-B$ for $A+(-B)$, when A, B have the same order.

We now list the **basic properties of matrix addition for $m \times n$ matrices over** F, some of which we have already investigated.

(1) $A+B$ exists for each pair of $m \times n$ matrices A, B;

(2) $A+(B+C) = (A+B)+C$ for all $m \times n$ matrices A, B, C [the associative property];

(3) $\exists$ an $m \times n$ matrix O such that, $\forall m \times n$ matrices A,

$$A+O = O+A = A;$$

(4) for each $m \times n$ matrix A, $\exists$ an $m \times n$ matrix $-A$ such that

$$A+(-A) = (-A)+A = O;$$

(5) $A+B = B+A$ for all $m \times n$ matrices A, B [the commutative property.]. We have already dealt with (1), (3) and (4).

Proof of (5). If $A = [a_{ij}]$ and $B = [b_{ij}]$, then

$$A+B = [a_{ij}]+[b_{ij}] = [a_{ij}+b_{ij}] = [b_{ij}+a_{ij}] = [b_{ij}]+[a_{ij}] = B+A.$$

Thus the commutative property of matrix addition depends on the commutative property of addition for F.

Proof of (2). If $A = [a_{ij}]$, $B = [b_{ij}]$ and $C = [c_{ij}]$, then

$$\begin{aligned}
A+(B+C) &= [a_{ij}]+([b_{ij}]+[c_{ij}]) \\
&= [a_{ij}]+[b_{ij}+c_{ij}] \\
&= [a_{ij}+(b_{ij}+c_{ij})] \\
&= [(a_{ij}+b_{ij})+c_{ij}] \\
&= [a_{ij}+b_{ij}]+[c_{ij}] \\
&= ([a_{ij}]+[b_{ij}])+[c_{ij}] \\
&= (A+B)+C.
\end{aligned}$$

Thus the associative property of matrix addition depends on the associative property of addition for F.

Note. We write $A+B+C$ for each of the equal matrices $A+(B+C)$ and $(A+B)+C$.

Exercise 1. If $A = \begin{bmatrix} 3 & 0 \\ 1 & -2 \end{bmatrix}$ and $B = \begin{bmatrix} -4 & 1 \\ 2 & 3 \end{bmatrix}$, verify that $B-A$ is the negative of $A-B$.

Prove for $m \times n$ matrices that $-(A-B) = B-A$.

2. If $A = \begin{bmatrix} 2 & 1 & 6 \\ 3 & 2 & 4 \\ 0 & 0 & 1 \end{bmatrix}$, $B = \begin{bmatrix} 3 & -1 & -2 \\ 0 & 1 & 5 \\ 3 & -2 & -1 \end{bmatrix}$, $C = \begin{bmatrix} 0 & 1 & 2 \\ 3 & 4 & 5 \\ 6 & 7 & 8 \end{bmatrix}$,

(a) verify that $A-B-C = A-(B+C)$;
(b) find the negative of $A+B+C$;
(c) find X such that $A+B+X = 0$;
(d) find X such that $A+B+X = C$.

Multiplication of a matrix by a scalar, i.e. by an element of F. We write $2A$ for $A+A$, $3A$ for $A+A+A$, etc., and nA for $A+A+ \ldots +A$ (n terms). Thus

$$nA = [a_{ij}+a_{ij}+\ldots+a_{ij}]$$
$$= [na_{ij}].$$

For example, $\quad 4\begin{bmatrix} 1 & 0 & -2 & 3 \\ 5 & -3 & 7 & 4 \end{bmatrix} = \begin{bmatrix} 4 & 0 & -8 & 12 \\ 20 & -12 & 28 & 16 \end{bmatrix}$.

We note that, in forming nA, each entry of A is multiplied by n. This suggests the following definition:

If $h \in F$, i.e. if h is a scalar, and if $A = [a_{ij}]$ is an $m \times n$ matrix over F, we define the **scalar multiple of A by h** to be the $m \times n$ matrix $[ha_{ij}]$, and denote it by hA; thus

$$hA = h[a_{ij}] = [ha_{ij}].$$

For example, $\quad \frac{3}{2}\begin{bmatrix} 4 & -2 \\ -1 & 0 \\ 6 & 3 \end{bmatrix} = \begin{bmatrix} 6 & -3 \\ -\frac{3}{2} & 0 \\ 9 & \frac{9}{2} \end{bmatrix}$.

Notes 1. $(-1)A = [-a_{ij}] = -A$, the negative of A.
2. $(0)A = [0a_{ij}] = [0] = O$, the zero $m \times n$ matrix.

Properties of multiplication of a matrix by a scalar.

(1) If $h, k \in F$, then $(h+k)A = hA + kA$.
(2) If $h \in F$ and A, B are $m \times n$ matrices over F, then

$$h(A+B) = hA + hB.$$

(3) If $h, k \in F$, then $h(kA) = (hk)A$.

Exercise 1. Write out proofs of (1), (2), and (3).

2. If $\quad 4\begin{bmatrix} 2 & 1 & 3 \\ -1 & 0 & 1 \end{bmatrix} + X = 6\begin{bmatrix} -1 & 3 & 4 \\ 5 & 1 & 0 \end{bmatrix}$, find X.

3. Multiplication of matrices

This is based on the definition of the product

$$[a_{i1}a_{i2} \ldots a_{in}] \begin{bmatrix} b_{1j} \\ b_{2j} \\ \vdots \\ b_{nj} \end{bmatrix},$$

i.e. of a row vector of dimension n by a column vector of the *same* dimension n. The product is called the **inner product** of the two vectors and is defined by:

$$[a_{i1}a_{i2} \ldots a_{in}] \begin{bmatrix} b_{1j} \\ b_{2j} \\ \vdots \\ b_{nj} \end{bmatrix} = a_{i1}b_{1j}+a_{i2}b_{2j}+\ldots+a_{in}b_{nj} = \sum_{r=1}^{n} a_{ir}b_{rj},$$

i.e. it is the sum of all the products of corresponding entries.

For example,

$$[2 \quad 1 \quad -2 \quad -3] \begin{bmatrix} 4 \\ 0 \\ 5 \\ -2 \end{bmatrix} = 2.4+1.0+(-2).5+(-3).(-2)$$

$$= 8+0-10+6 = 4.$$

If $A = [a_{ij}]$ and $B = [b_{ij}]$, then we say that the ordered pair of matrices A, B are **conformable for multiplication**, or that the product AB exists, if and only if *the number of columns of A equals the number of rows of B*, i.e. if and only if A is of order $m \times n$ and B is of order $n \times p$ for some positive integers m, n, p.

When A is of order $m \times n$ and B is of order $n \times p$, then AB is defined as an $m \times p$ matrix as follows:

$$AB = \begin{bmatrix} a_{11} & a_{12} \ldots a_{1n} \\ a_{21} & a_{22} \ldots a_{2n} \\ \cdot & \cdot \quad \cdot \quad \cdot \\ a_{m1} & a_{m2} \ldots a_{mn} \end{bmatrix} \begin{bmatrix} b_{11} & b_{12} \ldots b_{1p} \\ b_{21} & b_{22} \ldots b_{2p} \\ \cdot & \cdot \quad \cdot \quad \cdot \\ b_{n1} & b_{n2} \ldots b_{np} \end{bmatrix}$$

$$= \begin{bmatrix} x_{11} & x_{12} \ldots x_{1p} \\ x_{21} & x_{22} \ldots x_{2p} \\ \cdot & \cdot \quad \cdot \quad \cdot \\ x_{m1} & x_{m2} \ldots x_{mp} \end{bmatrix},$$

where x_{ij} is the inner product of the ith row of A by the jth column of B, and so

$$x_{ij} = [a_{i1} a_{i2} \dots a_{in}] \begin{bmatrix} b_{1j} \\ b_{2j} \\ \vdots \\ b_{nj} \end{bmatrix}$$

$$= a_{i1} b_{1j} + a_{i2} b_{2j} + \dots + a_{in} b_{nj}$$

$$= \sum_{r=1}^{n} a_{ir} b_{rj}.$$

For example,

(i) $\begin{bmatrix} 2 & 0 & -1 \\ -3 & 1 & 4 \end{bmatrix} \begin{bmatrix} 1 & 2 & 0 & -3 \\ -5 & 0 & 1 & 1 \\ 3 & 0 & 2 & -1 \end{bmatrix} = \begin{bmatrix} -1 & 4 & -2 & -5 \\ 4 & -6 & 9 & 6 \end{bmatrix};$

[Note the orders: $(2 \times 3).(3 \times 4)$ giving (2×4).]

(ii) $\begin{bmatrix} a_{11} & a_{12} \\ a_{21} & a_{22} \end{bmatrix} \begin{bmatrix} x_1 \\ x_2 \end{bmatrix} = \begin{bmatrix} a_{11} x_1 + a_{12} x_2 \\ a_{21} x_1 + a_{22} x_2 \end{bmatrix};$

[Note the orders: $(2 \times 2).(2 \times 1)$ giving (2×1).]

(iii) $\begin{bmatrix} 2 & 1 & -4 \\ -3 & 5 & 6 \end{bmatrix} \begin{bmatrix} 3 \\ -1 \end{bmatrix}$ is *not defined*.

[Here the orders are (2×3) and (2×1).]

Notes 1. When AB exists we say in forming AB that B is **premultiplied** by A and A is **postmultiplied** by B.

2. If A is of order $m \times n$ and B is of order $n \times p$, then AB exists; BA also exists if and only if $p = m$. Hence AB and BA both exist if and only if A is of order $m \times n$ and B is of order $n \times m$, for some positive integers m, n; then AB is of order $m \times m$ and BA is of order $n \times n$. It follows that the question of whether or not $AB = BA$ does not arise unless $m = n$; for, when AB and BA both exist, they are of the same order if and only if A and B are square matrices of the same order n for some positive integer n.

3. When A and B are both $n \times n$ matrices, examples show that, in general, $AB \neq BA$, i.e. **multiplication is not commutative.**

For example, if $A = \begin{bmatrix} 1 & 1 \\ 0 & 1 \end{bmatrix}$ and $B = \begin{bmatrix} 0 & -1 \\ 1 & 0 \end{bmatrix}$, then

$AB = \begin{bmatrix} 1 & -1 \\ 1 & 0 \end{bmatrix}$ and $BA = \begin{bmatrix} 0 & -1 \\ 1 & 1 \end{bmatrix}$, so that $AB \neq BA$.

Two given square matrices A, B such that $AB = BA$ are said to **commute**.

4. If $a, b \in \mathbf{R}$, the set of real numbers, and if $ab = 0$, then either $a = 0$ or $b = 0$. The corresponding result for matrices is *not* true, i.e. we can have $AB = O$ with $A \neq O$ and $B \neq O$.

For example, $\quad \begin{bmatrix} 1 & 0 \\ 1 & 0 \end{bmatrix} \begin{bmatrix} 0 & 0 \\ 1 & 1 \end{bmatrix} = \begin{bmatrix} 0 & 0 \\ 0 & 0 \end{bmatrix}.$

Exercise. If $h \in F$ and if AB exists, show that $h(AB) = (hA)B$.

Although the commutative property does not hold for matrix multiplication, it is an important fact that the associative property holds. This can be stated as follows:

Associative property of matrix multiplication.

If one of the products $A(BC)$ and $(AB)C$ exists, then so does the other, and

$$A(BC) = (AB)C.$$

Proof. It is easily checked that each product exists if and only if the orders of A, B and C are of the form $m \times n$, $n \times p$ and $p \times q$, respectively, for some positive integers m, n, p and q.

When the orders are of this form, then $A(BC)$ and $(AB)C$ are both of order $m \times q$.

Let $A = [a_{ij}]_{m \times n}$, $B = [b_{ij}]_{n \times p}$, $C = [c_{ij}]_{p \times q}$, and let $BC = [x_{ij}]_{n \times q}$ and $AB = [y_{ij}]_{m \times p}$.

The (i, j)th element of $A(BC)$

$$= \sum_{r=1}^{n} a_{ir} x_{rj}$$

$$= \sum_{r=1}^{n} a_{ir} \left\{ \sum_{s=1}^{p} b_{rs} c_{sj} \right\}$$

$$= a_{i1}(b_{11}c_{1j}+b_{12}c_{2j}+\ldots+b_{1p}c_{pj}) \qquad (3.1)$$
$$+a_{i2}(b_{21}c_{1j}+b_{22}c_{2j}+\ldots+b_{2p}c_{pj})$$
$$+\ldots\ldots\ldots\ldots\ldots$$
$$+a_{in}(b_{n1}c_{1j}+b_{n2}c_{2j}+\ldots+b_{np}c_{pj})$$

$$= (a_{i1}b_{11}+a_{i2}b_{21}+\ldots+a_{in}b_{n1})c_{1j}$$
$$+(a_{i1}b_{12}+a_{i2}b_{22}+\ldots+a_{in}b_{n2})c_{2j}$$
$$+\ldots\ldots\ldots\ldots\ldots$$
$$+(a_{i1}b_{1p}+a_{i2}b_{2p}+\ldots+a_{in}b_{np})c_{pj}$$

(on rearranging the summation in (3.1) from rows to columns)

$$= \sum_{s=1}^{p}\left\{\sum_{r=1}^{n} a_{ir}b_{rs}\right\}c_{sj}$$

$$= \sum_{s=1}^{p} y_{is}c_{sj}$$

$$= \text{the } (i,j)\text{th element of } (AB)C.$$

Since this is true for all i, j, it follows that $A(BC) = (AB)C$.

As usual when an associative property holds, we can omit brackets, writing in this case ABC for each of $A(BC)$ and $(AB)C$.

Similarly we write $ABCD$ unambiguously for the products arising when we insert brackets in all possible ways (assuming that the product exists); likewise for more than four matrices.

Powers of a square matrix. If A is of order $n \times n$ for some n, then the products $AA, AAA, AAAA, AAAAA, \ldots$ all exist, and are denoted by $A^2, A^3, A^4, A^5, \ldots$.

Exercise. If p, q are positive integers, show that

(i) $A^p.A^q = A^{p+q} = A^q.A^p$ and (ii) $(A^p)^q = A^{pq}$.

There are two other important results for matrices, namely the distributive properties. These can be stated as:

The distributive properties for matrices.

(1) $A(B+C) = AB+AC,$
(2) $(A+B)C = AC+BC,$

when the products on the left-hand side (or right-hand side) exist.

Proof of (1). It is easily checked that the matrices exist if and only if the orders of A, B and C are of the form $m \times n$, $n \times p$ and $n \times p$, respectively, for some positive integers m, n and p.

Let $A = [a_{ij}]$, $B = [b_{ij}]$ and $C = [c_{ij}]$. Then, the (i, j)th element of

$$A(B+C) = \sum_{r=1}^{n} a_{ir}(b_{rj} + c_{rj})$$

$$= \sum_{r=1}^{n} (a_{ir}b_{rj} + a_{ir}c_{rj})$$

$$= \sum_{r=1}^{n} a_{ir}b_{rj} + \sum_{r=1}^{n} a_{ir}c_{rj}$$

$$= \text{the } (i, j)\text{th element of } AB + AC.$$

Since this holds for all i, j, it follows that $A(B+C) = AB + AC$.

Exercise Prove similarly the distributive property (2).

Example 1. If A and B are square matrices of the same order, show that $A^2 - B^2 = (A - B)(A + B)$ if and only if A and B commute.

We have:
$$(A - B)(A + B) = A^2 + AB - BA - B^2$$
$$= A^2 - B^2 \Leftrightarrow AB - BA = O$$
$$\Leftrightarrow AB = BA$$
$$\Leftrightarrow A \text{ and } B \text{ commute.}$$

Notes 1. If A, B is any pair of square matrices of the same order, then the binomial theorem for the expansion of $(A + B)^n$, where n is a positive integer, does *not*, in general, hold; it does hold *if A and B commute*.

e.g.
$$(A + B)^2 = (A + B)(A + B)$$
$$= A^2 + AB + BA + B^2$$
$$= A^2 + 2AB + B^2 \Leftrightarrow AB = BA.$$

2. If A and B are both $n \times n$ matrices, then

$$(AB)^2 = ABAB,$$
$$A^2 B^2 = AABB,$$

and these are in general different.

The identity or unity $n \times n$ matrix. The $n \times n$ matrix

$$\begin{bmatrix} 1 & 0 & 0 \dots 0 \\ 0 & 1 & 0 \dots 0 \\ 0 & 0 & 1 \dots 0 \\ . & . & . \quad . \\ 0 & 0 & 0 \dots 1 \end{bmatrix},$$

with 1 down the so-called *main* (or *principal*) *diagonal* and 0 elsewhere, plays a special role in multiplication. For each $i = 1, 2, \ldots, n$, the ith row has 1 in the ith place and 0 elsewhere. This matrix is denoted by I_n, or simply by I if n is clearly understood, and called the **identity** or **unity** $n \times n$ **matrix**.

If A is any $m \times n$ matrix, then it is easily verified that

$$I_m A = A \quad \text{and} \quad AI_n = A.$$

In particular, if A is any $n \times n$ matrix, then

$$AI_n = I_n A = A.$$

Note. It is often useful to express I_n in terms of the so-called **Kronecker delta symbol** which is defined as follows:

$$\delta_{ij} = \begin{cases} 1 & \text{when} \quad i = j, \\ 0 & \text{when} \quad i \neq j. \end{cases}$$

Then $$I_n = [\delta_{ij}]_{n \times n}.$$

Exercise. Check that $I_n^m = I_n$, for each positive integer m.

4. The transpose of a matrix and some special forms of matrices.

If $\quad A = \begin{bmatrix} a_{11} & a_{12} \ldots a_{1n} \\ a_{21} & a_{22} \ldots a_{2n} \\ \cdot & \cdot \quad \cdot \quad \cdot \\ a_{m1} & a_{m2} \ldots a_{mn} \end{bmatrix}$, then the matrix $\begin{bmatrix} a_{11} & a_{21} \ldots a_{m1} \\ a_{12} & a_{22} \ldots a_{m2} \\ \cdot & \cdot \quad \cdot \quad \cdot \\ a_{1n} & a_{2n} \ldots a_{mn} \end{bmatrix}$,

which is obtained from A by interchanging rows and columns, is called the **transpose of A** and is denoted by A'. If A is of order $m \times n$, then A' is of order $n \times m$. Also, if $A = [a_{ij}]$, then $A' = [a'_{ij}]$, where $a'_{ij} = a_{ji}$ for all i, j.

For example, if $\quad A = \begin{bmatrix} 2 & 3 & -4 \\ -1 & 0 & 5 \end{bmatrix}$, then $A' = \begin{bmatrix} 2 & -1 \\ 3 & 0 \\ -4 & 5 \end{bmatrix}$.

If $\quad A = \begin{bmatrix} 3 & 5 & -2 \\ -1 & 0 & -1 \\ 2 & 1 & 6 \end{bmatrix}$, then $A' = \begin{bmatrix} 3 & -1 & 2 \\ 5 & 0 & 1 \\ -2 & -1 & 6 \end{bmatrix}$.

Also, if $\quad X = \begin{bmatrix} x_1 \\ x_2 \\ \vdots \\ x_n \end{bmatrix}$, then $X' = [x_1 \ x_2 \ldots x_n]$, i.e. the transpose of

a column matrix is the corresponding row matrix.

Exercise. Describe the transpose of a row matrix.

Properties of the transpose.

(1) $\qquad (A')' = A.$
(2) $\qquad (A+B)' = A' + B'.$ $\qquad$ (assuming $A+B$ exists)
(3) If $h \in F$, $\quad (hA)' = hA'.$
(4) $\qquad (AB)' = B'A'.$ $\qquad$ (assuming AB exists)

Proof of (4). Let $A = [a_{ij}]$ have order $m \times n$ and $B = [b_{ij}]$ have order $n \times p$, so that AB exists and is of order $m \times p$. Then $(AB)'$ has order $p \times m$; also B' has order $p \times n$ and A' has order $n \times m$ so that $B'A'$ has order $p \times m$. Hence $(AB)'$ and $B'A'$ have the same order. Also

$$A' = [a'_{ij}], \quad \text{where } a'_{ij} = a_{ji}, \quad \text{and} \quad B' = [b'_{ij}], \quad \text{where } b'_{ij} = b_{ji}.$$

The (i, j)th element of $(AB)' = $ the (j, i)th element of AB

$$= \sum_{r=1}^{n} a_{jr} b_{ri}$$

$$= \sum_{r=1}^{n} b'_{ir} a'_{rj}$$

$$= \text{the } (i, j)\text{th element of } B'A'.$$

Since this holds for all i, j, it follows that $(AB)' = B'A'$.

Extension. $\qquad (ABC)' = ((AB)C)' = C'(AB)' = C'B'A';$

$$(A_1 A_2 \ldots A_{n-1} A_n)' = A'_n A'_{n-1} \ldots A'_1;$$

in particular, if A is square, $\qquad (A^n)' = (A')^n.$

Exercise. Prove the properties (1), (2) and (3).

Some special forms of matrices.

I. Diagonal and scalar matrices. If the entries of a *square* matrix A, which do not lie on the main diagonal, are all 0, then A is called a **diagonal matrix**; such a matrix is of the form

$$A = \begin{bmatrix} a_{11} & 0 & 0 & \ldots 0 \\ 0 & a_{22} & 0 & \ldots 0 \\ 0 & 0 & a_{33} \ldots 0 \\ \cdot & \cdot & \cdot & \cdot \\ 0 & 0 & 0 & \ldots a_{nn} \end{bmatrix}.$$

A diagonal matrix for which $a_{11}, a_{22}, \ldots, a_{nn}$ are all equal, to a say, is called a **scalar matrix**; such a matrix is of the form

$$\begin{bmatrix} a & 0 & 0 \ldots 0 \\ 0 & a & 0 \ldots 0 \\ 0 & 0 & a \ldots 0 \\ \cdot & \cdot & \cdot \\ 0 & 0 & 0 \ldots a \end{bmatrix} = a \begin{bmatrix} 1 & 0 & 0 \ldots 0 \\ 0 & 1 & 0 \ldots 0 \\ 0 & 0 & 1 \ldots 0 \\ \cdot & \cdot & \cdot \\ 0 & 0 & 0 \ldots 1 \end{bmatrix},$$

and so of the form aI_n.

II. Polynomials in a square matrix A. For such a matrix, the powers A, A^2, A^3, A^4, ... exist; we often write A^0 for the identity matrix I of the same order as A.

The matrices p_0I, p_1A, $p_2A^2, \ldots, p_mA^m$, where $p_0, p_1, \ldots, p_m$ are elements of F, all exist, and hence so does the matrix

$$p_0I + p_1A + p_2A^2 + \ldots + p_mA^m.$$

Such a matrix is said to be a **polynomial in the square matrix A**. If $p_m \neq 0$, it is said to be of **degree** m.

Example 1. If $A = \begin{bmatrix} 1 & 2 \\ -3 & 0 \end{bmatrix}$, form the polynomial $A^2 + 3A + 5I$.

$$A^2 = \begin{bmatrix} 1 & 2 \\ -3 & 0 \end{bmatrix}\begin{bmatrix} 1 & 2 \\ -3 & 0 \end{bmatrix} = \begin{bmatrix} -5 & 2 \\ -3 & -6 \end{bmatrix}. \text{ Thus}$$

$$A^2 + 3A + 5I = \begin{bmatrix} -5 & 2 \\ -3 & -6 \end{bmatrix} + 3\begin{bmatrix} 1 & 2 \\ -3 & 0 \end{bmatrix} + 5\begin{bmatrix} 1 & 0 \\ 0 & 1 \end{bmatrix}$$

$$= \begin{bmatrix} -5+3+5 & 2+6+0 \\ -3-9+0 & -6+0+5 \end{bmatrix} = \begin{bmatrix} 3 & 8 \\ -12 & -1 \end{bmatrix}.$$

Example 2. Express $(2I+3A+4A^2)(5I-4A)$ as a polynomial in A.

$$(2I+3A+4A^2)(5I-4A) = 10I - 8A$$
$$+ 15A - 12A^2$$
$$+ 20A^2 - 16A^3$$
$$= 10I + 7A + 8A^2 - 16A^3.$$

Exercise. Expand $(5I-4A)(2I+3A+4A^2)$ as a polynomial in A, and compare with the result in Example 2.

Note. If $P(A)$ and $Q(A)$ are polynomials in A, then $P(A)$ and $Q(A)$ commute.

III. Symmetric and skew-symmetric matrices. A square matrix A is called **symmetric** if $A' = A$, i.e. if $a_{ji} = a_{ij}$ for all i, j.

Such a matrix is symmetrical about the main diagonal.

e.g. $\begin{bmatrix} 5 & 2 \\ 2 & 3 \end{bmatrix}$ and $\begin{bmatrix} a & h & g \\ h & b & f \\ g & f & c \end{bmatrix}$ are symmetric.

A square matrix A is called   if $A' = -A$, i.e. if $a_{ji} = -a_{ij}$ for i, j. It follows that, if A is skew-symmetric, then $a_{ii} = -a_{ii}$ and so $a_{ii} = 0$ for each i, i.e. a skew-symmetric matrix has 0 in each place of its main diagonal. For example, the matrices

$$\begin{bmatrix} 0 & a \\ -a & 0 \end{bmatrix}, \quad \begin{bmatrix} 0 & 1 & -2 \\ -1 & 0 & 5 \\ 2 & -5 & 0 \end{bmatrix} \text{ and } \begin{bmatrix} 0 & a & b & c \\ -a & 0 & d & e \\ -b & -d & 0 & f \\ -c & -e & -f & 0 \end{bmatrix}$$

are skew-symmetric.

Example 3. Show that, if A is any matrix, then the matrices AA' and $A'A$ are both symmetric.

If A is of order $m \times n$, then A' is of order $n \times m$ and so AA' is square of order m and $A'A$ is square of order n.

Now $(AA')' = (A')'A' = AA'$,   so that AA' is symmetric.
Also $(A'A)' = A'(A')' = A'A$,   and so $A'A$ is symmetric.

Example 4. Show that a square matrix A can be expressed in a unique way as a sum of a symmetric matrix and a skew-symmetric matrix.

Suppose in the first place, that $A = S + T$, where S is symmetric and T is skew-symmetric, so that $S' = S$ and $T' = -T$. Then

$$A = S + T,$$
$$A' = S' + T' = S - T,$$

and, by addition, $2S = A + A'$,   so that $S = \frac{1}{2}(A + A')$;
also, by subtraction, $2T = A - A'$,   so that $T = \frac{1}{2}(A - A')$.

Now
$$\tfrac{1}{2}(A+A')' = \tfrac{1}{2}(A'+A) = \tfrac{1}{2}(A+A')$$
$$\tfrac{1}{2}(A-A')' = \tfrac{1}{2}(A'-A) = -\tfrac{1}{2}(A-A')$$

so that $\tfrac{1}{2}(A+A')$ is in fact symmetric and $\tfrac{1}{2}(A-A')$ is in fact skew-symmetric. Since $\tfrac{1}{2}(A+A')+\tfrac{1}{2}(A-A') = A$, the result now follows.

Exercise 1. If A is a symmetric $n \times n$ matrix and B is an $m \times n$ matrix, show that BAB' is symmetric.

2. If A and B are symmetric $n \times n$ matrices, show that $AB-BA$ is skew-symmetric.

5. Some applications of matrices for compact notation.

I. Linear equations. The set of linear equations.

$$
\begin{aligned}
y_1 &= a_{11}x_1 + a_{12}x_2 + \ldots + a_{1n}x_n, \\
y_2 &= a_{21}x_1 + a_{22}x_2 + \ldots + a_{2n}x_n, \\
&\cdots \cdots \cdots \cdots \cdots \cdots \\
y_m &= a_{m1}x_1 + a_{m2}x_2 + \ldots + a_{mn}x_n
\end{aligned}
\tag{5.1}
$$

can be written in matrix form

$$Y = AX, \tag{5.2}$$

where Y is the $m \times 1$ column matrix $\begin{bmatrix} y_1 \\ y_2 \\ \vdots \\ y_m \end{bmatrix}$ and AX is the product of the

$m \times n$ matrix $A = \begin{bmatrix} a_{11} & a_{12} \ldots a_{1n} \\ a_{21} & a_{22} \ldots a_{2n} \\ \cdot & \cdot \quad \cdot \quad \cdot \\ a_{m1} & a_{m2} \ldots a_{mn} \end{bmatrix}$ and the $n \times 1$ column matrix

$$X = \begin{bmatrix} x_1 \\ x_2 \\ \vdots \\ x_n \end{bmatrix}.$$

The matrix A is called the **matrix of coefficients** of the set of linear equations (5.1)

Similarly, the set of linear equations

$$x_1 = b_{11}z_1 + b_{12}z_2 + \ldots + b_{1p}z_p,$$
$$x_2 = b_{21}z_1 + b_{22}z_2 + \ldots + b_{2p}z_p,$$
$$\ldots \ldots \ldots \ldots \ldots \ldots \ldots \quad (5.3)$$
$$x_n = b_{n1}z_1 + b_{n2}z_2 + \ldots + b_{np}z_p$$

can be written in matrix form

$$X = BZ, \quad (5.4)$$

where $\quad X = \begin{bmatrix} x_1 \\ x_2 \\ \vdots \\ x_n \end{bmatrix}, \quad Z = \begin{bmatrix} z_1 \\ z_2 \\ \vdots \\ z_p \end{bmatrix}$ and $B = \begin{bmatrix} b_{11} & b_{12} \ldots b_{1p} \\ b_{21} & b_{22} \ldots b_{2p} \\ \vdots & \vdots \quad \vdots \\ b_{n1} & b_{n2} \ldots b_{np} \end{bmatrix}$.

If we substitute for $x_1, \ldots, x_n$ in terms of $z_1, \ldots, z_p$ from equations (5.3) into equations (5.1), we obtain a new set of equations which express $y_1, \ldots, y_m$ in terms of $z_1, \ldots, z_p$. This process is very neatly expressed in matrix form by using (5.2) and (5.4). We have

$$Y = AX = A(BZ) = (AB)Z.$$

Consequently the matrix AB is the matrix of coefficients of the set of linear equations which express $y_1, \ldots, y_m$ in terms of $z_1, \ldots, z_p$.

This fact can be taken as a justification of the earlier definition of matrix multiplication.

Exercise. Express the sets of equations

$$(1) \quad \begin{aligned} y_1 &= 2x_1 - x_2 + 3x_3 \\ y_2 &= 5x_1 + x_2 - 2x_3 \end{aligned}, \quad (2) \quad \begin{aligned} x_1 &= z_1 - 2z_2 \\ x_2 &= 3z_1 + 5z_2 \\ x_3 &= 4z_1 + 7z_2 \end{aligned}$$

in matrix form. Determine the set of linear equations giving y_1, y_2 in terms of z_1, z_2, first by using matrices, and then by substituting from equations (2) into equations (1).

II. Quadratic forms. An expression of the form $ax^2 + 2bxy + cy^2$ is called a **quadratic form in the variables** x, y; the coefficients a, b, c are given elements of some number system F. Similarly an expression of the form $ax^2 + by^2 + cz^2 + 2fyz + 2gzx + 2hxy$, in which a, b, c, f, g, h are given elements of F, is called a **quadratic form in the variables** x, y, z. Each term is of second degree in the variables. For the general case of n variables $x_1, x_2, \ldots, x_n$ we use a double suffix notation for the coefficients of a **quadratic form in $x_1, x_2, \ldots, x_n$**; such a quadratic form is an expression of the form

$$\sum_{i=1}^{n} \sum_{j=1}^{n} a_{ij}x_ix_j,$$

in which the coefficients a_{ij} are given elements of F.

A quadratic form can be expressed in many ways as a product of matrices, but each form has a unique expression involving a symmetric matrix.

Consider the form $ax^2 + 2bxy + cy^2$, and verify that it equals the following product of matrices

$$[x \quad y] \begin{bmatrix} a & b+t \\ b-t & c \end{bmatrix} \begin{bmatrix} x \\ y \end{bmatrix},$$

for any $t \in F$.

If
$$X = \begin{bmatrix} x \\ y \end{bmatrix} \quad \text{and} \quad A_t = \begin{bmatrix} a & b+t \\ b-t & c \end{bmatrix},$$

then
$$ax^2 + 2bxy + cy^2 = X'A_tX.$$

The matrix A_t is symmetric if and only if $t = 0$. If we denote the corresponding matrix by S, so that

$$S = \begin{bmatrix} a & b \\ b & c \end{bmatrix},$$

then
$$ax^2 + 2bxy + cy^2 = X'SX,$$

and the symmetric matrix S is uniquely determined by the quadratic form.

Note. For the form $ax^2 + bxy + cy^2$, the matrix is $\begin{bmatrix} a & \frac{1}{2}b \\ \frac{1}{2}b & c \end{bmatrix}$.

The corresponding result holds for quadratic forms in n variables. Consider the form

$$\sum_{i=1}^{n} \sum_{j=1}^{n} a_{ij}x_ix_j.$$

If
$$A = [a_{ij}]_{n \times n} \quad \text{and} \quad X = \begin{bmatrix} x_1 \\ x_2 \\ \vdots \\ x_n \end{bmatrix},$$

then $\quad X'AX = [x_1x_2\ldots x_n]\begin{bmatrix} a_{11} & a_{12}\ldots a_{1n} \\ a_{21} & a_{22}\ldots a_{2n} \\ \cdot & \cdot\ \cdot\ \cdot\ \cdot\ \cdot \\ a_{n1} & a_{n2}\ \cdots\ a_{nn} \end{bmatrix}\begin{bmatrix} x_1 \\ x_2 \\ \vdots \\ x_n \end{bmatrix}$

$$= [x_1x_2\ldots x_n]\begin{bmatrix} a_{11}x_1+a_{12}x_2+\ldots+a_{1n}x_n \\ a_{21}x_1+a_{22}x_2+\ldots+a_{2n}x_n \\ \cdot\ \cdot\ \cdot\ \cdot\ \cdot\ \cdot\ \cdot\ \cdot\ \cdot\ \cdot \\ a_{n1}x_1+a_{n2}x_2+\ldots+a_{nn}x_n \end{bmatrix}$$

$$\begin{aligned} = \quad & x_1(a_{11}x_1+a_{12}x_2+\ldots+a_{1n}x_n) \\ + & x_2(a_{21}x_1+a_{22}x_2+\ldots+a_{2n}x_n) \\ + & \cdot\ \cdot\ \cdot\ \cdot\ \cdot\ \cdot\ \cdot\ \cdot\ \cdot \\ + & x_n(a_{n1}x_1+a_{n2}x_2+\ldots+a_{nn}x_n) \end{aligned}$$

$$= \sum_{i=1}^{n}\sum_{j=1}^{n} a_{ij}x_ix_j.$$

Now $A = S+K$, where S is the symmetric matrix $\frac{1}{2}(A+A')$ and K is the skew-symmetric matrix $\frac{1}{2}(A-A')$. Also by the distributive properties of matrix multiplication,

$$X'AX = X'(S+K)X = X'SX+X'KX.$$

We now show that $X'KX = 0$, the scalar zero. Since $X'KX$ is a 1×1 matrix and so a scalar, it follows that $X'KX = (X'KX)'$. Hence

$$X'KX = (X'KX)' = X'K'(X')' = X'(-K)X = -X'KX.$$

Thus $\quad\quad\quad\quad 2X'KX = 0,\quad \text{and so}\quad X'KX = 0.$

It follows that

$$\sum_{i=1}^{n}\sum_{j=1}^{n} a_{ij}x_ix_j = X'SX,$$

where S is the symmetric matrix $\frac{1}{2}(A+A')$ and A is the matrix $[a_{ij}]$ of coefficients.

For example,

$$ax^2+by^2+cz^2+2fyz+2gzx+2hxy = [xyz]\begin{bmatrix} a & h & g \\ h & b & f \\ g & f & c \end{bmatrix}\begin{bmatrix} x \\ y \\ z \end{bmatrix}.$$

Exercise 1. Express the quadratic form $2x^2-y^2+3z^2+4xz+5yz$ as a product of matrices with a symmetric coefficient matrix.

2. If $A = \begin{bmatrix} 1 & 3 & 2 \\ -1 & 0 & 5 \\ 4 & 1 & 1 \end{bmatrix}$ and $X = \begin{bmatrix} x \\ y \\ z \end{bmatrix}$, find a symmetric matrix

B such that the quadratic forms $X'AX$ and $X'BX$ are the same.

6. Some worked examples.

Example 1. *If* $A = \begin{bmatrix} 2 & -5 \\ 3 & 1 \end{bmatrix}$, *find scalars a, b, c, not all zero, such that*

$$aI + bA + cA^2 = 0.$$

$$A^2 = \begin{bmatrix} 2 & -5 \\ 3 & 1 \end{bmatrix}\begin{bmatrix} 2 & -5 \\ 3 & 1 \end{bmatrix} = \begin{bmatrix} -11 & -15 \\ 9 & -14 \end{bmatrix}.$$

Thus

$$aI + bA + cA^2 = \begin{bmatrix} a & 0 \\ 0 & a \end{bmatrix} + \begin{bmatrix} 2b & -5b \\ 3b & b \end{bmatrix} + \begin{bmatrix} -11c & -15c \\ 9c & -14c \end{bmatrix}$$

$$= \begin{bmatrix} a+2b-11c & -5b-15c \\ 3b+9c & a+b-14c \end{bmatrix}.$$

Hence $aI + bA + cA^2 = 0 \Leftrightarrow \begin{cases} a+2b-11c = 0 \\ -5b-15c = 0 \\ 3b+9c = 0 \\ a+b-14c = 0 \end{cases}$

$$\Leftrightarrow \begin{cases} a+2b-11c = 0 \\ b+3c = 0 \end{cases}$$

[since the other equations can be deduced from these]

$$\Leftrightarrow \begin{cases} a-17c = 0 \\ b+3c = 0 \end{cases} \Leftrightarrow \frac{a}{17} = \frac{b}{-3} = \frac{c}{1}.$$

Clearly $a = 17$, $b = -3$, $c = 1$ satisfy the conditions.

Example 2. *Show that a matrix X commutes with the matrix*

$$A = \begin{bmatrix} 0 & 0 & 0 \\ 1 & 0 & 0 \\ 0 & 1 & 0 \end{bmatrix}$$ *if and only if it is of the form* $\begin{bmatrix} x & 0 & 0 \\ y & x & 0 \\ z & y & x \end{bmatrix}$

for some scalars x, y, z.

A matrix X commuting with A must be of order 3×3. If $X = [x_{ij}]$, then

X commutes with $A \Leftrightarrow AX = XA$

$$\Leftrightarrow \begin{bmatrix} 0 & 0 & 0 \\ 1 & 0 & 0 \\ 0 & 1 & 0 \end{bmatrix} \begin{bmatrix} x_{11} & x_{12} & x_{13} \\ x_{21} & x_{22} & x_{23} \\ x_{31} & x_{32} & x_{33} \end{bmatrix}$$

$$= \begin{bmatrix} x_{11} & x_{12} & x_{13} \\ x_{21} & x_{22} & x_{23} \\ x_{31} & x_{32} & x_{33} \end{bmatrix} \begin{bmatrix} 0 & 0 & 0 \\ 1 & 0 & 0 \\ 0 & 1 & 0 \end{bmatrix}$$

$$\Leftrightarrow \begin{bmatrix} 0 & 0 & 0 \\ x_{11} & x_{12} & x_{13} \\ x_{21} & x_{22} & x_{23} \end{bmatrix} = \begin{bmatrix} x_{12} & x_{13} & 0 \\ x_{22} & x_{23} & 0 \\ x_{32} & x_{33} & 0 \end{bmatrix}$$

$$\Leftrightarrow x_{12} = 0, \ x_{13} = 0, \ x_{23} = 0,$$
$$x_{22} = x_{11}, \ x_{21} = x_{32}, \ x_{33} = x_{22}$$

$$\Leftrightarrow X = \begin{bmatrix} x_{11} & 0 & 0 \\ x_{21} & x_{11} & 0 \\ x_{31} & x_{21} & x_{11} \end{bmatrix}$$

$$\Leftrightarrow X = \begin{bmatrix} x & 0 & 0 \\ y & x & 0 \\ z & y & x \end{bmatrix} \text{ for some scalars } x, y, z.$$

Check that an X which commutes with A is of the form $xI + yA + zA^2$.

Example 3.

If
$$A = \begin{bmatrix} 2 & 1 \\ 1 & 2 \end{bmatrix} \quad and \quad B = \begin{bmatrix} 3 & 0 \\ 0 & 1 \end{bmatrix},$$

show that the most general 2×2 matrix X such that $AX = XB$ is of the form

$$\begin{bmatrix} x & y \\ x & -y \end{bmatrix}.$$

Let
$$X = \begin{bmatrix} x & y \\ z & t \end{bmatrix}.$$

Then
$$AX = XB \Leftrightarrow \begin{bmatrix} 2 & 1 \\ 1 & 2 \end{bmatrix}\begin{bmatrix} x & y \\ z & t \end{bmatrix} - \begin{bmatrix} x & y \\ z & t \end{bmatrix}\begin{bmatrix} 3 & 0 \\ 0 & 1 \end{bmatrix} = O$$

$$\Leftrightarrow \begin{bmatrix} 2x+z & 2y+t \\ x+2z & y+2t \end{bmatrix} - \begin{bmatrix} 3x & y \\ 3z & t \end{bmatrix} = O$$

$$\Leftrightarrow \begin{bmatrix} z-x & y+t \\ x-z & y+t \end{bmatrix} = O$$

$$\Leftrightarrow z = x \quad \text{and} \quad t = -y$$

$$\Leftrightarrow X = \begin{bmatrix} x & y \\ x & -y \end{bmatrix}.$$

Example 4.

If $\quad A = \begin{bmatrix} 1 & -1 & 0 \\ 2 & 3 & 4 \\ 0 & 1 & 2 \end{bmatrix} \quad and \quad B = \begin{bmatrix} 2 & 2 & -4 \\ -4 & 2 & -4 \\ 2 & -1 & 5 \end{bmatrix},$

verify that $BA = 6I$ *and hence solve the set of linear equations*

$$x - y = 3,$$
$$2x + 3y + 4z = 17,$$
$$y + 2z = 7.$$

It is easily checked that $\qquad BA = 6I.$

The matrix form of the set of equations is $AX = H$, where

$$X = \begin{bmatrix} x \\ y \\ z \end{bmatrix} \quad \text{and} \quad H = \begin{bmatrix} 3 \\ 17 \\ 7 \end{bmatrix}.$$

From $AX = H$, we deduce that

$$B(AX) = BH,$$
$$(BA)X = BH,$$
$$(6I)X = BH,$$

and so $\qquad X = \tfrac{1}{6}BH.$

Thus

$$\begin{bmatrix} x \\ y \\ z \end{bmatrix} = \frac{1}{6} \begin{bmatrix} 2 & 2 & -4 \\ -4 & 2 & -4 \\ 2 & -1 & 5 \end{bmatrix} \begin{bmatrix} 3 \\ 17 \\ 7 \end{bmatrix}$$

$$= \frac{1}{6} \begin{bmatrix} 12 \\ -6 \\ 24 \end{bmatrix} = \begin{bmatrix} 2 \\ -1 \\ 4 \end{bmatrix},$$

and so the only possible solution of the equations is $x = 2, y = -1, z = 4$. It is easily verified that this is in fact a solution.

7. Matrices with inverses

In dealing with inverses we shall assume that the entries in the matrices are elements of a field F, i.e. an algebraic system in which we can add, subtract, multiply and divide (except by zero); we can take F to be the rational numbers, or real numbers or complex numbers.

If $a \in F$ and $a \neq 0$, then there is an element $a^{-1} \in F$ (the **inverse** or **reciprocal** of a) such that $aa^{-1} = a^{-1}a = 1$. We are thus led to ask the question: When, for a given matrix A, can we find a matrix X such that $AX = XA = I$?

In the first place we note that such a matrix A must be square.

We note also that the condition $A \neq 0$ is not enough to ensure the existence of an X such that $AX = XA = I$. For example, if

$$A = \begin{bmatrix} 1 & 1 \\ 0 & 0 \end{bmatrix} \quad \text{and} \quad X = \begin{bmatrix} x & y \\ z & t \end{bmatrix}, \quad \text{then}$$

$$AX = I \Rightarrow \begin{bmatrix} 1 & 1 \\ 0 & 0 \end{bmatrix}\begin{bmatrix} x & y \\ z & t \end{bmatrix} = \begin{bmatrix} 1 & 0 \\ 0 & 1 \end{bmatrix} \Rightarrow \begin{bmatrix} x+z & y+t \\ 0 & 0 \end{bmatrix} = \begin{bmatrix} 1 & 0 \\ 0 & 1 \end{bmatrix}$$

$$\Rightarrow 0 = 1, \text{ which is impossible.}$$

We shall answer the above question completely in the case of 2×2 and 3×3 matrices and sketch the answer for $n \times n$ matrices. We shall show that it is enough to satisfy only one of the equations $AX = I$, $XA = I$, by proving that the other is then automatically satisfied.

Definitions: right inverse, left inverse, inverse. If A is an $n \times n$ matrix and I is the unity $n \times n$ matrix, then an $n \times n$ matrix X such that $AX = I$ is called a **right inverse** of A. Similarly, a matrix Y such that $YA = I$ is called a **left inverse** of A. A matrix X which is both a right and a left inverse of A, i.e. such that $AX = XA = I$ is called **an inverse** of A. A matrix with an inverse is said to be **invertible**.

Theorem 7.1. *If A has a right inverse X and a left inverse Y then $X = Y$, and this inverse of A is unique.*

Proof. Since $AX = I$ and $YA = I$ we have

$$X = IX = (YA)X = Y(AX) = YI = Y.$$

Also, if $AX_1 = I$, then, as above, $Y = X_1$, and so $X_1 = X$. Hence X is unique.

Corollary. If a square matrix A has an inverse, then that inverse is unique. It is denoted by A^{-1}, and satisfies the equations

$$AA^{-1} = A^{-1}A = I.$$

Note. Since matrix multiplication is noncommutative, the notation $\dfrac{B}{A}$

cannot be used for a product involving B and A^{-1}; in general the matrices $A^{-1}B$ and BA^{-1} are *not* equal.

Theorem 7.2. *If a square matrix A has a right inverse X, i.e. if $AX = I$, then X is also a left inverse of A and so A has the inverse X. [Similarly, with left and right interchanged.]*

We shall prove this result later for 2×2 and 3×3 matrices, and obtain a necessary and sufficient condition for such a matrix to have an inverse. The method will involve the use of determinants and these will be discussed in the next section.

We complete this section with some general results involving invertible matrices.

Theorem 7.3.

(1) *If A^{-1} exists, then $(A^{-1})^{-1} = A$; also $(A')^{-1}$ exists and*

$$(A')^{-1} = (A^{-1})'.$$

(2) *If A^{-1} and B^{-1} exist, where A, B are both $n \times n$ matrices, then $(AB)^{-1}$ exists and*

$$(AB)^{-1} = B^{-1}A^{-1}.$$

Proof. (1) Since $AA^{-1} = A^{-1}A = I$, it follows that A^{-1} has A as an inverse. But an inverse is unique; so $(A^{-1})^{-1} = A$.

Also $\qquad\qquad (AA^{-1})' = (A^{-1}A)' = I',$

and so $\qquad\qquad (A^{-1})'A' = A'(A^{-1})' = I.$

Thus A' has $(A^{-1})'$ as an inverse, and so $(A')^{-1} = (A^{-1})'$, again using the uniqueness of an inverse.

(2) $\qquad (AB)(B^{-1}A^{-1}) = A(BB^{-1})A^{-1} = AIA^{-1} = AA^{-1} = I,$

and $\qquad (B^{-1}A^{-1})(AB) = B^{-1}(A^{-1}A)B = B^{-1}IB = B^{-1}B = I.$

Hence AB has $B^{-1}A^{-1}$ as an inverse, so that $(AB)^{-1} = B^{-1}A^{-1}$, by uniqueness of an inverse.

Extension of (2). If $A_1, A_2, \ldots, A_m$ are all invertible $n \times n$ matrices, then $A_1 A_2 \ldots A_m$ is invertible and

$$(A_1 A_2 \ldots A_m)^{-1} = A_m^{-1} \ldots A_2^{-1} A_1^{-1}.$$

In particular, if A is an invertible $n \times n$ matrix, then

$$(A^m)^{-1} = (A^{-1})^m; \text{ each is denoted by } A^{-m}.$$

Example 1. Show that, if $AX = O$ and if A^{-1} exists, then $X = O$.

$$AX = O \Rightarrow A^{-1}(AX) = A^{-1}O \Rightarrow (A^{-1}A)X = O \Rightarrow IX = O \Rightarrow X = O.$$

Example 2. Show that, if A is symmetric and invertible, then A^{-1} is symmetric.

$$(A^{-1})' = (A')^{-1} = A^{-1},$$

since A' is symmetric; thus A^{-1} is symmetric.

Example 3. If A and B are invertible $n \times n$ matrices, show that

$$A^{-1} + B^{-1} = A^{-1}(A+B)B^{-1},$$

and hence find, if $A + B$ is *also* invertible, an expression for $(A^{-1} + B^{-1})^{-1}$.

$$\begin{aligned} A^{-1}(A+B)B^{-1} &= (A^{-1}A)B^{-1} + A^{-1}(BB^{-1}) \\ &= B^{-1} + A^{-1} = A^{-1} + B^{-1}. \end{aligned}$$

Hence $(A^{-1} + B^{-1})^{-1} = \{A^{-1}(A+B)B^{-1}\}^{-1} = B(A+B)^{-1}A.$

8. Determinants, inverse matrices

I. Determinant of a 2×2 matrix and 2×2 inverse matrices.

If A is the 2×2 matrix $\begin{bmatrix} a_{11} & a_{12} \\ a_{21} & a_{22} \end{bmatrix}$, then the **determinant** of A is defined to be the

scalar $a_{11}a_{22} - a_{12}a_{21}$, and is denoted by $|A|$ or $\det A$ or $\begin{vmatrix} a_{11} & a_{12} \\ a_{21} & a_{22} \end{vmatrix}$,

so that $|A| = a_{11}a_{22} - a_{12}a_{21}$. Clearly, $|A'| = |A|$.

For example, if $A = \begin{bmatrix} 2 & -3 \\ 1 & 2 \end{bmatrix}$, then $|A| = 4 + 3 = 7$,

and, if $A = \begin{bmatrix} 1 & 0 \\ 0 & -1 \end{bmatrix}$, then $|A| = -1$.

Also $|I| = 1$ and $|O| = 0$.

Multiplicative property of determinants of 2×2 matrices. *If A, B are 2×2 matrices, then*

$$|AB| = |A||B|.$$

Proof. Let $A = \begin{bmatrix} a & b \\ c & d \end{bmatrix}$ and $B = \begin{bmatrix} x & y \\ z & t \end{bmatrix}$. Then

$$AB = \begin{bmatrix} ax+bz & ay+bt \\ cx+dz & cy+dt \end{bmatrix}, \quad \text{and so}$$

$$\begin{aligned}
|AB| &= (ax+bz)(cy+dt) - (ay+bt)(cx+dz) \\
&= ad(xt-yz) - bc(xt-yz) \\
&= (ad-bc)(xt-yz) \\
&= |A||B|.
\end{aligned}$$

Nonsingular matrices. A 2×2 matrix A is called **nonsingular** if $|A| \neq 0$, and **singular** if $|A| = 0$.

In the following theorem we show that a 2×2 matrix A is invertible if and only if it is nonsingular and at the same time prove Theorem 7.2 in the case of 2×2 matrices.

Theorem 8.1. *The 2×2 matrix $A = \begin{bmatrix} a & b \\ c & d \end{bmatrix}$ has a right inverse if and only if A is nonsingular. If A is nonsingular, i.e. if $|A| \neq 0$, then A has right inverse*

$$\frac{1}{|A|} \begin{bmatrix} d & -b \\ -c & a \end{bmatrix}, \quad \text{i.e.} \quad \begin{bmatrix} \dfrac{d}{|A|} & \dfrac{-b}{|A|} \\[2mm] \dfrac{-c}{|A|} & \dfrac{a}{|A|} \end{bmatrix}.$$

This matrix is also a left inverse of A and therefore is the inverse of A.

Proof. Suppose first that A has a right inverse X, so that $AX = I$. Then $|AX| = |I|$, and so $|A||X| = 1$. It follows that $|A| \neq 0$, and hence that A is nonsingular.

[Similarly, if A has a left inverse or an inverse, then A is nonsingular.]
Suppose now that A is nonsingular, so that $|A| \neq 0$.

Then the matrix $\quad X = \dfrac{1}{|A|} \begin{bmatrix} d & -b \\ -c & a \end{bmatrix}$ exists.

Also, $AX = \dfrac{1}{|A|}\begin{bmatrix} a & b \\ c & d \end{bmatrix}\begin{bmatrix} d & -b \\ -c & a \end{bmatrix} = \dfrac{1}{|A|}\begin{bmatrix} ad-bc & 0 \\ 0 & ad-bc \end{bmatrix}$

$\qquad = \dfrac{1}{|A|}\begin{bmatrix} |A| & 0 \\ 0 & |A| \end{bmatrix} = \begin{bmatrix} 1 & 0 \\ 0 & 1 \end{bmatrix} = I.$

Thus X is a right inverse of A.

Proceeding similarly we can check that $XA = I$, and so X is also a left inverse of A.

It follows that X is the inverse of A. If we denote this inverse by A^{-1}, we have

$$A^{-1} = \begin{bmatrix} a & b \\ c & d \end{bmatrix}^{-1} = \frac{1}{|A|}\begin{bmatrix} d & -b \\ -c & a \end{bmatrix} = \begin{bmatrix} \dfrac{d}{ad-bc} & \dfrac{-b}{ad-bc} \\ \dfrac{-c}{ad-bc} & \dfrac{a}{ad-bc} \end{bmatrix}.$$

Exercise. Determine which of the following matrices have inverses and write down the inverse when it exists.

(i) $\begin{bmatrix} 2 & -1 \\ -4 & 3 \end{bmatrix}$, (ii) $\begin{bmatrix} 1 & -1 \\ -1 & 1 \end{bmatrix}$, (iii) $\begin{bmatrix} \cos\theta & \sin\theta \\ -\sin\theta & \cos\theta \end{bmatrix}$.

II. Determinant of a 3×3 matrix, and 3×3 inverse matrices. If A is the 3×3 matrix $[a_{ij}]$, then the **determinant** of A is denoted by $|A|$ or det A or

$$\begin{vmatrix} a_{11} & a_{12} & a_{13} \\ a_{21} & a_{22} & a_{23} \\ a_{31} & a_{32} & a_{33} \end{vmatrix},$$

and is defined to be the scalar

$$a_{11}\begin{vmatrix} a_{22} & a_{23} \\ a_{32} & a_{33} \end{vmatrix} - a_{12}\begin{vmatrix} a_{21} & a_{23} \\ a_{31} & a_{33} \end{vmatrix} + a_{13}\begin{vmatrix} a_{21} & a_{22} \\ a_{31} & a_{32} \end{vmatrix}. \tag{8.1}$$

Here $\begin{vmatrix} a_{22} & a_{23} \\ a_{32} & a_{33} \end{vmatrix}$ is the 2×2 determinant obtained from A by omitting the row and column in which a_{11} lies, $\begin{vmatrix} a_{21} & a_{23} \\ a_{31} & a_{33} \end{vmatrix}$ is the 2×2 determinant obtained from A by omitting the row and column in which a_{12} lies, and $\begin{vmatrix} a_{21} & a_{22} \\ a_{31} & a_{32} \end{vmatrix}$ is the 2×2 determinant obtained from A by omitting the row and column in which a_{13} lies.

The expression for $|A|$ given by (8.1) is called the **expansion of** $|A|$ **from the first row**.

We show that $|A|$ is given by similar expansions from the other two rows. From (8.1),

$$|A| = a_{11}(a_{22}a_{33} - a_{23}a_{32}) - a_{12}(a_{21}a_{33} - a_{23}a_{31}) + a_{13}(a_{21}a_{32} - a_{22}a_{31})$$

$$= -a_{21}(a_{12}a_{32} - a_{13}a_{33}) + a_{22}(a_{11}a_{33} - a_{13}a_{31}) - a_{23}(a_{11}a_{32} - a_{12}a_{31})$$

$$= -a_{21}\begin{vmatrix} a_{12} & a_{13} \\ a_{32} & a_{33} \end{vmatrix} + a_{22}\begin{vmatrix} a_{11} & a_{13} \\ a_{31} & a_{33} \end{vmatrix} - a_{23}\begin{vmatrix} a_{11} & a_{12} \\ a_{31} & a_{32} \end{vmatrix}, \qquad (8.2)$$

giving the expansion of $|A|$ from the second row.

Similarly, we can check that

$$|A| = a_{31}\begin{vmatrix} a_{12} & a_{13} \\ a_{22} & a_{23} \end{vmatrix} - a_{32}\begin{vmatrix} a_{11} & a_{13} \\ a_{21} & a_{23} \end{vmatrix} + a_{33}\begin{vmatrix} a_{11} & a_{12} \\ a_{21} & a_{22} \end{vmatrix}, \qquad (8.3)$$

giving the expansion of $|A|$ from the third row.

In the three expansions (8.1), (8.2) and (8.3), the factor multiplying a given a_{ij} is the 2×2 determinant obtained from A by omitting the row and column in which a_{ij} lies, together with a sign $+$ or $-$ according to the chess-board pattern

$$\begin{bmatrix} + & - & + \\ - & + & - \\ + & - & + \end{bmatrix}.$$

This signed factor of a_{ij} is denoted by A_{ij} and called the **cofactor** of a_{ij} in $|A|$, so that

$$A_{11} = \begin{vmatrix} a_{22} & a_{23} \\ a_{32} & a_{33} \end{vmatrix}, \quad A_{12} = -\begin{vmatrix} a_{21} & a_{23} \\ a_{31} & a_{33} \end{vmatrix},$$

etc., and, from (8.1), (8.2) and (8.3),

$$|A| = a_{11}A_{11} + a_{12}A_{12} + a_{13}A_{13},$$
$$|A| = a_{21}A_{21} + a_{22}A_{22} + a_{23}A_{23}, \qquad (8.4)$$
$$|A| = a_{31}A_{31} + a_{32}A_{32} + a_{33}A_{33}.$$

The *transpose* of the matrix of cofactors, namely $[A_{ij}]'$, is called the **adjugate** of A and denoted by adj A. We shall see shortly that this matrix plays an important role in determining inverses of invertible 3×3 matrices; but first we obtain some properties of 3×3 determinants that we shall require for matrix inverses, and also some additional properties.

Some properties of 3×3 determinants. Let $A = [a_{ij}]_{3 \times 3}$.

(1) $$|A'| = |A|.$$

Proof. $|A'| = \begin{vmatrix} a_{11} & a_{21} & a_{31} \\ a_{12} & a_{22} & a_{32} \\ a_{13} & a_{23} & a_{33} \end{vmatrix}$

$$= a_{11}(a_{22}a_{33} - a_{23}a_{32}) - a_{21}(a_{12}a_{33} - a_{13}a_{32})$$
$$+ a_{31}(a_{12}a_{23} - a_{13}a_{22})$$
$$= a_{11}(a_{22}a_{33} - a_{23}a_{32}) - a_{12}(a_{21}a_{33} - a_{23}a_{31})$$
$$+ a_{13}(a_{21}a_{32} - a_{22}a_{31})$$
$$= |A|.$$

It follows from (1) that $|A|$ can be expanded from columns in a manner similar to that for rows. For example,

$$|A| = a_{11} \begin{vmatrix} a_{22} & a_{23} \\ a_{32} & a_{33} \end{vmatrix} - a_{21} \begin{vmatrix} a_{12} & a_{13} \\ a_{32} & a_{33} \end{vmatrix} + a_{31} \begin{vmatrix} a_{12} & a_{13} \\ a_{22} & a_{23} \end{vmatrix}$$
$$= a_{11}A_{11} + a_{21}A_{21} + a_{31}A_{31}.$$

(2) *If two rows (or columns) of a 3×3 matrix A are equal, then $|A| = 0$.*

Proof. If the second and third rows are identical, then clearly $A_{11} = 0$, $A_{12} = 0$, $A_{13} = 0$ and so, by (8.1), $|A| = 0$.

Similarly for the two other possibilities. Also, the result for columns follows from (1).

(3) If $i \neq j$, then $a_{i1}A_{j1} + a_{i2}A_{j2} + a_{i3}A_{j3} = 0$. \hfill (8.5)

Proof. Consider the case $i = 2, j = 1$.

$$a_{21}A_{11} + a_{22}A_{12} + a_{23}A_{13} = a_{21} \begin{vmatrix} a_{22} & a_{23} \\ a_{32} & a_{33} \end{vmatrix} - a_{22} \begin{vmatrix} a_{21} & a_{23} \\ a_{31} & a_{33} \end{vmatrix} + a_{23} \begin{vmatrix} a_{21} & a_{22} \\ a_{31} & a_{32} \end{vmatrix}$$
$$= \begin{vmatrix} a_{21} & a_{22} & a_{23} \\ a_{21} & a_{22} & a_{23} \\ a_{31} & a_{32} & a_{33} \end{vmatrix}$$
$$= 0, \quad \text{by (2)}.$$

Similarly for the other possible cases.

(4) $$\begin{vmatrix} a_{11}+b_{11} & a_{12}+b_{12} & a_{13}+b_{13} \\ a_{21} & a_{22} & a_{23} \\ a_{31} & a_{32} & a_{33} \end{vmatrix} = \begin{vmatrix} a_{11} & a_{12} & a_{13} \\ a_{21} & a_{22} & a_{23} \\ a_{31} & a_{32} & a_{33} \end{vmatrix} + \begin{vmatrix} b_{11} & b_{12} & b_{13} \\ a_{21} & a_{22} & a_{23} \\ a_{31} & a_{32} & a_{33} \end{vmatrix}.$$

Proof. The left-hand side

$$= (a_{11}+b_{11})A_{11}+(a_{12}+b_{12})A_{12}+(a_{13}+b_{13})A_{13}$$
$$= (a_{11}A_{11}+a_{12}A_{12}+a_{13}A_{13})+(b_{11}A_{11}+b_{12}A_{12}+b_{13}A_{13})$$
$$= \text{right-hand side.}$$

Similar results hold when other rows or columns are sums of terms.

(5)
$$\begin{vmatrix} pa_{11} & pa_{12} & pa_{13} \\ a_{21} & a_{22} & a_{23} \\ a_{31} & a_{32} & a_{33} \end{vmatrix} = p \begin{vmatrix} a_{11} & a_{12} & a_{13} \\ a_{21} & a_{22} & a_{23} \\ a_{31} & a_{32} & a_{33} \end{vmatrix}.$$

This follows at once by expansion from the first rows, and similar results hold when other rows or columns are multiplied by a scalar.

As a consequence of (5),
$$|pA| = \begin{vmatrix} pa_{11} & pa_{12} & pa_{13} \\ pa_{21} & pa_{22} & pa_{23} \\ pa_{31} & pa_{32} & pa_{33} \end{vmatrix} = p^3 |A|.$$

Note. The analogous result for a 2×2 matrix A is $|pA| = p^2 |A|$.

(6)
$$\begin{vmatrix} a_{11} & a_{12} & a_{13} \\ a_{21} & a_{22} & a_{23} \\ a_{31} & a_{32} & a_{33} \end{vmatrix}$$

$$= \begin{vmatrix} a_{11}+pa_{21}+qa_{31} & a_{12}+pa_{22}+qa_{32} & a_{13}+pa_{23}+qa_{33} \\ a_{21} & a_{22} & a_{23} \\ a_{31} & a_{32} & a_{33} \end{vmatrix}$$

Proof. The right-hand side, by using (4) and (5), equals

$$\begin{vmatrix} a_{11} & a_{12} & a_{13} \\ a_{21} & a_{22} & a_{23} \\ a_{31} & a_{32} & a_{33} \end{vmatrix} + p \begin{vmatrix} a_{21} & a_{22} & a_{23} \\ a_{21} & a_{22} & a_{23} \\ a_{31} & a_{32} & a_{33} \end{vmatrix} + q \begin{vmatrix} a_{31} & a_{32} & a_{33} \\ a_{21} & a_{22} & a_{23} \\ a_{31} & a_{32} & a_{33} \end{vmatrix}$$
$$= |A|+0+0 \; [\text{by (2)}] = |A|.$$

This is a typical case of the general result that a determinant is not altered by adding to any row multiples of any of the other rows, and similarly for columns.

(7) *If two rows (or columns) of a determinant are interchanged, then only the sign of the determinant is changed.*

Proof. Let $A = [a_{ij}]_{3 \times 3}$. We have, when rows 1 and 2 are interchanged,

$$\begin{vmatrix} a_{21} & a_{22} & a_{23} \\ a_{11} & a_{12} & a_{13} \\ a_{31} & a_{32} & a_{33} \end{vmatrix} = -a_{11} \begin{vmatrix} a_{22} & a_{23} \\ a_{32} & a_{33} \end{vmatrix} + a_{12} \begin{vmatrix} a_{21} & a_{23} \\ a_{31} & a_{33} \end{vmatrix} - a_{13} \begin{vmatrix} a_{21} & a_{22} \\ a_{31} & a_{32} \end{vmatrix}$$
$$= -|A|.$$

Similarly for the other possible interchanges of rows (and similarly for columns).

For the remaining properties (8) and (9) we require some information about the six permutations of the integers 1, 2, 3. As noted in Chapter **5**, Section **3**, these can be listed as: 123, 231, 312, 132, 213, 321.

A change of position of two integers is called an **interchange**. A permutation is changed to a new permutation by an interchange of integers. For example, 231 is transformed to 321 by the interchange of the integers 2 and 3 and we write $231 \rightarrow 321$; similarly $321 \rightarrow 123$ by the interchange of 1 and 3. Each of the six permutations can be transformed to the permutation 123 by zero, one or two interchanges. Check that 231 and 312 need two interchanges and that 132, 213 and 321 each require one interchange; 123 of course requires no interchange.

A permutation *rst* of 123 is called **even** or **odd** according as an *even* or *odd* number of interchanges are required to change the order *rst* to the order 123.

For many applications of permutations it is convenient to introduce the **signature** of a permutation. For the permutation *rst* of 123 this is denoted by the symbol ε_{rst} and is simply $+1$ or -1 according as the permutation is even or odd; e.g.

$$\varepsilon_{123} = +1, \quad \varepsilon_{231} = +1, \quad \varepsilon_{132} = -1, \text{ etc.}$$

Expression for $|A|$ involving signatures of permutations. Let $A = [a_{ij}]_{3 \times 3}$. We have, on expanding $|A|$ from the first row,

$$
\begin{aligned}
|A| &= a_{11} \begin{vmatrix} a_{22} & a_{23} \\ a_{32} & a_{33} \end{vmatrix} - a_{12} \begin{vmatrix} a_{21} & a_{23} \\ a_{31} & a_{33} \end{vmatrix} + a_{13} \begin{vmatrix} a_{21} & a_{22} \\ a_{31} & a_{32} \end{vmatrix} \\
&= \quad a_{11}a_{22}a_{33} + a_{12}a_{23}a_{31} + a_{13}a_{21}a_{32} \\
&\quad - a_{11}a_{23}a_{32} - a_{12}a_{21}a_{33} - a_{13}a_{22}a_{31} \\
&= \quad \varepsilon_{123}a_{11}a_{22}a_{33} + \varepsilon_{231}a_{12}a_{23}a_{31} + \varepsilon_{312}a_{13}a_{21}a_{32} \\
&\quad + \varepsilon_{132}a_{11}a_{23}a_{32} + \varepsilon_{213}a_{12}a_{21}a_{33} + \varepsilon_{321}a_{13}a_{22}a_{31} \\
&= \sum_{i,j,k} \varepsilon_{ijk} a_{1i}a_{2j}a_{3k},
\end{aligned}
\tag{8.6}
$$

where the summation is over the six distinct permutations *ijk* of 123.

We now continue with properties (8) and (9) of 3×3 determinants.

(8) *If rst is a permutation of* 123, *then*

$$
\begin{vmatrix} a_{r1} & a_{r2} & a_{r3} \\ a_{s1} & a_{s2} & a_{s3} \\ a_{t1} & a_{t2} & a_{t3} \end{vmatrix} = \varepsilon_{rst} \begin{vmatrix} a_{11} & a_{12} & a_{13} \\ a_{21} & a_{22} & a_{23} \\ a_{31} & a_{32} & a_{33} \end{vmatrix}.
\tag{8.7}
$$

Proof. Suppose that m interchanges are required to transform rst to 123; then these m interchanges move the rows of the determinant on the left-hand side of equation (8.7) to those of the determinant on the right-hand side. Hence, using property (7) and the definition of ε_{rst},

$$\begin{vmatrix} a_{r1} & a_{r2} & a_{r3} \\ a_{s1} & a_{s2} & a_{s3} \\ a_{t1} & a_{t2} & a_{t3} \end{vmatrix} = (-1)^m \begin{vmatrix} a_{11} & a_{12} & a_{13} \\ a_{21} & a_{22} & a_{23} \\ a_{31} & a_{32} & a_{33} \end{vmatrix} = \varepsilon_{rst} \begin{vmatrix} a_{11} & a_{12} & a_{13} \\ a_{21} & a_{22} & a_{23} \\ a_{31} & a_{32} & a_{33} \end{vmatrix}.$$

(9) *If A and B are both 3×3 matrices, then*

$$|AB| = |A||B|.$$

Proof. Let $A = [a_{ij}]$, $B = [b_{ij}]$ and $AB = [x_{ij}]$. Now, from (8.6),

$$|AB| = \sum_{i,j,k} \varepsilon_{ijk} x_{1i} x_{2j} x_{3k}$$

$$= \sum_{i,j,k} \varepsilon_{ijk} \left(\sum_{r=1}^{3} a_{1r} b_{ri} \right) \left(\sum_{s=1}^{3} a_{2s} b_{sj} \right) \left(\sum_{t=1}^{3} a_{3t} b_{tk} \right)$$

$$= \sum_{r,s,t} \left(\sum_{i,j,k} \varepsilon_{ijk} b_{ri} b_{sj} b_{tk} \right) a_{1r} a_{2s} a_{3t}.$$

Again using (8.6),

$$\sum_{i,j,k} \varepsilon_{ijk} b_{ri} b_{sj} b_{tk} = \begin{vmatrix} b_{r1} & b_{r2} & b_{r3} \\ b_{s1} & b_{s2} & b_{s3} \\ b_{t1} & b_{t2} & b_{t3} \end{vmatrix},$$

and so, by property (2), the sum has the value 0 if at least two of r, s, t are equal, and, by (8), has the value $\varepsilon_{rst}|B|$, when rst is a permutation of 123.

Hence
$$|AB| = |B| \sum_{r,s,t} \varepsilon_{rst} a_{1r} a_{2s} a_{3t},$$

where the summation is over the six distinct permutations rst of 123.

Consequently, $$|AB| = |A||B|.$$

[Clearly this proof is not easy!]

Note. In general, $|A + B| \neq |A| + |B|$.

[Check this by suitable examples.]

We can now prove the 3×3 forms of Theorems 8.1 and 7.2. A 3×3 matrix A is called **nonsingular** or **singular** according as $|A| \neq 0$ or $|A| = 0$.

Theorem 8.2. *The* 3×3 *matrix* $A = [a_{ij}]$ *has a right inverse if and only if* A *is nonsingular. If* A *is nonsingular, i.e. if* $|A| \neq 0$*, then* A *has right inverse*

$\dfrac{1}{|A|}$ adj A. *This matrix is also a left inverse of* A *and therefore is the inverse*

of A.

Proof. Suppose first that A has a right inverse X, so that $AX = I$. Then $|AX| = |I|$, and so $|A| \, |X| = 1$. It follows that $|A| \neq 0$, and hence that A is nonsingular.

[Similarly, if A has a left inverse or an inverse, then A is nonsingular.]

Suppose now that A is nonsingular, so that $|A| \neq 0$. Then the matrix

$$X = \frac{1}{|A|} \text{ adj } A = \frac{1}{|A|} [A_{ij}]' = \frac{1}{|A|} \begin{bmatrix} A_{11} & A_{21} & A_{31} \\ A_{12} & A_{22} & A_{32} \\ A_{13} & A_{23} & A_{33} \end{bmatrix} \text{ exists.}$$

Also,

$$AX = \frac{1}{|A|} \begin{bmatrix} a_{11} & a_{12} & a_{13} \\ a_{21} & a_{22} & a_{23} \\ a_{31} & a_{32} & a_{33} \end{bmatrix} \begin{bmatrix} A_{11} & A_{21} & A_{31} \\ A_{12} & A_{22} & A_{32} \\ A_{13} & A_{23} & A_{33} \end{bmatrix}$$

$$= \frac{1}{|A|} \begin{bmatrix} a_{11}A_{11}+a_{12}A_{12}+a_{13}A_{13} & a_{11}A_{21}+a_{12}A_{22}+a_{13}A_{23} & a_{11}A_{31}+a_{12}A_{32}+a_{13}A_{33} \\ a_{21}A_{11}+a_{22}A_{12}+a_{23}A_{13} & a_{21}A_{21}+a_{22}A_{22}+a_{23}A_{23} & a_{21}A_{31}+a_{22}A_{32}+a_{23}A_{33} \\ a_{31}A_{11}+a_{32}A_{12}+a_{33}A_{13} & a_{31}A_{21}+a_{32}A_{22}+a_{33}A_{23} & a_{31}A_{31}+a_{32}A_{32}+a_{33}A_{33} \end{bmatrix}$$

$$= \frac{1}{|A|} \begin{bmatrix} |A| & 0 & 0 \\ 0 & |A| & 0 \\ 0 & 0 & |A| \end{bmatrix}, \text{ by using (8.4) and (8.5),}$$

$$= I, \quad \text{so that} \quad X = \frac{1}{|A|} \text{ adj } A \text{ is a right inverse of } A.$$

Proceeding similarly and using the results corresponding to (8.4) and (8.5) for expansions by columns, we can check that $XA = I$, so that

$$X = \frac{1}{|A|} \text{ adj } A \text{ is also a left inverse of } A.$$

It follows that $\dfrac{1}{|A|}$ adj A is the inverse of A. If we denote this inverse

by A^{-1}, then $A^{-1} = \dfrac{1}{|A|}$ adj A.

Note. If $|A| = 0$, then $A(\text{adj}A) = (\text{adj}A)A = O$.

Example 1. Find the inverse of the matrix $A = \begin{bmatrix} 2 & 1 & 4 \\ 1 & 0 & 2 \\ 2 & 3 & 1 \end{bmatrix}$, and hence

solve the set of equations

$$\begin{aligned} 2x + y + 4z &= 2, \\ x \quad\; + 2z &= 3, \\ 2x + 3y + z &= -6. \end{aligned}$$

Here, $[A_{ij}] = \begin{bmatrix} \begin{vmatrix} 0 & 2 \\ 3 & 1 \end{vmatrix} & -\begin{vmatrix} 1 & 2 \\ 2 & 1 \end{vmatrix} & \begin{vmatrix} 1 & 0 \\ 2 & 3 \end{vmatrix} \\[6pt] -\begin{vmatrix} 1 & 4 \\ 3 & 1 \end{vmatrix} & \begin{vmatrix} 2 & 4 \\ 2 & 1 \end{vmatrix} & -\begin{vmatrix} 2 & 1 \\ 2 & 3 \end{vmatrix} \\[6pt] \begin{vmatrix} 1 & 4 \\ 0 & 2 \end{vmatrix} & -\begin{vmatrix} 2 & 4 \\ 1 & 2 \end{vmatrix} & \begin{vmatrix} 2 & 1 \\ 1 & 0 \end{vmatrix} \end{bmatrix} = \begin{bmatrix} -6 & 3 & 3 \\ 11 & -6 & -4 \\ 2 & 0 & -1 \end{bmatrix}$$

Thus $\quad$ adj $A = [A_{ij}]' = \begin{bmatrix} -6 & 11 & 2 \\ 3 & -6 & 0 \\ 3 & -4 & -1 \end{bmatrix}$.

We can now find $|A|$ by expanding by its first row or by forming the first entry in A adj A, since A adj $A = |A|I$. We have:

$$|A| = \begin{bmatrix} 2 & 1 & 4 \end{bmatrix} \begin{bmatrix} -6 \\ 3 \\ 3 \end{bmatrix} \quad \text{(1st row of } A \times \text{1st column of adj } A\text{)}$$

$$= -12 + 3 + 12 = 3.$$

Hence $\quad A^{-1} = \dfrac{1}{|A|}$ adj $A = \dfrac{1}{3} \begin{bmatrix} -6 & 11 & 2 \\ 3 & -6 & 0 \\ 3 & -4 & -1 \end{bmatrix}$.

Now the matrix form of the given set of equations is

$$AX = H, \quad \text{where} \quad X = \begin{bmatrix} x \\ y \\ z \end{bmatrix} \quad \text{and} \quad H = \begin{bmatrix} 2 \\ 3 \\ -6 \end{bmatrix}.$$

The solution is $X = A^{-1}H = \dfrac{1}{3}\begin{bmatrix} -6 & 11 & 2 \\ 3 & -6 & 0 \\ 3 & -4 & -1 \end{bmatrix}\begin{bmatrix} 2 \\ 3 \\ -6 \end{bmatrix} = \begin{bmatrix} 3 \\ -4 \\ 0 \end{bmatrix}$,

so that $\qquad x = 3, y = -4$ and $z = 0$.

Notes 1. $AB = I$ with A and B square matrices (2×2 or 3×3) implies that $B = A^{-1}$ and $A = B^{-1}$. This is an immediate consequence of Theorems 8.1 and 8.2.

2. If A^{-1} exists, then $|A^{-1}| = \dfrac{1}{|A|}$; for, $AA^{-1} = I \Rightarrow |A||A^{-1}| = 1$.

Example 2. Show that, if $A^2 - A + I = O$, then A has an inverse, and find this inverse.

We have $A - A^2 = I$, so that $A(I - A) = I$. Thus A^{-1} exists and $A^{-1} = I - A$.

III. Determinant of an $n \times n$ matrix, and $n \times n$ inverse matrices. If A is the 4×4 matrix $[a_{ij}]$, then the determinant of A is denoted by $|A|$ or $\det A$ or

$$\begin{vmatrix} a_{11} & a_{12} & a_{13} & a_{14} \\ a_{21} & a_{22} & a_{23} & a_{24} \\ a_{31} & a_{32} & a_{33} & a_{34} \\ a_{41} & a_{42} & a_{43} & a_{44} \end{vmatrix},$$

and is the scalar given by

$$|A| = a_{11}\begin{vmatrix} a_{22} & a_{23} & a_{24} \\ a_{32} & a_{33} & a_{34} \\ a_{42} & a_{43} & a_{44} \end{vmatrix} - a_{12}\begin{vmatrix} a_{21} & a_{23} & a_{24} \\ a_{31} & a_{33} & a_{34} \\ a_{41} & a_{43} & a_{44} \end{vmatrix} + a_{13}\begin{vmatrix} a_{21} & a_{22} & a_{24} \\ a_{31} & a_{32} & a_{34} \\ a_{41} & a_{42} & a_{44} \end{vmatrix}$$
$$- a_{14}\begin{vmatrix} a_{21} & a_{22} & a_{23} \\ a_{31} & a_{32} & a_{33} \\ a_{41} & a_{42} & a_{43} \end{vmatrix};$$

this expression is called the **expansion of $|A|$ from the first row**.

If we introduce a **cofactor** notation, similar to that for 3×3 determinants, we have

$$|A| = a_{11}A_{11} + a_{12}A_{12} + a_{13}A_{13} + a_{14}A_{14}.$$

It can be shown that 4×4 determinants have the same properties as 3×3 determinants; in particular, there is a cofactor A_{ij} corresponding to each a_{ij} in A, this being $(-1)^{i+j}$ times the 3×3 determinant obtained from A by omitting the row and column in which a_{ij} lies. The signs arising from $(-1)^{i+j}$ follow the chess-board pattern

$$\begin{bmatrix} + & - & + & - \\ - & + & - & + \\ + & - & + & - \\ - & + & - & + \end{bmatrix}.$$

Also, a 4×4 matrix $A = [a_{ij}]$ has an inverse if and only if it is non-singular (i.e. $|A| \neq 0$) and, if $|A| \neq 0$, this inverse is

$$A^{-1} = \frac{1}{|A|} \text{ adj } A,$$

where adj $A = [A_{ij}]'$, the transpose of the 4×4 matrix of cofactors.

Further,

$$|A| = \sum_{i,j,k,l} \varepsilon_{ijkl} a_{1i} a_{2j} a_{3k} a_{4l},$$

where ε_{ijkl} is $+1$ or -1 according as $ijkl$ is an even or odd permutation of 1234 (*even* and *odd* being defined in terms of the number of interchanges needed to transform $ijkl$ to 1234) and where the summation is over the $4!(=24)$ distinct permutations of 1234.

The corresponding definitions and results can be set up by induction for $n \times n$ matrices $(n>4)$. Such a matrix $A = [a_{ij}]_{n \times n}$ has an inverse if and only if $|A| \neq 0$, and, if $|A| \neq 0$, this inverse is

$$A^{-1} = \frac{1}{|A|} \text{ adj } A,$$

where adj $A = [A_{ij}]'$, the transpose of the matrix of cofactors. $[A_{ij}$ is $(-1)^{i+j}$ times the $(n-1) \times (n-1)$ determinant obtained from A by omitting the row and column in which a_{ij} lies.]

Further,

$$|A| = \sum_{i_1, i_2, \dots i_n} \varepsilon_{i_1 i_2 \dots i_n} a_{1 i_1} a_{2 i_2} \dots a_{n i_n},$$

where $\varepsilon_{i_1 i_2 \dots i_n}$ is $+1$ or -1 according as $i_1 i_2 \dots i_n$ is an even or an odd permutation of $12 \dots n$, and where the summation is over the $n!$ distinct permutations of $12 \dots n$.

Note. If A, B are $n \times n$ matrices such that $AB = I$, then

$$B = A^{-1} \quad \text{and} \quad A = B^{-1}. \tag{8.8}$$

Orthogonal matrices. These form a special class of nonsingular matrices. An $n \times n$ matrix A is called **orthogonal** if $AA' = I$. Then $A' = A^{-1}$ and so $A'A = I$. Also, if A is orthogonal, then

$$|AA'| = |I|, \quad \text{and so} \quad |A|^2 = 1; \quad \text{thus} \quad |A| = 1 \text{ or } -1.$$

Example 3. Show that the matrix $A = \begin{bmatrix} \cos \theta & \sin \theta \\ -\sin \theta & \cos \theta \end{bmatrix}$ is orthogonal.

$$AA' = \begin{bmatrix} \cos \theta & \sin \theta \\ -\sin \theta & \cos \theta \end{bmatrix} \begin{bmatrix} \cos \theta & -\sin \theta \\ \sin \theta & \cos \theta \end{bmatrix} = \begin{bmatrix} 1 & 0 \\ 0 & 1 \end{bmatrix} = I;$$

thus A is orthogonal.

Note that $|A| = 1$.

Exercise. Verify that $\begin{bmatrix} 1 & 0 \\ 0 & -1 \end{bmatrix}$ is an orthogonal matrix of determinant -1.

Example 4. A geometrical transformation in space maps a point P with coordinate vector $X = \begin{bmatrix} x_1 \\ x_2 \\ x_3 \end{bmatrix}$ to the point P' with coordinate vector $Y = \begin{bmatrix} y_1 \\ y_2 \\ y_3 \end{bmatrix}$ by means of the matrix equation $Y = AX$ where A is an orthogonal 3×3 matrix. Show that $OP = OP'$, where O is the origin.

We have $(OP')^2 = y_1^2 + y_2^2 + y_3^2 = Y'Y = (AX)'(AX) = X'(A'A)X$
$$= X'IX = X'X = x_1^2 + x_2^2 + x_3^2 = OP^2,$$

and the result follows.

9. Inverse matrices by elementary row operations

We illustrate this work with a 3×3 matrix, but the method applies to square matrices of any order.

Let $A = [a_{ij}]_{3 \times 3}$ and let $E_{12} = \begin{bmatrix} 0 & 1 & 0 \\ 1 & 0 & 0 \\ 0 & 0 & 1 \end{bmatrix}$, i.e. the matrix obtained from I by interchanging rows 1 and 2.

$$E_{12}A = \begin{bmatrix} 0 & 1 & 0 \\ 1 & 0 & 0 \\ 0 & 0 & 1 \end{bmatrix} \begin{bmatrix} a_{11} & a_{12} & a_{13} \\ a_{21} & a_{22} & a_{23} \\ a_{31} & a_{32} & a_{33} \end{bmatrix} = \begin{bmatrix} a_{21} & a_{22} & a_{23} \\ a_{11} & a_{12} & a_{13} \\ a_{31} & a_{32} & a_{33} \end{bmatrix},$$

so that *premultiplying* A by E_{12} has the effect of producing a new matrix from A by interchanging rows 1 and 2 of A. We say that E_{12} has *transformed* the matrix A to the new matrix.

Such a matrix E_{12} is called an **elementary matrix**; it is clearly nonsingular. There is such a matrix E_{ij} corresponding to each pair of rows i, j.

Now let $E_1(c) = \begin{bmatrix} c & 0 & 0 \\ 0 & 1 & 0 \\ 0 & 0 & 1 \end{bmatrix}$, where $c \neq 0$. It is easily checked that

$$E_1(c)A = \begin{bmatrix} ca_{11} & ca_{12} & ca_{13} \\ a_{21} & a_{22} & a_{23} \\ a_{31} & a_{32} & a_{33} \end{bmatrix}.$$

Thus, premultiplying A by this nonsingular matrix $E_1(c)$ (for $c \neq 0$) multiplies the first row of A by c.

Similarly we can define $E_2(c)$ and $E_3(c)$.

Again we call these matrices $E_i(c)$ **elementary matrices**.

Finally, let $E_{12}(k) = \begin{bmatrix} 1 & k & 0 \\ 0 & 1 & 0 \\ 0 & 0 & 1 \end{bmatrix}$; it is easy to check that

$$E_{12}(k)A = \begin{bmatrix} a_{11}+ka_{21} & a_{12}+ka_{22} & a_{13}+ka_{23} \\ a_{21} & a_{22} & a_{23} \\ a_{31} & a_{32} & a_{33} \end{bmatrix},$$

so that premultiplying A by this nonsingular matrix $E_{12}(k)$ adds k times the second row to the first row.

Similarly we can define $E_{ij}(k)$ for each ordered pair of rows i, j.

Again we list these matrices $E_{ij}(k)$ as **elementary matrices**.

Premultiplying A by an elementary matrix is called an **elementary row operation**; we say that A has been **transformed** by the row operation to the new matrix produced.

Clearly the work described can be extended to $n \times n$ matrices.

Theorem 9.1. *If a sequence of elementary row operations transforms an $n \times n$ matrix A into I, then A is invertible and the same sequence transforms I into A^{-1}.*

Proof. Let the sequence of operations be $E_m E_{m-1} \ldots E_2 E_1$, where $E_1, \ldots, E_m$ are, in a simplified notation, the corresponding elementary matrices, and where E_1 is applied first, E_2 second, etc. Then

$$(E_m E_{m-1} \ldots E_2 E_1) A = I,$$

and so

$$((E_m E_{m-1} \ldots E_2 E_1) I) A = I,$$

and therefore A^{-1} exists and, by (8.8),

$$(E_m E_{m-1} \ldots E_2 E_1) I = A^{-1}.$$

We take advantage of this result by considering the matrix $[A \; I]$ formed by adjoining the rows of I to those of A and we proceed by elementary row operations to transform A to the matrix I; by Theorem 9.1, I will at the same time be transformed to A^{-1}.

Example 1. If $A = \begin{bmatrix} 2 & -1 & 4 \\ 4 & 0 & 2 \\ 3 & -2 & 7 \end{bmatrix}$, find A^{-1}.

We consider $[A \; I]$ and indicate elementary row operations needed to transform A to I.

$$[A \; I] = \begin{bmatrix} 2 & -1 & 4 & 1 & 0 & 0 \\ 4 & 0 & 2 & 0 & 1 & 0 \\ 3 & -2 & 7 & 0 & 0 & 1 \end{bmatrix}$$

$$\rightarrow \begin{bmatrix} 1 & -\frac{1}{2} & 2 & \frac{1}{2} & 0 & 0 \\ 4 & 0 & 2 & 0 & 1 & 0 \\ 3 & -2 & 7 & 0 & 0 & 1 \end{bmatrix} \quad \begin{array}{l} [\frac{1}{2} \times \text{row 1}] \\ [\text{The idea is to place a 1 in the first} \\ \text{position of the first row.}] \end{array}$$

$$\rightarrow \begin{bmatrix} 1 & -\frac{1}{2} & 2 & \frac{1}{2} & 0 & 0 \\ 0 & 2 & -6 & -2 & 1 & 0 \\ 0 & -\frac{1}{2} & 1 & -\frac{3}{2} & 0 & 1 \end{bmatrix} \quad \begin{array}{l} [\text{row } 2 + (-4) \times \text{row 1;} \\ \text{row } 3 + (-3) \times \text{row 1}] \\ [\text{The idea is to produce the} \\ \text{first column of } I.] \end{array}$$

$$\rightarrow \begin{bmatrix} 1 & -\frac{1}{2} & 2 & \frac{1}{2} & 0 & 0 \\ 0 & 1 & -3 & -1 & \frac{1}{2} & 0 \\ 0 & -\frac{1}{2} & 1 & -\frac{3}{2} & 0 & 1 \end{bmatrix} \quad \begin{array}{l} [\frac{1}{2} \times \text{row 2}] \\ [\text{The aim is to produce a 1 in the} \\ \text{second place of the second row.}] \end{array}$$

$$\rightarrow \begin{bmatrix} 1 & -\frac{1}{2} & 2 & \frac{1}{2} & 0 & 0 \\ 0 & 1 & -3 & -1 & \frac{1}{2} & 0 \\ 0 & 0 & -\frac{1}{2} & -2 & \frac{1}{4} & 1 \end{bmatrix} \quad \begin{array}{l} [\text{row } 3 + (\frac{1}{2}) \times \text{row 2}] \\ [\text{The aim is to produce a 0 under} \\ \text{the 1 of the second row.}] \end{array}$$

$$\rightarrow \begin{bmatrix} 1 & -\frac{1}{2} & 2 & \frac{1}{2} & 0 & 0 \\ 0 & 1 & -3 & -1 & \frac{1}{2} & 0 \\ 0 & 0 & 1 & 4 & -\frac{1}{2} & -2 \end{bmatrix} \quad \begin{array}{l} [(-2) \times \text{row 3}] \\ [\text{The aim is to produce the} \\ \text{last row of } I.] \end{array}$$

$$\rightarrow \begin{bmatrix} 1 & -\frac{1}{2} & 0 & -\frac{15}{2} & 1 & 4 \\ 0 & 1 & 0 & 11 & -1 & -6 \\ 0 & 0 & 1 & 4 & -\frac{1}{2} & -2 \end{bmatrix} \begin{array}{l} [\text{row } 1+(-2)\times \text{row } 3; \\ \text{row } 2+(3)\times \text{row } 3] \\ [\text{The aim is to produce the} \\ \text{last column of } I.] \end{array}$$

$$\rightarrow \begin{bmatrix} 1 & 0 & 0 & -2 & \frac{1}{2} & 1 \\ 0 & 1 & 0 & 11 & -1 & -6 \\ 0 & 0 & 1 & 4 & -\frac{1}{2} & -2 \end{bmatrix} \quad [\text{row } 1+(\tfrac{1}{2})\times \text{row } 2]$$

$$= [I \ A^{-1}].$$

Hence
$$A^{-1} = \begin{bmatrix} -2 & \frac{1}{2} & 1 \\ 11 & -1 & -6 \\ 4 & -\frac{1}{2} & -2 \end{bmatrix}.$$

Check this result by computing $\dfrac{1}{|A|}$ adj A.

Also try to obtain the result by using different row operations.

10. Linear equations

We restrict ourselves in the first place to square matrices.
Consider the set of linear equations given by the matrix equation
$$AX = H,$$

where A is a given $n\times n$ matrix $[a_{ij}]$, $X = \begin{bmatrix} x_1 \\ x_2 \\ \vdots \\ x_n \end{bmatrix}$ and H is a given column

matrix $\begin{bmatrix} h_1 \\ h_2 \\ \vdots \\ h_n \end{bmatrix}$.

If A is nonsingular, then the system has a unique solution for X, which can be obtained as follows:
$$AX = H \Rightarrow A^{-1}(AX) = A^{-1}H \Rightarrow (A^{-1}A)X = A^{-1}H$$
$$\Rightarrow IX = A^{-1}H$$
$$\Rightarrow X = A^{-1}H$$
$$\Rightarrow X = \frac{1}{|A|}(\text{adj } A)H. \qquad (10.1)$$

But, since $A(A^{-1}H) = H$, $X = A^{-1}H$ is in fact a solution of $AX = H$, and the result follows.

This result, when the product (adj A)H is written out explicitly, is often called **Cramer's rule** for the solution when $|A| \neq 0$.

If A is singular (i.e. $|A| = 0$), then the set of equations may or may not have a solution. If they have *no* solution, we say that the equations are **inconsistent**. If they have a solution, we say that they are **consistent**; the solution then involves one or more parameters.

Special case in which $H = O$. In this case we say that we have a **homogeneous** set of equations

$$AX = O.$$

$X = O$ is always a solution of this equation. It is the only solution when $|A| \neq 0$, since then A^{-1} exists and

$$A^{-1}(AX) = A^{-1}O \quad \text{gives} \quad X = O.$$

It follows that, if there is a non-zero solution for X, then $|A| = 0$.

It can in fact be shown that a non-zero solution for X exists if and only if $|A| = 0$.

Example 1. Show that the system of linear equations

$$\begin{aligned}
x + y + z &= 1, \\
x + ay + z &= a, \\
2x + 3y + az &= -1
\end{aligned} \tag{10.2}$$

has a unique solution provided $a \neq 1$ and $a \neq 2$, and find this solution.

Investigate fully the cases $a = 1$ and $a = 2$, giving the most general solution when the system is consistent.

We consider the matrix

$$\begin{bmatrix} 1 & 1 & 1 & 1 \\ 1 & a & 1 & a \\ 2 & 3 & a & -1 \end{bmatrix}$$

consisting of the matrix of coefficients of the terms in x, y, z in the equations and the column of numbers on the right-hand sides of the equations. This is often called the **augmented** matrix for the system. We apply row operations to this matrix in order to transform the system to a simpler equivalent system.

$$\begin{bmatrix} 1 & 1 & 1 & 1 \\ 1 & a & 1 & a \\ 2 & 3 & a & -1 \end{bmatrix}$$

$$\rightarrow \begin{bmatrix} 1 & 1 & 1 & 1 \\ 0 & a-1 & 0 & a-1 \\ 0 & 1 & a-2 & -3 \end{bmatrix} \quad \begin{array}{l} [\text{row } 2+(-1)\times\text{row } 1; \\ \text{row } 3+(-2)\times\text{row } 1] \end{array}$$

$$\rightarrow \begin{bmatrix} 1 & 1 & 1 & 1 \\ 0 & 1 & a-2 & -3 \\ 0 & a-1 & 0 & a-1 \end{bmatrix} \quad [\text{interchange row 2 and row 3}]$$

$$\rightarrow \begin{bmatrix} 1 & 1 & 1 & 1 \\ 0 & 1 & a-2 & -3 \\ 0 & 0 & -(a-1)(a-2) & 4(a-1) \end{bmatrix} \quad [\text{row } 3+(-(a-1))\times\text{row } 2]$$

It follows that the given system of equations is equivalent to the following system:

$$\begin{aligned} x+y+z &= 1, \\ y+(a-2)z &= -3, \\ -(a-1)(a-2)z &= 4(a-1). \end{aligned} \tag{10.3}$$

If $a \neq 1$ and $a \neq 2$, then $z = -\dfrac{4}{a-2}$, $y = 1$ and $x = \dfrac{4}{a-2}$ is the

corresponding unique solution.

Case $a = 1$: The system is equivalent to

$$\begin{aligned} x+y+z &= 1, \\ y-z &= -3, \end{aligned}$$

and this has solution $\begin{cases} y = z-3 \\ x = 4-2z \end{cases}$, with z a parameter.

The solution can be written in the form $x = 4-2t$, $y = t-3$, $z = t$, with t a parameter; in matrix form the solution is

$$\begin{bmatrix} x \\ y \\ z \end{bmatrix} = \begin{bmatrix} 4 \\ -3 \\ 0 \end{bmatrix} + t\begin{bmatrix} -2 \\ 1 \\ 1 \end{bmatrix}.$$

Check that $\begin{bmatrix} -2 \\ 1 \\ 1 \end{bmatrix}$ is a solution of the corresponding homogeneous equations.

Case $a = 2$: From the third equation in (10.3), $0 = 4$, which is impossible. Hence when $a = 2$ the system of equations is inconsistent.

Note. In transforming system (10.2) to the equivalent system (10.3), we say that we have reduced (10.2) to an **echelon form**.

Exercise. Solve Example 1 by forming the adjugate of the matrix

$$A = \begin{bmatrix} 1 & 1 & 1 \\ 1 & a & 1 \\ 2 & 3 & a \end{bmatrix}$$

of coefficients, and using (10.1) when $|A| \neq 0$.

Example 2. Find the values of a for which the set of equations in x, y, z

$$\begin{aligned}(a+2)x \quad -y+(a+1)z &= 0, \\ 2x+ay \qquad\quad +z &= 0, \\ ax-3y \qquad\quad +az &= 0\end{aligned}$$

has a non-zero solution, and discuss the solutions for these values of a.

The set of equations is $AX = O$, where

$$A = \begin{bmatrix} a+2 & -1 & a+1 \\ 2 & a & 1 \\ a & -3 & a \end{bmatrix} \quad \text{and} \quad X = \begin{bmatrix} x \\ y \\ z \end{bmatrix}.$$

The set has a non-zero solution

$$\begin{aligned}&\Leftrightarrow |A| = 0 \\ &\Leftrightarrow (a+2)(a^2+3)+a+(a+1)(-6-a^2) = 0 \\ &\Leftrightarrow a^2-2a = 0 \\ &\Leftrightarrow a = 0 \quad \text{or} \quad a = 2.\end{aligned}$$

Case $a = 0$: The set of equations is

$$\begin{aligned} 2x-y+z &= 0, \\ 2x+z &= 0, \\ -3y &= 0,\end{aligned}$$

and so $\begin{cases} y = 0 \\ z = -2x \end{cases}$, with x a parameter, is the general solution.

This can be written as $x = t, y = 0, z = -2t$, with t a parameter.

Case $a = 2$: The set of equations is:

$$4x - y + 3z = 0,$$
$$2x + 2y + z = 0,$$
$$2x - 3y + 2z = 0.$$

This is equivalent to $\begin{cases} 4x - y + 3z = 0 \\ \quad\ 5y - z = 0, \end{cases}$ $\begin{array}{l} [(2) \times \text{row } 2 - \text{row } 1; \\ \text{row } 3 - \text{row } 2.] \end{array}$

and so has general solution

$$\begin{cases} z = 5y \\ 4x = y - 15y = -14y \end{cases}, \quad \text{with } y \text{ a parameter.}$$

This can be written as $x = -\frac{7}{2}t$, $y = t$, $z = 5t$, with t a parameter.

Exercise. Try to solve this problem by reducing the system to an echelon form.

Systems of linear equations with non-square matrices. We shall illustrate the echelon treatment for such systems with one example.

Example 3. Show that the set of equations

$$x - y + 5z - 3t = p,$$
$$2x + y + 4z + t = q,$$
$$2x + 7y - 8z + 15t = r$$

in the variables x, y, z, t is consistent if and only if $r = 3q - 4p$, and find the general solution when this condition is satisfied.

Here, $\begin{bmatrix} 1 & -1 & 5 & -3 & p \\ 2 & 1 & 4 & 1 & q \\ 2 & 7 & -8 & 15 & r \end{bmatrix}$

$\rightarrow \begin{bmatrix} 1 & -1 & 5 & -3 & p \\ 0 & 3 & -6 & 7 & q-2p \\ 0 & 9 & -18 & 21 & r-2p \end{bmatrix}$ $\begin{array}{l} [\text{row } 2 + (-2) \times \text{row } 1; \\ \text{row } 3 + (-2) \times \text{row } 1] \end{array}$

$\rightarrow \begin{bmatrix} 1 & -1 & 5 & -3 & p \\ 0 & 3 & -6 & 7 & q-2p \\ 0 & 0 & 0 & 0 & r-3q+4p \end{bmatrix}$ $[\text{row } 3 + (-3) \times \text{row } 2]$

Hence the system is equivalent to:

$$x - y + 5z - 3t = p,$$
$$3y - 6z + 7t = q - 2p,$$
$$0 = r - 3q + 4p,$$

and so is consistent if and only if $r = 3q - 4p$. When the condition is satisfied, the solution is:

$$\begin{cases} y = 2z - \frac{7}{3}t + \frac{1}{3}q - \frac{2}{3}p, \\ x = -3z + \frac{2}{3}t + \frac{1}{3}q + \frac{1}{3}p, \end{cases} \quad \text{with } z \text{ and } t \text{ parameters.}$$

EXERCISE 8

1. Let $A = \begin{bmatrix} 2 & 0 \\ 3 & -1 \\ -5 & 3 \\ 1 & 4 \end{bmatrix}$, $B = \begin{bmatrix} -1 & 1 \\ 0 & -3 \\ 3 & 0 \\ -2 & 1 \end{bmatrix}$, $C = \begin{bmatrix} 0 & 1 \\ -1 & 0 \end{bmatrix}$, $D = \begin{bmatrix} 0 & 1 & 2 \\ -1 & 0 & 3 \end{bmatrix}$,

$E = \begin{bmatrix} 0 & 1 & -1 & 1 \\ 0 & 0 & 1 & -1 \\ 0 & 0 & 0 & 1 \\ 0 & 0 & 0 & 0 \end{bmatrix}$.

(i) Determine the following sums and products when they exist:
$$A - 2B, \ 2A + 3B, \ A + C, \ AC, \ AD, \ (A + B)D, \ DE, \ A'(BC), \ D'B'.$$

(ii) Find the matrices E^n $(n = 2, 3, 4, \ldots)$.

(iii) Describe the effect of postmultiplying any 4×4 matrix by E.

2. If $P = \begin{bmatrix} 3 & 4 \\ -1 & -1 \end{bmatrix}$, show, by induction, that for all $n \geqslant 1$,

$$P^n = \begin{bmatrix} 1 + 2n & 4n \\ -n & 1 - 2n \end{bmatrix}.$$

3. If $A = \begin{bmatrix} a + 2b & 3b \\ -b & a \end{bmatrix}$, $B = \begin{bmatrix} x + 2y & 3y \\ -y & x \end{bmatrix}$, show that AB is of the form

$$\begin{bmatrix} p + 2q & 3q \\ -q & p \end{bmatrix}.$$

4. If $A = \begin{bmatrix} 1 & 0 \\ 1 & 2 \end{bmatrix}$ and $B = \begin{bmatrix} 2 & -1 \\ -1 & 2 \end{bmatrix}$, show that the matrix $X = \begin{bmatrix} x & y \\ z & t \end{bmatrix}$ satisfies the

equation $AX = XB$ if and only if X is a scalar multiple of the matrix

$$\begin{bmatrix} 1 & 1 \\ -1 & -1 \end{bmatrix}.$$

5. (i) Find the set of all 2×2 matrices X such that $X^2 = O$.

(ii) Find the set of all 2×2 matrices X such that $X^2 = I$.

6. If $A = \begin{bmatrix} a & b \\ c & d \end{bmatrix}$ and if there exist matrices X, Y such that $XY - YX = A$, prove that

$$a + d = 0.$$

Find suitable matrices X, Y for the particular case in which

$$A = \begin{bmatrix} 1 & 2 \\ 0 & -1 \end{bmatrix}.$$

7. If $A = \begin{bmatrix} 3 & 2 \\ -2 & -1 \end{bmatrix}$ and $I = \begin{bmatrix} 1 & 0 \\ 0 & 1 \end{bmatrix}$, show that $(A - I)^2 = O$.

8. Find all the matrices X of the form $\begin{bmatrix} x & 0 \\ y & z \end{bmatrix}$ such that $X^2 + 3X + 2I = O$.

9. If $I = \begin{bmatrix} 1 & 0 \\ 0 & 1 \end{bmatrix}$ and $J = \begin{bmatrix} 0 & 1 \\ -1 & 0 \end{bmatrix}$, show that $J^2 = -I$.

Prove that the sum and product of two matrices of the form $aI + bJ$, where a, b are real scalars, are of the same form.

Prove that $aI + bJ$ (a and b real) has an inverse if and only if $a^2 + b^2 \neq 0$.

10. If $A = \begin{bmatrix} 2 & -1 \\ -1 & 2 \end{bmatrix}$ and $B = \begin{bmatrix} 1 & 0 \\ 0 & 3 \end{bmatrix}$, show that the most general 2×2 matrix X such

that $AX = XB$ is of the form $\begin{bmatrix} x & y \\ x & -y \end{bmatrix}$.

Hence find a 2×2 matrix P such that $P'AP = B$ and $P'P = I$, where I is the unity 2×2 matrix.

11. If $P = \begin{bmatrix} \cos\theta & \sin\theta \\ -\sin\theta & \cos\theta \end{bmatrix}$ and $Q = \begin{bmatrix} \cos\phi & \sin\phi \\ -\sin\phi & \cos\phi \end{bmatrix}$, show that PQ can be expressed in

the same form as P and Q.

If $X = \begin{bmatrix} \frac{1}{2}\sqrt{3} & \frac{1}{2} \\ -\frac{1}{2} & \frac{1}{2}\sqrt{3} \end{bmatrix}$, find the least positive integer n for which $X^n = I$.

12. Show that any 2×2 real orthogonal matrix can be expressed in the form

$$\begin{bmatrix} \cos\theta & \sin\theta \\ -\sin\theta & \cos\theta \end{bmatrix} \quad \text{or} \quad \begin{bmatrix} \cos\theta & \sin\theta \\ \sin\theta & -\cos\theta \end{bmatrix}$$

for some real θ.

13. Evaluate P^2, where $P = \begin{bmatrix} a & b & c & 0 \\ b & -a & 0 & c \\ c & 0 & -a & -b \\ 0 & c & -b & a \end{bmatrix}$,

and hence show, if a, b, c are *real* numbers not all zero, that P^{-1} exists and find it.

14. If A is an $m \times n$ *real* matrix such that $AA' = O$, prove, by considering the entries in the main diagonal of AA', that $A = O$.

15. If $U = \begin{bmatrix} 0 & 1 & 0 \\ 0 & 0 & 1 \\ 0 & 0 & 0 \end{bmatrix}$, evaluate U^2 and U^3.

If $A = aI + bU + cU^2$ and $B = a^2I - abU + (b^2 - ac)U^2$, where a, b, c are scalars and I is the unity 3×3 matrix, evaluate the product AB. Deduce that, if $a \neq 0$, A has an inverse, and find this inverse.

16. Show that a matrix X commutes with the matrix

$$A = \begin{bmatrix} 0 & 1 & 0 \\ 0 & 0 & 1 \\ 1 & 0 & 0 \end{bmatrix} \text{ if and only if } X \text{ is of the form } \begin{bmatrix} a & b & c \\ c & a & b \\ b & c & a \end{bmatrix}.$$

17. If $A = \begin{bmatrix} 1 & 0 & 0 \\ -1 & -2 & -1 \\ 2 & 3 & 2 \end{bmatrix}$, show that $A^3 - A = A^2 - I$.

Prove by induction that, for every integer $n \geqslant 3$,
$$A^n - A^{n-2} = A^2 - I.$$

Hence find A^{100}.

18. (i) Find the inverse of the matrix $\begin{bmatrix} 2 & 1 & 1 \\ 1 & -1 & 2 \\ 3 & 2 & -1 \end{bmatrix}$, and hence solve the set of equations

$$\begin{aligned} 2x + y + z &= 1, \\ x - y + 2z &= -1, \\ 3x + 2y - z &= 4. \end{aligned}$$

(ii) Find the inverse of the matrix $\begin{bmatrix} 1 & -1 & 1 \\ 1 & 1 & 2 \\ 2 & -1 & 3 \end{bmatrix}$, and hence solve the set of equations

$$\begin{aligned} x - y + z &= 1, \\ x + y + 2z &= 0, \\ 2x - y + 3z &= 2. \end{aligned}$$

19. If A is an $m \times n$ matrix such that the $m \times m$ matrix AA' is nonsingular and if B is the $n \times n$ matrix $A'(AA')^{-1}A$, show that B is symmetric and that $B^2 = B$.

20. A square matrix A is such that $A^2 = A$ and $(A - A')^2 = O$. Prove that $AA' + A'A = A + A'$, and deduce that $(AA')^2 = AA'$.

21. The non-zero matrix A is nilpotent of order $k (\geqslant 2)$, i.e. $A^k = O$ and $A^{k-1} \neq O$. Verify that $I - A$ has inverse $I + A + A^2 + \ldots + A^{k-1}$.

Show that the matrix $A = \begin{bmatrix} -1 & -5 & -4 \\ 0 & -2 & -1 \\ 1 & 3 & 3 \end{bmatrix}$ is nilpotent of order 3, and hence find the

inverse of the matrix $\begin{bmatrix} 2 & 5 & 4 \\ 0 & 3 & 1 \\ -1 & -3 & -2 \end{bmatrix}$.

22. (i) If A and B are $n \times n$ matrices, A is nonsingular and $A^{-1} = A + B$, show that
$$A^3 - B^3 = (A - B)(A^2 + AB + B^2).$$

(ii) If X and Y are $n \times n$ matrices such that $X + Y = XY$ and if X is nonsingular, show that Y is nonsingular and that $X^{-1} + Y^{-1} = I$.

(iii) The $n \times n$ matrices A, B are such that $I - AB$ is nonsingular. Show that $I - BA$ is nonsingular and that
$$(I - BA)^{-1} = I + B(I - AB)^{-1}A.$$

23. Express each of the quadratic forms
  (i) $x^2 + 2y^2 - z^2 - 2xy + 3xz$,
  (ii) $2x^2 + 4y^2 - 3xz + 6yz$
in matrix form $X'AX$ with A a symmetric matrix.

24. (i) Find a, b, c so that the matrix $\dfrac{1}{3}\begin{bmatrix} a & 2 & 2 \\ 2 & 1 & b \\ 2 & c & 1 \end{bmatrix}$ is orthogonal.

(ii) Show that, if P and Q are orthogonal $n \times n$ matrices, then PQ, PQ' and $P'Q$ are orthogonal.

25. Let $A = \begin{bmatrix} a+2 & 1 & 1 \\ a & -1 & a+1 \\ 2 & a & -a \end{bmatrix}$ and $X = \begin{bmatrix} x \\ y \\ z \end{bmatrix}$. Find the real value of a for which the system of equations $AX = O$ has a non-zero solution.

26. Show that the set of equations
$$\begin{aligned} x+y+z &= 1, \\ ax+ay+z &= a+1, \\ ax+2y+2z &= 2 \end{aligned}$$
has a unique solution if $a \neq 1$ and $a \neq 2$, and find the solution.
Discuss the cases $a = 1$ and $a = 2$.

27. Show that the set of equations
$$\begin{aligned} x+y-z &= 1, \\ x+2y+az &= -1, \\ x+ay-z &= 1 \end{aligned}$$
has a unique solution if $a \neq \pm 1$, and find this solution.
Discuss the cases $a = 1$ and $a = -1$.

28. Find the real value of a for which the system of equations
$$\begin{aligned} x+ay+2z &= 0, \\ ax-3y+(a+1)z &= 0, \\ -x+2y+az &= 0 \end{aligned}$$
has a non-zero solution, and find this solution.

29. Show that the system of equations in x, y, z, t,
$$\begin{aligned} x+5y-2z+6t &= p, \\ 4x-3y+7z+12t &= q, \\ 5x-44y+35z-6t &= r, \end{aligned}$$
has a solution if and only if $r = 3q - 7p$, and find the solution set in terms of the parameters y and t when this condition is satisfied.

30. Show that the system of equations
$$\begin{aligned} x+y+2z+t &= 5, \\ 3x+2y-z+3t &= 6, \\ 4x+3y+z+4t &= 11, \\ 2x+y-3z+2t &= 1 \end{aligned}$$
is consistent, and find its general solution.

31. Prove that, if the set of equations in x, y
$$a_1x+b_1y = c_1, \quad a_2x+b_2y = c_2, \quad a_3x+b_3y = c_3$$
is consistent, then $\begin{vmatrix} a_1 & b_1 & c_1 \\ a_2 & b_2 & c_2 \\ a_3 & b_3 & c_3 \end{vmatrix} = 0$. Show, by considering the system
$$x+y = 1, \quad 2x+2y = 1, \quad 3x+3y = 1,$$
that the converse is not true.

Algebraic Structures 1
groups

1. Operations in and on a set

We begin with a general definition which enables us to tie together a great many ideas that have appeared in earlier chapters of this book.

Let S be a given non-empty set; a rule which assigns to certain *ordered* pairs (a, b) of elements a, b of S a unique element of S is called a **binary operation** (i.e. **two-variable operation**) **in** S. We shall usually omit the word "binary" when it is clear that two elements of S are involved. The elements a, b *need not be distinct*; from now on we shall assume this without further mention.

If the rule assigns to *every* ordered pair of elements of S a unique element of S, the rule is called an **operation on** S; we then say that S is **closed under (or with respect to) the operation**.

From the definition, it is clear that a binary operation **on** S is a mapping from $S^2 = S \times S$ to S and a binary operation **in** S is a mapping from a *subset* of S^2 to S.

That this rather abstract definition does cover many earlier ideas is verified by examining the following examples:

Example 1. Take the set $\mathbf{N}$ of natural numbers. Then
 - (i) *addition* is an operation *on* $\mathbf{N}$, the element $a+b$ corresponding to pair $(a, b) \in \mathbf{N}^2$;
 [$\mathbf{N}$ is closed under addition.]
 - (ii) *multiplication* is an operation *on* $\mathbf{N}$, the element ab corresponding to pair $(a, b) \in \mathbf{N}^2$;
 [$\mathbf{N}$ is closed under multiplication.]
 - (iii) *subtraction* is an operation *in* $\mathbf{N}$, *but not on* $\mathbf{N}$, the element $a-b$ corresponding to the pair $(a, b) \in \mathbf{N}^2$ if and only if $a > b$;
 [$\mathbf{N}$ is *not closed* under subtraction.]
 - (iv) *raising a to the bth power* is an operation *on* $\mathbf{N}$, the element a^b corresponding to the pair $(a, b) \in \mathbf{N}^2$;
 [$\mathbf{N}$ is closed under this operation.]

(v) *forming* $\sqrt{(ab)}$ is an operation *in, but not on,* **N**; e.g. $\sqrt{(1.2)} \notin \mathbf{N}$, and, in fact, $\sqrt{(ab)} \in \mathbf{N}$ for those pairs $(a, b) \in \mathbf{N}^2$ such that $ab = x^2$ for some $x \in \mathbf{N}$.
[**N** is not closed under this operation.]

Example 2. *Let* $P(X)$ *be the power set of a non-empty set* X, i.e. *the set of all subsets of* X. Then

(i) *union* is an operation *on* $P(X)$, the element $A \cup B$ corresponding to pair $(A, B) \in P(X) \times P(X)$;
[$P(X)$ is closed under union.]

(ii) *intersection* is an operation on $P(X)$, the element $A \cap B$ corresponding to pair $(A, B) \in P(X) \times P(X)$.
[$P(X)$ is closed under intersection.]

Similarly, $P(X)$ is closed under the operations which assign to the pair $(A, B) \in (P(X))^2$ the elements (iii) $A - B$, (iv) $B - A$ and (v) $A + B$, respectively.

Example 3. *Let* S *be the set of all real* 2×2 *matrices.* Then

(i) *addition of matrices* is an operation *on* S, the matrix $A + B$ corresponding to pair $(A, B) \in S^2$;
[S is closed under addition of matrices.]

(ii) *multiplication of matrices* is an operation *on* S, the matrix AB corresponding to pair $(A, B) \in S^2$;
[S is closed under multiplication of matrices.]

(iii) *forming* $A - 2B$ is an operation *on* S, the matrix $A - 2B$ corresponding to pair $(A, B) \in S^2$;
[S is closed under this operation.]

(iv) forming $A^{-1}B$ is an operation *in, but not on,* S, since the rule assigns an element of S to pair $(A, B) \in S^2$ if and only if A is nonsingular.
[S is not closed under this operation.]

Example 4. *Let* M *be the set of all mappings:* $X \to X$, *where* X *is a given set.* Then

(i) *composition of mappings* in which $f \circ g$ corresponds to $(f, g) \in M^2$ is an operation *on* M, and

(ii) *composition of mappings* in which $g \circ f$ corresponds to $(f, g) \in M^2$ is another operation *on* M;
[M is closed under each of these operations.]

(iii) *forming* $f^{-1} \circ g$ is an operation *in, but not on*, M, since the rule assigns an element of M to pair $(f, g) \in M^2$ if and only if f is a bijection.
[M is not closed under this operation.]

Example 5. *Let* $\mathbf{Z}_5$ *be the set of residues* $\{0, 1, 2, 3, 4\}$ *modulo* 5. Then

(i) *addition modulo* 5 is an operation *on* $\mathbf{Z}_5$ [see Example **1** of Section **6**, Chapter **6**], and

(ii) *multiplication modulo* 5 is also an operation *on* $\mathbf{Z}_5$; this is best shown by the following multiplication table (mod 5):

(mod 5)	0	1	2	3	4
0	0	0	0	0	0
1	0	1	2	3	4
2	0	2	4	1	3
3	0	3	1	4	2
4	0	4	3	2	1

Example 6. *Let* V *be the set of all geometrical vectors in* 3*-dimensional space.* Then

(i) *addition of vectors* is an operation *on* V, the vector $\mathbf{a} + \mathbf{b}$ corresponding to the pair $(\mathbf{a}, \mathbf{b}) \in V^2$;
[V is closed under addition of vectors.]

(ii) forming the vector product $\mathbf{a} \times \mathbf{b}$ is an operation *on* V, since a vector denoted by $\mathbf{a} \times \mathbf{b}$ is assigned to each pair of vectors $\mathbf{a}, \mathbf{b}$;
[V is closed under vector product.]

(iii) forming the scalar product $\mathbf{a} . \mathbf{b}$ is *not an operation in or on* V since $\mathbf{a} . \mathbf{b} \notin V$ for any pair of vectors $\mathbf{a}, \mathbf{b}$.

We shall meet other examples later. A study of the examples given shows that the concept of an operation on a set involves many core ideas in mathematics.

EXERCISE 9.1

1. For the set of **N** of natural numbers, determine which of the following expressions define operations in or on **N**:

(i) $a^2 b^2$, (ii) ab^3, (iii) $a - 2b$, (iv) $a^2 + b^2$, (v) $\dfrac{a}{b}$, (vi) $|a - b|$,
(vii) set $\{a, b\}$, (viii) (a, b), the g.c.d. of a, b.

2. Let $S = \left\{ \begin{bmatrix} 1 & a \\ 0 & 1 \end{bmatrix} : a \in \mathbf{R} \right\}$. Determine whether or not S is closed under matrix addition

and under matrix multiplication.

$\left[\textbf{Hint.} \text{ Consider the pair of elements } A = \begin{bmatrix} 1 & a \\ 0 & 1 \end{bmatrix}, B = \begin{bmatrix} 1 & b \\ 0 & 1 \end{bmatrix} \text{ of } S, \text{ where } a, b \in \mathbf{R}, \right.$

and form $A + B$ and AB. $\Big]$

If, in the definition of S the condition $a \in \mathbf{R}$ was replaced by the condition $a \in \mathbf{Z}$, what would your answers be?

3. Let $S = \left\{ \begin{bmatrix} 1 & a\sqrt{2} \\ 0 & 1 \end{bmatrix} : a \in \mathbf{Q}, a \neq 0 \right\}$. Determine whether or not S is closed under matrix

addition and under matrix multiplication.
   What would your answers be if the condition $a \neq 0$ were omitted from the description of S?

4. If $S = \{1, 2, 3, 4, 5\}$, determine whether or not S is closed under addition (mod 6) and under multiplication (mod 6).
   What would your answers be if (mod 6) were replaced by (mod 5)?

2. Properties for an operation, definition of a group

An operation on a set is by itself of little interest; it must have some suitable properties likely to lead to interesting and useful results. For the choice of such properties we look at properties of the operations of addition and multiplication on the integers, rational numbers, real numbers, 2×2 matrices, 3-dimensional vectors, the bijective mappings of a set onto itself, etc. We shall list the basic properties for six such examples.

Example 1. **R,** *the set of real numbers under addition:*

(1) For each pair a, b in **R**,
$$\exists a + b \in \mathbf{R}$$
    ; closure property (Cl.)

(2) $\forall a, b, c \in \mathbf{R}$,
$$a + (b + c) = (a + b) + c$$
    ; associative property (Ass.)

(3) $\exists$ element $0 \in \mathbf{R}$ such that,
$$\forall a \in \mathbf{R}, a + 0 = 0 + a = a$$
    ; existence of "identity" (Id.)

(4) For each $a \in \mathbf{R}, \exists -a \in \mathbf{R}$
          such that
$$a + (-a) = (-a) + a = 0$$
    ; existence of "inverse" (In.)

(5) $\forall a, b \in \mathbf{R}, a+b = b+a$; commutative property (Comm.)

The number 0 (zero) is called the **additive identity for R** and $-a$ the **additive inverse** of $a \in \mathbf{R}$.

In the following examples we shall use the abbreviations Cl., Ass., Id., In., Comm. as indicated above for Closure property, etc.

Example 2. R $- \{0\}$, *the set of non-zero real numbers under multiplication:*

(1) For each pair a, b in $\mathbf{R} - \{0\}$, $\exists ab \in \mathbf{R} - \{0\}$; Cl.

(2) $\forall a, b, c \in \mathbf{R} - \{0\}$, $a(bc) = (ab)c$; Ass.

(3) $\exists$ element $1 \in \mathbf{R} - \{0\}$ such that, $\forall a \in \mathbf{R} - \{0\}$,
$$a1 = 1a = a$$; Id.

(4) For each $a \in \mathbf{R} - \{0\}$, $\exists a^{-1} \in \mathbf{R} - \{0\}$ such that
$$aa^{-1} = a^{-1}a = 1$$; In.

(5) $\forall a, b \in \mathbf{R} - \{0\}$, $ab = ba$; Comm.

The number 1 is the **multiplicative identity** for $\mathbf{R} - \{0\}$ and $a^{-1}\left(= \dfrac{1}{a}\right)$, the **multiplicative inverse** of $a \in \mathbf{R} - \{0\}$.

Example 3. *S, the set of all 2×2 real matrices under matrix addition:*
This has properties identical to those listed for Example 1 when the real numbers $a, b, c, 0, -a$ are replaced by 2×2 matrices $A, B, C, O, -A$, respectively.

Example 4. *V, the set of all vectors in three-dimensional space under vector addition:*
Again this has properties identical to those listed for Example 1 when the real numbers $a, b, c, 0, -a$ are replaced by vectors $\mathbf{a, b, c, 0, -a}$, respectively.

Example 5. *S_0, the set of all non-singular 2×2 real matrices under matrix multiplication:*

(1) For each pair $A, B \in S_0$, $\exists AB \in S_0$
[**Note.** $|A| \neq 0$ and $|B| \neq 0 \Rightarrow |AB| \neq 0$.]   ; Cl.

(2) $\forall A, B, C \in S_0, A(BC) = (AB)C$;   Ass.

(3) $\exists$ element $I \in S_0$ such that, $\forall A \in S_0$,
$$AI = IA = A$$
;   Id.

(4) For each $A \in S_0, \exists A^{-1} \in S_0$ such that
$$AA^{-1} = A^{-1}A = I$$
;   In.

For this example, the commutative property does not hold.

Example 6. M_0, *the set of all bijective mappings from a set S onto itself under composition of mappings, $f \circ g$ corresponding to the pair f, g:*

(1) For each pair $f, g \in M_0, \exists f \circ g \in M_0$;   Cl.
   [See Problem 11 (with $S = T = U$) of Exercise 3, Chapter 3.]

(2) $\forall f, g, h \in M_0, f \circ (g \circ h) = (f \circ g) \circ h$;   Assoc.
   [See Theorem 3.1 (with $S = T = U$) of Chapter 3.]

(3) $\exists$ element $i_S \in M_0$ such that, $\forall f \in M_0$,
$$f \circ i_S = i_S \circ f = f$$
;   Id.
   [See Problem 9 (with $S = T$) of Exercise 3, Chapter 3.]

(4) For each $f \in M_0, \exists f^{-1} \in M_0$ such that
$$f \circ f^{-1} = f^{-1} \circ f = i_S$$
;   In.
   [See Theorem 3.2 (with $S = T$) of Chapter 3.]

For this example, the commutative property does not in general hold. [See worked examples **14** and **15** of Chapter **3**.]

Common to these six examples (and many others) we have the following structure:

A set together with an operation on the set satisfying

    (1) the closure property,
    (2) the associative property,
    (3) the property of existence of an identity,
    (4) the property of existence of an inverse for each element.

Such a structure is called a **group** and the study of the properties of groups is called **Group Theory**. These structures appear in many branches of mathematics and its applications.

As we saw in the six examples, a group may or may not possess the **commutative property**, listed as (5) in examples **1, 2, 3** and **4**.

We now present the formal definition of a group using a commonly accepted general notation.

Definition of a group.

Let S be a given set and let $\circ$ (read as "circle") denote an operation in S, $a \circ b$ denoting the element of S corresponding by the operation to a pair of elements $a, b \in S$. Then we say that S *forms a group under the operation* $\circ$ or that *the pair $\{S, \circ\}$ is a group* if the following four axioms (called the **group axioms**) are satisfied:

$\mathbf{G}_1$: S is closed under $\circ$ (i.e. $\circ$ is an operation *on* S); (*closure property*).

$\mathbf{G}_2$: $\forall a, b, c \in S$, $a \circ (b \circ c) = (a \circ b) \circ c$; (*associative property*).

$\mathbf{G}_3$: $\exists$ element $e \in S$ such that, $\forall a \in S$,
$$a \circ e = e \circ a = a.$$
[We show later that e is *unique*; e is called *the identity element*.]

$\mathbf{G}_4$: For each $a \in S$, $\exists a^{-1} \in S$ such that
$$a \circ a^{-1} = a^{-1} \circ a = e.$$
[We show later that a^{-1} is *unique*; a^{-1} is called *the inverse of a*.]

If, in addition to $\mathbf{G}_1$, $\mathbf{G}_2$, $\mathbf{G}_3$ and $\mathbf{G}_4$, the following axiom also holds, namely,

$\mathbf{G}_5$: $\forall a, b \in S$, $a \circ b = b \circ a$; (*commutative property*),

then the group is said to be **commutative** or **abelian** (after the mathematician Abel).

If it is known that $\mathbf{G}_5$ does *not* hold, we often say that the group is **non-abelian**.

Note. When $\circ$ is denoted by $+$, we use 0 and $-a$ in place of e and a^{-1} for the identity and inverse, and call the group an **additive group** or say that it is expressed in **additive notation**.

When $\circ$ is denoted by $.$, we sometimes use 1 for the identity element, and call the group a **multiplicative group** or say that it is expressed in **multiplicative notation**. In this case we usually omit the $.$ and write ab for $a \circ b$ (instead of $a . b$).

A group is said to be **finite** or **infinite** according as the set S has a *finite* or *infinite* number of elements. The number of elements in a finite group is called the **order of the group**.

The examples **1**, **2**, **3**, **4** and **5** of this section are all infinite groups, the first four being abelian and the fifth, the multiplicative group of non-singular 2×2 real matrices, being non-abelian.

The example **6**, the group of bijections of a set S onto itself under composition of mappings, is infinite if S is infinite and finite if S is finite.

[In fact, if S has n elements, then a bijection: $S \to S$ is a permutation mapping of the elements of S; it follows that the group is of order $n!$ since there are $n!$ distinct permutations of n different objects. The group is abelian if $n = 1$ or 2 and non-abelian for each $n \geqslant 3$. It is called the **symmetric group of degree** n or **of order** $n!$.]

Note 1. When $a \circ (b \circ c) = (a \circ b) \circ c$, we can, without ambiguity, denote the (common) element by $a \circ b \circ c$; we can insert brackets in $a \circ b \circ c$ in the two ways possible without affecting the element. As an extension of this simplified notation we can show, by induction, that $a_1 \circ a_2 \circ a_3 \circ \ldots$ $\circ a_n$, where the a_i belong to a given group, can be taken as the common value of all the elements obtained from the expression $a_1 \circ a_2 \circ a_3 \circ \ldots \circ a_n$ by inserting brackets in all possible ways.

Note 2. The element $a \circ a \circ \ldots \circ a$ (m such terms) is denoted by a^m. If m and n are positive integers,

$$a^m \circ a^n = (a \circ a \circ \ldots \circ a) \circ (a \circ a \circ \ldots \circ a)$$
$$\text{(m terms)} \qquad \text{(n terms)}$$
$$= a \circ a \circ \ldots \circ a = a^{m+n}.$$
$$\text{($m+n$ terms)}$$

Also, $a^n \circ a^m = a^{m+n}$, and so $a^m \circ a^n = a^n \circ a^m$; we say that "positive powers of a commute".

Also, it is easy to show that $(a^m)^n = a^{mn}$.

In additive notation, $a + a + \ldots + a$ (m terms) is, as usual, denoted by ma; clearly $ma + na = (m+n)a = na + ma$, where m and n are positive integers; also $n(ma) = (nm)a$.

Note 3. Let set G form a group under an operation $\circ$, and let H be a subset of G. Then, if H itself forms a group under this same operation $\circ$, we say that $\{H, \circ\}$ **is a subgroup of the group** $\{G, \circ\}$ or, simply, that H **is a subgroup of the group** G, when the operation is clearly understood.

Note 4. If elements a, b in a group are such that $a \circ b = b \circ a$, we say that a and b **commute.**

3. Some examples of groups

I. Groups from number systems.

(1) **Z**, the set of all integers, forms an abelian group under $+$.

Q, **R** and **C** also form abelian groups under $+$.

Z, **Q** and **R** are subgroups of the additive group of **C**.

N and **W** do not form groups under $+$; explain why.

(2) $\mathbf{C}-\{0\}$ forms an abelian group under multiplication and $\mathbf{Q}-\{0\}$, $\mathbf{R}-\{0\}$ are both subgroups of this multiplicative group.

$\mathbf{C}$, $\mathbf{Q}$, $\mathbf{R}$ themselves do not form groups under multiplication since 0 has no multiplicative inverse.

Explain why $\mathbf{N}-\{0\}$, $\mathbf{W}-\{0\}$ and $\mathbf{Z}-\{0\}$ do not form groups under multiplication.

II. Groups of matrices.

(1) *The set of all 2×2 real matrices forms an abelian group under matrix addition,* the identity being the zero 2×2 matrix O and the additive inverse of A being the matrix $-A$.

(2) *The set M_0 of all non-singular 2×2 real matrices forms a group under matrix multiplication* [See Example **5** of Section **2**.], the matrix $I = \begin{bmatrix} 1 & 0 \\ 0 & 1 \end{bmatrix}$ being the identity and the inverse matrix A^{-1} being the group inverse of A. If $A = \begin{bmatrix} a & b \\ c & d \end{bmatrix}$, then

$$A^{-1} = \begin{bmatrix} d/\Delta & -b/\Delta \\ -c/\Delta & a/\Delta \end{bmatrix}, \text{ where } \Delta = ad - bc \neq 0.$$

This group is non-abelian, since e.g.

$$\begin{bmatrix} 1 & -1 \\ 0 & 1 \end{bmatrix}\begin{bmatrix} 0 & 1 \\ -1 & 1 \end{bmatrix} \neq \begin{bmatrix} 0 & 1 \\ -1 & 1 \end{bmatrix}\begin{bmatrix} 1 & -1 \\ 0 & 1 \end{bmatrix}.$$

(3) *The set M_1 of all 2×2 real matrices of determinant $+1$ forms a non-abelian group under matrix multiplication.*

Proof. We verify that the group axioms $\mathbf{G}_1$, $\mathbf{G}_2$, $\mathbf{G}_3$ and $\mathbf{G}_4$ hold.

$\mathbf{G}_1$: Take $A, B \in M_1$, so that A, B are 2×2 real matrices with $|A| = 1$ and $|B| = 1$. Then AB is a 2×2 real matrix and $|AB| = |A||B| = 1$. It follows that, if $A, B \in M_1$, then $AB \in M_1$, and so M_1 is closed under matrix multiplication.

$\mathbf{G}_2$: The associative property holds for M_1 since it holds for all 2×2 matrices.

$\mathbf{G}_3$: The matrix $I = \begin{bmatrix} 1 & 0 \\ 0 & 1 \end{bmatrix}$ belongs to M_1 since it is a 2×2 real matrix of determinant 1. Also, for any $A \in M_1$, $IA = AI = A$, so that M_1 has an identity element.

G_4: If $A \in M_1$, then $|A| = 1$, so that the matrix A^{-1} exists. Also A^{-1} is a 2×2 real matrix and $|A^{-1}| = 1/|A| = 1$. Hence $A^{-1} \in M_1$ and so is an inverse for A in M_1 (since $AA^{-1} = A^{-1}A = I$).

It follows that M_1 forms a group under matrix multiplication.

That the group is non-abelian is easily checked by finding two elements A, B of M_1 for which $AB \neq BA$, i.e. two elements which do not commute. The elements

$$A = \begin{bmatrix} 1 & -1 \\ 0 & 1 \end{bmatrix} \quad \text{and} \quad B = \begin{bmatrix} 0 & 1 \\ -1 & 1 \end{bmatrix} \quad \text{will do.}$$

M_1 is a subgroup of the group M_0 of (2).

(4) **Example.** Show that the set S of all matrices of the form $\begin{bmatrix} a & b \\ -b & a \end{bmatrix}$,

where a and b are real numbers not both zero, forms a group under matrix multiplication. Determine whether S is an abelian group.

State which of the following subsets of S are subgroups:

$$\text{(i)} \quad U = \left\{ \begin{bmatrix} 0 & b \\ -b & 0 \end{bmatrix} : b \in \mathbf{R} - \{0\} \right\},$$

$$\text{(ii)} \quad V = \left\{ \begin{bmatrix} a & 0 \\ 0 & a \end{bmatrix} : a \in \mathbf{R} - \{0\} \right\},$$

and justify your statement in each case.

We again consider G_1, G_2, G_3 and G_4 in turn.

G_1: Take $A = \begin{bmatrix} a & b \\ -b & a \end{bmatrix}$ and $B = \begin{bmatrix} x & y \\ -y & x \end{bmatrix}$ both in S so that a, b

are real numbers not both zero and x, y are real numbers not both zero. Then

$$AB = \begin{bmatrix} a & b \\ -b & a \end{bmatrix} \begin{bmatrix} x & y \\ -y & x \end{bmatrix} = \begin{bmatrix} ax-by & ay+bx \\ -(ay+bx) & ax-by \end{bmatrix} = \begin{bmatrix} p & q \\ -q & p \end{bmatrix},$$

say, where p and q are both real. Thus AB is of the right form and will belong to S provided p, q are not both zero. To show that this is the case we use determinants.

Since $\qquad\qquad |AB| = |A||B|,$ we have:
$$p^2 + q^2 = (a^2 + b^2)(x^2 + y^2).$$

But $a^2 + b^2 \neq 0$ since a and b are not both zero, and $x^2 + y^2 \neq 0$ since x and y are not both zero. Hence $p^2 + q^2 \neq 0$, and so p and q are not both zero.

It follows that S is closed under matrix multiplication.

G_2: The associative property holds in S since it holds for all 2×2 matrices.

G_3: The matrix $I = \begin{bmatrix} 1 & 0 \\ 0 & 1 \end{bmatrix}$ belongs to S since it is given by $a = 1$, $b = 0$. Since $IA = AI = A$ for any $A \in S$, it follows that S has an identity.

G_4: If $A = \begin{bmatrix} a & b \\ -b & a \end{bmatrix} \in S$, then $|A| = a^2 + b^2 \neq 0$, and so the matrix A^{-1} exists. Also

$$A^{-1} = \frac{1}{a^2 + b^2} \begin{bmatrix} a & -b \\ b & a \end{bmatrix}, \text{ i.e.}$$

$$A^{-1} = \begin{bmatrix} \dfrac{a}{a^2 + b^2} & \left(\dfrac{-b}{a^2 + b^2}\right) \\ -\left(\dfrac{-b}{a^2 + b^2}\right) & \dfrac{a}{a^2 + b^2} \end{bmatrix}.$$

Clearly the entries in A^{-1} are real numbers which are not both zero since a and b are not both zero. Hence $A^{-1} \in S$ and is an inverse for A in S.

It follows that S forms a group under matrix multiplication.

To check whether or not S is abelian we form, in the notation used for G_1, BA. We have:

$$BA = \begin{bmatrix} x & y \\ -y & x \end{bmatrix} \begin{bmatrix} a & b \\ -b & a \end{bmatrix} = \begin{bmatrix} ax - by & ay + bx \\ -(ay + bx) & ax - by \end{bmatrix}.$$

It follows that, $\forall A, B \in S$, $AB = BA$ and so S is abelian.

Consider now the subset U of S: We can see in several ways that U is not a subgroup of S. For example, taking

$$A = \begin{bmatrix} 0 & 1 \\ -1 & 0 \end{bmatrix} \text{ in } U, \text{ we have: } A^2 = \begin{bmatrix} 0 & 1 \\ -1 & 0 \end{bmatrix} \begin{bmatrix} 0 & 1 \\ -1 & 0 \end{bmatrix} = \begin{bmatrix} -1 & 0 \\ 0 & -1 \end{bmatrix},$$

and so $A^2 \notin U$. Thus U is not closed under multiplication, i.e. axiom $\mathbf{G_1}$ is not satisfied.

Consider now the subset V of S: Using the work that we have already done for S it is easily verified that V is a subgroup of S. We merely put $b = 0$ and $y = 0$ in the work for $\mathbf{G_1}$ and $\mathbf{G_4}$.

III. A finite group of order 4.

(1) *The set $S = \{1, i, -1, -i\}$ forms an abelian group under multiplication of complex numbers.*

In a case like this it is useful to draw up a multiplication table as follows:

.	1	i	-1	$-i$
1	1	i	-1	$-i$
i	i	-1	$-i$	1
-1	-1	$-i$	1	i
$-i$	$-i$	1	i	-1

$\mathbf{G_1}$: If $a, b \in S$, then clearly, from the table, $ab \in S$, and so S is closed under multiplication.

$\mathbf{G_2}$: For any complex numbers, $a(bc) = (ab)c$, and so the associative property holds.

$\mathbf{G_3}$: $1 \in S$ and 1 is an identity for S since, $\forall a \in S$, $1a = a1 = a$.

$\mathbf{G_4}$: If $a \in S$, $\exists a^{-1} \in S$ such that $aa^{-1} = a^{-1}a = 1$; this is checked at once from the table which shows that $1^{-1} = 1$, $i^{-1} = -i$, $(-1)^{-1} = -1$ and $(-i)^{-1} = i$.

Thus S forms a group under multiplication.

Since $ab = ba$ for any complex numbers a, b, it follows that the group is abelian.

The set S can be written as $\{1, i, i^2, i^3\}$, where $i^4 = 1$.

This group of order 4 is a subgroup of the multiplicative group of all non-zero complex numbers.

(2) *The set $S = \left\{ \begin{bmatrix} 1 & 0 \\ 0 & 1 \end{bmatrix}, \begin{bmatrix} 0 & 1 \\ -1 & 0 \end{bmatrix}, \begin{bmatrix} -1 & 0 \\ 0 & -1 \end{bmatrix}, \begin{bmatrix} 0 & -1 \\ 1 & 0 \end{bmatrix} \right\}$*

forms an abelian group under matrix multiplication.

We can check this statement step by step by considering the properties G_1, G_2, G_3, G_4 or we can proceed as follows:

$$\text{If } A = \begin{bmatrix} 0 & 1 \\ -1 & 0 \end{bmatrix}, \text{ then } A^2 = \begin{bmatrix} 0 & 1 \\ -1 & 0 \end{bmatrix}\begin{bmatrix} 0 & 1 \\ -1 & 0 \end{bmatrix} = \begin{bmatrix} -1 & 0 \\ 0 & -1 \end{bmatrix},$$

$$A^3 = \begin{bmatrix} 0 & 1 \\ -1 & 0 \end{bmatrix}\begin{bmatrix} -1 & 0 \\ 0 & -1 \end{bmatrix} = \begin{bmatrix} 0 & -1 \\ 1 & 0 \end{bmatrix},$$

$$A^4 = \begin{bmatrix} 0 & 1 \\ -1 & 0 \end{bmatrix}\begin{bmatrix} 0 & -1 \\ 1 & 0 \end{bmatrix} = \begin{bmatrix} 1 & 0 \\ 0 & 1 \end{bmatrix} = I.$$

Hence the set $S = \{I, A, A^2, A^3\}$, where A is such that $A^4 = I$. The multiplication table (putting $A^4 = I$ where necessary) for S is:

.	I	A	A^2	A^3
I	I	A	A^2	A^3
A	A	A^2	A^3	I
A^2	A^2	A^3	I	A
A^3	A^3	I	A	A^2

The group properties can now be easily checked, and that the group is abelian follows from the fact that $A^m A^n = A^n A^m$ for each pair of non-negative integers m, n (A^0 being I).

We now show that the two groups just considered in (1) and (2) are similar in structure. This becomes clearer if we rewrite the multiplication table for (1) as follows:

.	1	i	i^2	i^3
1	1	i	i^2	i^3
i	i	i^2	i^3	1
i^2	i^2	i^3	1	i
i^3	i^3	1	i	i^2

The group in (1) is the set $\{1, i, i^2, i^3\}$, with $i^4 = 1$, and multiplication of complex numbers;

the group in (2) is the set $\{I, A, A^2, A^3\}$, with $A^4 = I$, and multiplication of 2×2 matrices.

The two groups are clearly identical in structure; if we define a mapping from (1) to (2) by mapping i^n to A^n, for $n = 0, 1, 2, 3$, then the mapping is a bijection and any product of elements in (1) is mapped to the corresponding product in (2).

Two such groups are said to be **isomorphic** to each other.

In general, two groups $\{G, \circ\}$ and $\{G', *\}$, where $\circ$ denotes the operation on G and $*$ the operation on G', are said to be **isomorphic** to each other if $\exists$ a bijection $f : G \to G'$ such that,

$$\left.\begin{array}{l} \text{if } f \text{ maps } a \text{ to } a' \\ \text{and } f \text{ maps } b \text{ to } b' \end{array}\right\}, \text{ then } f \text{ maps } a \circ b \text{ to } a' * b'.$$

This condition can also be expressed in the form:

$$\forall a, b \in G, f(a \circ b) = f(a) * f(b).$$

Two isomorphic groups have the same structure although they may differ in notation and in the nature of their elements.

Note. A group whose elements can be written in the form

$$\{e, a, a^2, \ldots, a^{n-1}\},$$

where $a^n = e$, the identity element, and where n is the smallest positive integer such that $a^n = e$, is called a **finite cyclic group of order** n. The element a is called a **generator** of the cyclic group. If we go on to form the powers $a^n, a^{n+1}, a^{n+2}, \ldots, a^{2n-1}, \ldots$, they keep occurring in the cycle $e, a, a^2, \ldots, a^{n-1}$, repeatedly.

A cyclic group is necessarily abelian; for, in any group, $a^m \circ a^n = a^n \circ a^m = a^{m+n}$ for any non-negative integers m, n, where we write a^0 for e, the identity element.

Any cyclic group of order 4 can be expressed in the form $\{e, a, a^2, a^3\}$, where e is the identity element, and $a^4 = e$. This is called the **abstract cyclic group** C_4 of order 4, since the nature of the elements is immaterial. The two examples (1) and (2) above are said to be **representations** of this abstract cyclic group of order 4, the one involving complex numbers and the other 2×2 matrices.

Every abstract group can be represented in many ways. We now give three other representations of the cyclic group of order 4.

(3) *Check that the set $\{1, 2, 3, 4\}$ forms an abelian group under multiplication* (mod 5). [Draw up a multiplication table (mod 5).]

Since $3 \equiv 2^3$ (mod 5) and $2^4 = 16 \equiv 1$ (mod 5), the group is formed by the set $\{1, 2, 2^2, 2^3\}$ under multiplication (mod 5), and $2^4 \equiv 1$ (mod 5). [Note the rearrangement in the listing of the elements 3, 4.]

Thus the group is a representation of the cyclic group of order 4.

(4) *Check that the set*

$$S = \left\{ \begin{bmatrix} 1 & 1 \\ 0 & 0 \end{bmatrix}, \begin{bmatrix} i & i \\ 0 & 0 \end{bmatrix}, \begin{bmatrix} -1 & -1 \\ 0 & 0 \end{bmatrix}, \begin{bmatrix} -i & -i \\ 0 & 0 \end{bmatrix} \right\}$$

forms an abelian group under matrix multiplication.

If we write $E = \begin{bmatrix} 1 & 1 \\ 0 & 0 \end{bmatrix}$ and $A = \begin{bmatrix} i & i \\ 0 & 0 \end{bmatrix}$, then

$$S = \{E, A, A^2, A^3\} \quad \text{and} \quad A^4 = E.$$

Thus the group is another representation of the cyclic group of order 4.

Note. Although the group in (4) is a multiplicative group of 2×2 matrices, the identity element of the group is *not* the identity matrix, and the inverse of an element in the group is *not* the inverse matrix.

(5) *Let S be the following set of permutation mappings of the set* $\{1, 2, 3, 4\}$ *onto itself:*

$$\overset{I}{\begin{pmatrix} 1 & 2 & 3 & 4 \\ 1 & 2 & 3 & 4 \end{pmatrix}} \quad \overset{A}{\begin{pmatrix} 1 & 2 & 3 & 4 \\ 2 & 3 & 4 & 1 \end{pmatrix}} \quad \overset{B}{\begin{pmatrix} 1 & 2 & 3 & 4 \\ 3 & 4 & 1 & 2 \end{pmatrix}} \quad \overset{C}{\begin{pmatrix} 1 & 2 & 3 & 4 \\ 4 & 1 & 2 & 3 \end{pmatrix}}.$$

Check that S forms an abelian group under composition of mappings.

[As usual, the second row in each case gives the permutation of 1234 produced by the permutation mapping listed above. For example, the mapping A maps 1 to 2, 2 to 3, 3 to 4 and 4 to 1.]

Now $A^2(1\ 2\ 3\ 4) = A(2\ 3\ 4\ 1) = (3\ 4\ 1\ 2) = B(1\ 2\ 3\ 4)$
$$\therefore A^2 = B.$$

Also $A^3(1\ 2\ 3\ 4) = A(3\ 4\ 1\ 2) = (4\ 1\ 2\ 3) = C(1\ 2\ 3\ 4)$
$$\therefore A^3 = C.$$

Further, $A^4(1\ 2\ 3\ 4) = A(4\ 1\ 2\ 3) = (1\ 2\ 3\ 4) = I(1\ 2\ 3\ 4)$
$$\therefore A^4 = I, \quad \text{the identity mapping.}$$

Thus $S = \{I, A, A^2, A^3\} \quad \text{and} \quad A^4 = I.$

Again it is clear that we have another representation of the cyclic group of order 4.

IV. A non-abelian group of order 6. Let $X = \{1, 2, 3\}$. There are $3! = 6$ permutations of the numbers 1, 2, 3 and so 6 permutation mappings, i.e.

bijections, of the set X onto itself. These can be represented as follows:

$$\begin{pmatrix} 1 & 2 & 3 \\ 1 & 2 & 3 \end{pmatrix}, \begin{pmatrix} 1 & 2 & 3 \\ 2 & 3 & 1 \end{pmatrix}, \begin{pmatrix} 1 & 2 & 3 \\ 3 & 1 & 2 \end{pmatrix}, \begin{pmatrix} 1 & 2 & 3 \\ 1 & 3 & 2 \end{pmatrix}, \begin{pmatrix} 1 & 2 & 3 \\ 2 & 1 & 3 \end{pmatrix}, \begin{pmatrix} 1 & 2 & 3 \\ 3 & 2 & 1 \end{pmatrix},$$

where, for example, $\begin{pmatrix} 1 & 2 & 3 \\ 2 & 3 & 1 \end{pmatrix}$ denotes as usual the permutation mapping

of X onto itself which maps $\begin{cases} 1 \to 2, \\ 2 \to 3, \\ 3 \to 1. \end{cases}$

$\begin{pmatrix} 1 & 2 & 3 \\ 1 & 2 & 3 \end{pmatrix}$ is the identity mapping I. If we denote $\begin{pmatrix} 1 & 2 & 3 \\ 2 & 3 & 1 \end{pmatrix}$ by A

and $\begin{pmatrix} 1 & 2 & 3 \\ 1 & 3 & 2 \end{pmatrix}$ by B, then

$$A^2(1\ 2\ 3) = A(2\ 3\ 1) = (3\ 1\ 2) \qquad \therefore\ A \circ A = A^2 = \begin{pmatrix} 1 & 2 & 3 \\ 3 & 1 & 2 \end{pmatrix}.$$

$$A^3(1\ 2\ 3) = A(3\ 1\ 2) = (1\ 2\ 3) \qquad \therefore\ A^3 = I.$$

$$B^2(1\ 2\ 3) = B(1\ 3\ 2) = (1\ 2\ 3) \qquad \therefore\ B^2 = I.$$

$$A \circ B(1\ 2\ 3) = A(1\ 3\ 2) = (2\ 1\ 3) \qquad \therefore\ A \circ B = \begin{pmatrix} 1 & 2 & 3 \\ 2 & 1 & 3 \end{pmatrix}.$$

$$A^2 \circ B(1\ 2\ 3) = A(2\ 1\ 3) = (3\ 2\ 1) \qquad \therefore\ A^2 \circ B = \begin{pmatrix} 1 & 2 & 3 \\ 3 & 2 & 1 \end{pmatrix}.$$

$$(A \circ B)^2(1\ 2\ 3) = A \circ B(2\ 1\ 3) = (1\ 2\ 3) \qquad \therefore\ (A \circ B)^2 = I.$$

Now, we have seen in Example 6 of Section 2 that the set of all bijections of a given set onto itself forms a group under composition of mappings. Hence the above set of permutation mappings of $X = \{1, 2, 3\}$ onto itself forms a group of order 6 under composition of mappings. The group can be expressed in the form $G = \{I, A, A^2, B, AB, A^2B\}$, where $A^3 = I$, $B^2 = I$ and $(AB)^2 = I$, and where, for simplicity of notation, we have written AB for $A \circ B$ and A^2B for $A^2 \circ B$.

Now $BA(1\ 2\ 3) = B(2\ 3\ 1) = (3\ 2\ 1) = A^2B(1\ 2\ 3)$.

Hence $BA = A^2B \neq AB$, so that the group is non-abelian.

A further representation for this same group can be obtained in terms of the **permutation matrices** for the set X.

We note first that

$$\begin{bmatrix} 0 & 1 & 0 \\ 0 & 0 & 1 \\ 1 & 0 & 0 \end{bmatrix} \begin{bmatrix} 1 \\ 2 \\ 3 \end{bmatrix} = \begin{bmatrix} 2 \\ 3 \\ 1 \end{bmatrix}.$$

Thus premultiplying the column vector $\begin{bmatrix} 1 \\ 2 \\ 3 \end{bmatrix}$ by the matrix $\begin{bmatrix} 0 & 1 & 0 \\ 0 & 0 & 1 \\ 1 & 0 & 0 \end{bmatrix}$

produces the same permutation of $\{1, 2, 3\}$ as the permutation mapping

$\begin{pmatrix} 1 & 2 & 3 \\ 2 & 3 & 1 \end{pmatrix}$. The matrix $\begin{bmatrix} 0 & 1 & 0 \\ 0 & 0 & 1 \\ 1 & 0 & 0 \end{bmatrix}$ is obtained from the identity 3×3

matrix $I = \begin{bmatrix} 1 & 0 & 0 \\ 0 & 1 & 0 \\ 0 & 0 & 1 \end{bmatrix}$ by permuting the *rows* of I using the same permu-

tation mapping that maps $\begin{bmatrix} 1 \\ 2 \\ 3 \end{bmatrix} \rightarrow \begin{bmatrix} 2 \\ 3 \\ 1 \end{bmatrix}$. The matrix $\begin{bmatrix} 0 & 1 & 0 \\ 0 & 0 & 1 \\ 1 & 0 & 0 \end{bmatrix}$ is called

the **permutation matrix** corresponding to the permutation mapping

$$\begin{pmatrix} 1 & 2 & 3 \\ 2 & 3 & 1 \end{pmatrix}.$$

The set of all the permutation matrices for the permutation mappings on a set of three elements is:

$$\left\{ \begin{bmatrix} 1 & 0 & 0 \\ 0 & 1 & 0 \\ 0 & 0 & 1 \end{bmatrix}, \begin{bmatrix} 0 & 1 & 0 \\ 0 & 0 & 1 \\ 1 & 0 & 0 \end{bmatrix}, \begin{bmatrix} 0 & 0 & 1 \\ 1 & 0 & 0 \\ 0 & 1 & 0 \end{bmatrix}, \begin{bmatrix} 1 & 0 & 0 \\ 0 & 0 & 1 \\ 0 & 1 & 0 \end{bmatrix}, \begin{bmatrix} 0 & 1 & 0 \\ 1 & 0 & 0 \\ 0 & 0 & 1 \end{bmatrix}, \begin{bmatrix} 0 & 0 & 1 \\ 0 & 1 & 0 \\ 1 & 0 & 0 \end{bmatrix} \right\}, \quad (3.1)$$

i.e. $\{ \quad I \quad , \quad A \quad , \quad A^2 \quad , \quad B \quad , \quad A^2B \quad , \quad AB \quad \}$, say,

corresponding to the permutations

$$\left\{ \begin{pmatrix} 1 & 2 & 3 \\ 1 & 2 & 3 \end{pmatrix}, \begin{pmatrix} 1 & 2 & 3 \\ 2 & 3 & 1 \end{pmatrix}, \begin{pmatrix} 1 & 2 & 3 \\ 3 & 1 & 2 \end{pmatrix}, \begin{pmatrix} 1 & 2 & 3 \\ 1 & 3 & 2 \end{pmatrix}, \begin{pmatrix} 1 & 2 & 3 \\ 2 & 1 & 3 \end{pmatrix}, \begin{pmatrix} 1 & 2 & 3 \\ 3 & 2 & 1 \end{pmatrix} \right\}.$$

The set of matrices (4.1) forms a group under matrix multiplication, another representation of the group of order 6 just described. [Note the interchange of the last two matrices from the corresponding products in the representation by permutation mappings.]

4. Some properties of a group $\{G, \circ\}$

We prove some of the basic properties of any group.

I. The identity element e is unique.

Proof. Let e_1 be any identity element. Then we have:

$$\forall a \in G, \quad a \circ e = e \circ a = a \quad (e \text{ an identity})$$

and
$$a \circ e_1 = e_1 \circ a = a \quad (e_1 \text{ an identity}).$$

Taking $a = e_1$ in the first row gives $e_1 = e_1 \circ e$, and taking $a = e$ in the second row gives $e = e_1 \circ e$.

Hence $e_1 = e$, and the result follows.

II. For each $a \in G$, the inverse element a^{-1} is unique.

Proof. Let a_1 in G be any inverse for a. Then

$$a \circ a_1 = a_1 \circ a = e.$$

But
$$a \circ a^{-1} = a^{-1} \circ a = e.$$

Hence $a^{-1} = a^{-1} \circ e = a^{-1} \circ (a \circ a_1) = (a^{-1} \circ a) \circ a_1 = e \circ a_1 = a_1$.
Thus a^{-1} is unique.

III. The equations $a \circ x = b$ and $y \circ a = b$ have unique solutions in G for x and y, for each given pair of elements a, b in G.

Proof. We show in corresponding columns the steps involved for the two equations. In each case, we start by *assuming* that a solution exists and deduce that it must be a certain element; then we check that this element is in fact a solution.

$a \circ x = b \qquad (4.1)$	$y \circ a = b \qquad (4.2)$
$\Rightarrow a^{-1} \circ (a \circ x) = a^{-1} \circ b$	$\Rightarrow (y \circ a) \circ a^{-1} = b \circ a^{-1}$
$\Rightarrow (a^{-1} \circ a) \circ x = a^{-1} \circ b$	$\Rightarrow y \circ (a \circ a^{-1}) = b \circ a^{-1}$
$\Rightarrow \qquad e \circ x = a^{-1} \circ b$	$\Rightarrow \qquad y \circ e = b \circ a^{-1}$
$\Rightarrow \qquad x = a^{-1} \circ b.$	$\Rightarrow \qquad y = b \circ a^{-1}.$

But $x = a^{-1} \circ b$ does satisfy (4.1); for,	But $y = b \circ a^{-1}$ does satisfy (4.2); for,
$a \circ (a^{-1} \circ b) = (a \circ a^{-1}) \circ b$	$(b \circ a^{-1}) \circ a = b \circ (a^{-1} \circ a)$
$= e \circ b = b.$	$= b \circ e = b.$
Thus (1) has unique solution	Thus (2) has unique solution
$x = a^{-1} \circ b.$	$y = b \circ a^{-1}.$

Note. If $a \circ x = e$, then x is called a **right inverse** of a. Then **III** shows that, in a group, a right inverse is unique and is in fact $a^{-1} \circ e$, i.e. a^{-1}, the group inverse itself.

Similarly, if $y \circ a = e$, then y is called a **left inverse** of a. Again **III** shows that, in a group, a left inverse is unique and is in fact the group inverse a^{-1}.

IV. The cancellation laws.

$$x \circ y = x \circ z \Rightarrow y = z.$$
(left cancellation law)

$$y \circ x = z \circ x \Rightarrow y = z.$$
(right cancellation law)

Proof.

$$x \circ y = x \circ z$$
$$\Rightarrow x^{-1} \circ (x \circ y) = x^{-1} \circ (x \circ z)$$
$$\Rightarrow (x^{-1} \circ x) \circ y = (x^{-1} \circ x) \circ z$$
$$\Rightarrow \qquad e \circ y = e \circ z$$
$$\Rightarrow \qquad\qquad y = z.$$

Proof.

$$y \circ x = z \circ x$$
$$\Rightarrow (y \circ x) \circ x^{-1} = (z \circ x) \circ x^{-1}$$
$$\Rightarrow y \circ (x \circ x^{-1}) = z \circ (x \circ x^{-1})$$
$$\Rightarrow \qquad y \circ e = z \circ e$$
$$\Rightarrow \qquad\qquad y = z.$$

V. $(a^{-1})^{-1} = a$.

Proof. $a \circ a^{-1} = e = a^{-1} \circ a$.

These two equations imply that a is an inverse for a^{-1}. But an inverse is unique; hence $(a^{-1})^{-1} = a$.

In additive notation, property **V** becomes: $-(-a) = a$.

VI. $(a \circ b)^{-1} = b^{-1} \circ a^{-1}$.

Proof.

$$(a \circ b) \circ (b^{-1} \circ a^{-1}) = a \circ (b \circ b^{-1}) \circ a^{-1}$$
$$= (a \circ e) \circ a^{-1} = a \circ a^{-1} = e.$$

Also

$$(b^{-1} \circ a^{-1}) \circ (a \circ b) = b^{-1} \circ (a^{-1} \circ a) \circ b$$
$$= b^{-1} \circ (e \circ b) = b^{-1} \circ b = e.$$

Hence $b^{-1} \circ a^{-1}$ is an inverse for $a \circ b$. But an inverse is unique; hence $(a \circ b)^{-1} = b^{-1} \circ a^{-1}$.

In additive notation, property **VI** becomes: $-(a+b) = (-b)+(-a)$.

Extension of VI: $(a_1 \circ a_2 \circ \ldots \circ a_n)^{-1} = a_n^{-1} \circ a_{n-1}^{-1} \circ \ldots \circ a_1^{-1}$.

This can be proved $\forall n \geqslant 2$ by induction on n.

In particular, if n is a positive integer,

$$(a^n)^{-1} = a^{-1} \circ a^{-1} \circ \ldots \circ a^{-1} \ (n \text{ such terms})$$
$$= (a^{-1})^n.$$

We denote this element by a^{-n}.

Example 1. A group has precisely three elements e, a, b, where e is the identity element. If o denotes the operation of the group, show that $a o b = e$ and deduce that the group is the cyclic group of order 3.

Since $a o b \in$ the group, $a o b = e$, a or b.

Now $a o b = a \Rightarrow a o b = a o e$

                     $\Rightarrow b = e$ (by left cancellation law).

But $b \neq e$, and so $a o b \neq a$.

Similarly, $a o b = b \Rightarrow a o b = e o b$

                       $\Rightarrow a = e$ (by right cancellation law).

But $a \neq e$, and so $a o b \neq b$.

It follows that $a o b = e$.   (4.3)

To show that the group is cyclic we want to show that it is $\{e, a, a^2\}$, where $a^3 = e$; in particular we want to show that $a^2 = b$.

Now $a^2 \in$ the group, and so $a^2 = e$, a or b.

     $a^2 = e \Rightarrow a o a = a o b$ (by (4.3))

                $\Rightarrow a = b$ (by left cancellation law).

But $a \neq b$, and so $a^2 \neq e$.

Also $a^2 = a \Rightarrow a o a = a o e$

                $\Rightarrow a = e$ (by left cancellation law).

It follows that $a^2 = b$ and that $a^3 = a o b = e$, and so that the group is cyclic (and so abelian) of order 3.

Example 2. A set S with an operation o satisfies the axioms G_1, G_2, G_3 for a group. If it also satisfies

    G_4': For each $a \in S$, $\exists x \in S$ such that

$$a o x = e,$$

i.e. if *each* $a \in S$ has a *right* inverse in S, show that S forms a group under o.

Proof. It is enough to show that the x in G_4' satisfies also

$$x o a = e,$$

i.e., that x is also a *left* inverse of a. For, then we would have

$$a o x = x o a = e,$$

and so have axiom G_4 also satisfied.

Now
$$a \circ x = e$$
$$\Rightarrow x \circ (a \circ x) = x \circ e$$
$$\Rightarrow (x \circ a) \circ x = x.$$

But, by $\mathbf{G}'_4$, $\exists c \in S$ such that $x \circ c = e$.

Thus
$$a \circ x = e \Rightarrow (x \circ a) \circ (x \circ c) = x \circ c$$
$$\Rightarrow (x \circ a) \circ e = e$$
$$\Rightarrow x \circ a = e.$$

Exercise. State and prove the similar result with right and left inverses interchanged.

Note on the group table for a group $\{G, \circ\}$ of order n.

If $G = \{a_1, a_2, \ldots, a_n\}$, we draw up the group table in the usual way:

In the square marked x we put $a_i \circ a_j$, the element corresponding by the operation $\circ$ to the pair of elements a_i, a_j in G. Then:

(i) the closure property holds if and only if the rows (and columns) involve only $a_1, a_2, \ldots, a_n$;

(ii) an identity exists if and only if some row and the corresponding column are identical to the row heading $a_1, a_2, \ldots, a_n$ and the column heading, respectively;

(iii) since, in a group, $a_i \circ a_j = a_i \circ a_k \Leftrightarrow a_j = a_k$, it follows that, in the row of a group table determined by the element a_i, the n entries $a_i \circ a_1, a_i \circ a_2, \ldots, a_i \circ a_n$ are distinct and so form a permutation of the elements $a_1, a_2, \ldots, a_n$ of the group. It follows that each row and similarly each column in a group table contains a permutation of the elements of the group.

The conditions involved in (i), (ii) and (iii) hold in any group table, but, conversely, a table satisfying the conditions in (i), (ii) and (iii) is not necessarily the table of a group. Consider the following table:

·	e	a	b	c	d
e	e	a	b	c	d
a	a	e	d	b	c
b	b	c	e	d	a
c	c	d	a	e	b
d	d	b	c	a	e

Verify that the conditions in (i), (ii) and (iii) are satisfied. Show that the table is not a group table by showing that the associative law does not hold, and, in particular, that $a(bc) \neq (ab)c$.

The associative law is not easily checked from a table, and is best verified by other methods.

A finite group is abelian if and only if its group table is symmetrical about its main diagonal.

5. Some groups in plane geometry

We shall assume that the reader has some knowledge of plane geometrical transformations, and, in particular, of **isometries**. For completeness we include a description of some of the most important facts.

An **isometry** T of the plane $\mathbf{R}^2$ (we can regard the plane as a representation of $\mathbf{R}^2$ by using a coordinate system) is a bijection of $\mathbf{R}^2$ onto itself which preserves distance; this means that, if $A, B \in \mathbf{R}^2$ and if T maps A to A' and B to B', then

$$|A'B'| = |AB|,$$

where $|AB|$ means the distance from A to B in terms of the chosen unit of length.

It is easily shown that an isometry is a linear mapping in the sense that it maps collinear points onto collinear points; also it preserves angles in magnitude and maps a figure onto a congruent figure. Further, an isometry is uniquely determined by three pairs of corresponding points (A, A'), (B, B'), (C, C'), where A, B, C are non-collinear and the triangles ABC and $A'B'C'$ are congruent to each other.

A point P is called a fixed point of a transformation T if $T(P) = P$, i.e. if T maps P onto itself.

An isometry is called **direct** (or **even**) if it preserves the direction round every triangle, and called **opposite** (or **odd**) if it maps each triangle onto a congruent triangle described in the opposite direction.

It can be shown that every *direct* isometry is a composition of two reflections and is either

(i) the identity transformation, for which every point is a fixed point,

or (ii) a rotation, which has one fixed point,

or (iii) a translation, which has no fixed point.

Also every *opposite* isometry is a composition of one or three reflections and is either

(iv) a reflection, which has a line of fixed points,

or (v) a glide reflection, which has no fixed point.

The effect of a glide reflection T is the composition of a reflection and a translation as shown in the figure below, where P_1 is the mirror image of P in the line m and the distance $|P, T(P)|$ is constant.

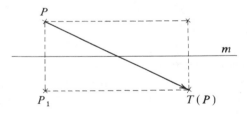

The composition of two isometries obeys the following pattern:

$$\text{direct} \times \text{direct} = \text{direct},$$
$$\text{direct} \times \text{opposite} = \text{opposite},$$
$$\text{opposite} \times \text{direct} = \text{opposite},$$
$$\text{opposite} \times \text{opposite} = \text{direct}.$$

We first establish the following result.

Theorem 5.1. *The set $\mathscr{I}$ of all isometries of the plane forms a group under composition of mappings.*

Proof. G_1. Let T, $U \in \mathscr{I}$ and let P, Q be any two points in $\mathbf{R}^2$. Now, since U is an isometry of $\mathbf{R}^2$,

$$|U(P), U(Q)| = |PQ|$$

Also, since T is an isometry of $\mathbf{R}^2$,

$$|T(U(P)), T(U(Q))| = |U(P), U(Q)| = |PQ|.$$

Hence the mapping $T \circ U$ preserves distance in $\mathbf{R}^2$. Since $T \circ U$ is also a bijection of $\mathbf{R}^2$ onto itself, it is an isometry of $\mathbf{R}^2$.

Hence $\mathscr{I}$ is closed under composition of mappings.

$\mathbf{G_2}$. The associative property holds in $\mathscr{I}$ since it holds for all mappings of $\mathbf{R}^2$ onto $\mathbf{R}^2$.

$\mathbf{G_3}$. The identity mapping $I \in \mathscr{I}$ (since it is a bijection and preserves distance) and acts as an identity for $\mathscr{I}$.

$\mathbf{G_4}$. If $T \in \mathscr{I}$, then the inverse mapping T^{-1} is a bijection of $\mathbf{R}^2$ onto itself. Also, since

$$|T^{-1}(P), T^{-1}(Q)| = |T(T^{-1}(P)), T(T^{-1}(Q))| = |PQ|,$$

T^{-1} preserves distance and so is an isometry of $\mathbf{R}^2$, i.e. $T^{-1} \in \mathscr{I}$ and acts in $\mathscr{I}$ as an inverse for T.

Hence $\mathscr{I}$ forms a group under composition of mappings.

The group $\mathscr{I}$ has many important subgroups such as

(i) the set of all *direct* isometries of $\mathbf{R}^2$,
(ii) the set of all isometries of $\mathbf{R}^2$ which leave a given point O fixed,
(iii) the set of all rotations about a given point O,
(iv) the set of all translations of the plane (including the identity transformation),
(v) the set of all isometries which leave a given point O fixed and which map a given geometrical figure onto itself; this group is called the **symmetry group of the figure about** O.

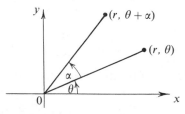

Proof of (iii). We use polar coordinates with respect to fixed axes Ox, Oy.

The rotation R_α about O through an angle of measure α radians is defined by:

$$R_\alpha : (r, \theta) \to (r, \theta + \alpha).$$

$\mathbf{G_1}$. Take rotations $R_{\alpha_1} : (r, \theta) \to (r, \theta + \alpha_1)$ and
$$R_{\alpha_2} : (r, \theta) \to (r, \theta + \alpha_2).$$

Then $R_{\alpha_1} \circ R_{\alpha_2} = R_{\alpha_2} \circ R_{\alpha_1}$ is the mapping which maps (r, θ) to $(r, \theta + \alpha_1 + \alpha_2)$, and so is the rotation about O through an angle of measure $\alpha_1 + \alpha_2$; it follows that the set of rotations about O is closed under composition of mappings.

$$[\text{Note that } R_{\alpha_1} \circ R_{\alpha_2} = R_{\alpha_1 + \alpha_2}.]$$

G$_2$. The associative property holds since it holds for all mappings of $\mathbf{R}^2$ onto itself.

G$_3$. The identity mapping $I:(r, \theta) \to (r, \theta)$ clearly is an identity for the set of rotations about O.

G$_4$. The inverse mapping of $R_\alpha:(r, \theta) \to (r, \theta + \alpha)$ is the mapping $R_\alpha^{-1}:(r, \theta) \to (r, \theta - \alpha)$, i.e. is the rotation $R_{-\alpha}$.

It follows that the set of rotations about O forms a group under composition of mappings and, as indicated in the proof of **G$_1$**, that this group is abelian. It is clearly a subgroup of $\mathscr{I}$.

Note 1. If we map $R_{\alpha_1} \to \alpha_1$, then, since $R_{\alpha_1} \circ R_{\alpha_2} = R_{\alpha_1 + \alpha_2}$, it follows that this group of rotations about O is isomorphic to the additive group of real numbers.

Note 2. We can represent the group of rotations about O as a group of 2×2 real matrices as follows:

In cartesian coordinates, if R_α maps point (x, y) to point (x', y'), and if (x, y) has polar coordinates (r, θ), then

$$x' = r \cos(\theta + \alpha) = r(\cos \theta \cos \alpha - \sin \theta \sin \alpha) = x \cos \alpha - y \sin \alpha,$$

and $y' = r \sin(\theta + \alpha) = r(\sin \theta \cos \alpha + \cos \theta \sin \alpha) = x \sin \alpha + y \cos \alpha.$

Thus,
$$\begin{bmatrix} x' \\ y' \end{bmatrix} = \begin{bmatrix} \cos \alpha & -\sin \alpha \\ \sin \alpha & \cos \alpha \end{bmatrix} \begin{bmatrix} x \\ y \end{bmatrix}.$$

It follows that the set of rotations about O can be represented by the set of matrices

$$S = \left\{ \begin{bmatrix} \cos \alpha & -\sin \alpha \\ \sin \alpha & \cos \alpha \end{bmatrix} : \alpha \in \mathbf{R} \right\}.$$

Exercise. Show that S forms a group under matrix multiplication.

Exercise. Prove (i) and (ii).

Exercise. Prove (iv) using vectors. Note that a translation T can be represented as a shift through a vector $\mathbf{\alpha}$, i.e. by a mapping:

$$\mathbf{r} \to \mathbf{r} + \boldsymbol{\alpha},$$

where $\mathbf{r}$ is the position vector of a point P with respect to an origin O.

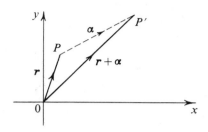

Proof of (v); the group of symmetries about O of a given figure.

Let S denote the set of all isometries with O as fixed point which map a given geometrical configuration V onto itself.

$\mathbf{G_1}$. The composition of two such isometries is itself in S, since it is an isometry, keeps O fixed and maps V onto itself. Thus S is closed under composition of mappings.

$\mathbf{G_2}$. The associative property holds in S since it holds for all isometries.

$\mathbf{G_3}$. Clearly the identity mapping is in S and acts as an identity for S.

$\mathbf{G_4}$. The inverse of an element T of S is the usual inverse isometry T^{-1}.

[For, $T^{-1}(O) = T^{-1}(T(O)) = O$, so that T^{-1} keeps O fixed; also $\quad T^{-1}(V) = T^{-1}(T(V)) = V$, so that T^{-1} maps V onto itself.]

Hence S forms a group under composition of mappings, called the **group of symmetries** (or the **symmetry group**) of V about O.

This group reflects the symmetry of V with respect to the point O. Since each isometry involved keeps O fixed, it is either

    (i) the identity transformation I,

or (ii) a rotation about O,

or (iii) a reflection about a line through O.

When V has no special symmetry about O the group reduces to a group of order 1, consisting of the identity transformation I alone.

We shall always take O to be the centroid of the geometrical configuration V or a point about which V has "most symmetry".

Examples. Some figures and their symmetry groups.

(1) **Letter L.** Clearly the group is $\{I\}$ for any choice of O.

(2) **Letter E.** The symmetry group is $\{I, R\}$, where $R^2 = I$, R being the reflection in the line of symmetry. This is a representation of the cyclic group C_2 of order 2.

Note that the exact position of O on the line of symmetry is not significant.

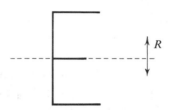

(3) **Letter N.** Here the group is $\{I, H\}$, where $H^2 = I$, H being the half-turn about O, i.e. a rotation through an angle of measure π radians. The group is a representation of C_2.

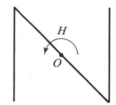

(4) **Swastika.** Here the group is $\{I, A, A^2, A^3\}$, where $A^4 = I$, A being the rotation about O through an angle of measure $\frac{1}{2}\pi$ radians.

The group is a representation of C_4, the cyclic group of order 4.

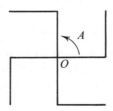

(5) **Rectangle.** The rectangle has four symmetries:

(i) I, corresponding to the permutation mapping

$$I = \begin{pmatrix} 1 & 2 & 3 & 4 \\ 1 & 2 & 3 & 4 \end{pmatrix};$$

(ii) R_1, corresponding to the permutation mapping

$$R_1 = \begin{pmatrix} 1 & 2 & 3 & 4 \\ 2 & 1 & 4 & 3 \end{pmatrix};$$

(iii) R_2, corresponding to the permutation mapping

$$R_2 = \begin{pmatrix} 1 & 2 & 3 & 4 \\ 4 & 3 & 2 & 1 \end{pmatrix};$$

(iv) H, corresponding to the permutation mapping

$$H = \begin{pmatrix} 1 & 2 & 3 & 4 \\ 3 & 4 & 1 & 2 \end{pmatrix}.$$

Here, R_1 is reflection in the axis of symmetry $X'X$, so that

$$R_1^2 = I;$$

R_2 is reflection in the axis of symmetry $Y'Y$, so that

$$R_2^2 = I;$$

H is the half-turn about O, so that $H^2 = I$.

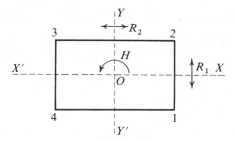

It is convenient to use the same notation for a geometrical transformation and the corresponding permutation mapping which arises from the way in which the set of four vertices $\{1, 2, 3, 4\}$ is mapped onto itself. For example, under reflection R_1, the vertices 1 and 2 are interchanged and 3 and 4 are interchanged, so that the corresponding permutation mapping is $\begin{pmatrix} 1 & 2 & 3 & 4 \\ 2 & 1 & 4 & 3 \end{pmatrix}.$

Now, using the product notation,

$$R_1R_2(1\ 2\ 3\ 4) = R_1(4\ 3\ 2\ 1) = (3\ 4\ 1\ 2) = H(1\ 2\ 3\ 4)$$

Hence $R_1R_2 = H$;   similarly, $R_2R_1 = H$.

It follows that the symmetry group of a rectangle is $\{I, R_1, R_2, R_1R_2\}$, with $R_1^2 = I$ and $R_2^2 = I$, and so it is a group of order 4; it is abelian since

$$R_1R_2 = R_2R_1.$$

It is a representation of the Klein **four-group**, which can be expressed in the form $\{e, a, b, a \circ b\}$, where e is the identity element and $a^2 = e$ and $b^2 = e$. This group is abelian.

Note 1. The four-group is not isomorphic to C_4, the cyclic group of order 4; for, $C_4 = \{e, a, a^2, a^3\}$ and $a^2 \neq e$, whereas in the four-group every element has square e.

Note 2. The group table for the symmetry group of the rectangle is:

	I	R_1	R_2	R_1R_2
I	I	R_1	R_2	R_1R_2
R_1	R_1	I	R_1R_2	R_2
R_2	R_2	R_1R_2	I	R_1
R_1R_2	R_1R_2	R_2	R_1	I

Check that this is correct. [Note that $R_2R_1R_2 = R_2R_2R_1 = IR_1 = R_1$.]

Exercise. Show that the set $\{1, 3, 5, 7\}$ forms a group under multiplication (mod 8) and that this group is a representation of the four-group.

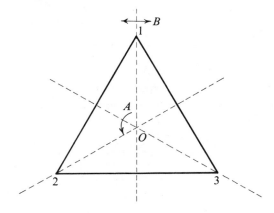

(6) **Equilateral triangle.** Three isometries which map the equilateral triangle onto itself are the rotations through angles of measures 0, $2\pi/3$ and $4\pi/3$ about the centroid O. These three transformations can be denoted by I, A, A^2, where A represents a rotation through $2\pi/3$ and $A^3 = I$, the identity transformation.

Under A, the vertices $1, 2, 3$ are mapped to $2, 3, 1$, and under A^2, the vertices $1, 2, 3$ are mapped to $3, 1, 2$, respectively. Thus, the permutation mappings corresponding to I, A, A^2 are

$$I = \begin{pmatrix} 1 & 2 & 3 \\ 1 & 2 & 3 \end{pmatrix}, \quad A = \begin{pmatrix} 1 & 2 & 3 \\ 2 & 3 & 1 \end{pmatrix}, \quad A^2 = \begin{pmatrix} 1 & 2 & 3 \\ 3 & 1 & 2 \end{pmatrix},$$

in which the same letter is used for the corresponding mapping.

Three further isometries which map the triangle onto itself are the reflections in the three medians of the triangle. If we denote the reflection in the median through vertex 1 by B, then $B^2 = I$, and B maps the vertices $1, 2, 3$ to $1, 3, 2$, respectively. Thus B corresponds to the permutation mapping $B = \begin{pmatrix} 1 & 2 & 3 \\ 1 & 3 & 2 \end{pmatrix}$.

Similarly, the reflection B_2 in the median through vertex 2 corresponds to the permutation mapping $B_2 = \begin{pmatrix} 1 & 2 & 3 \\ 3 & 2 & 1 \end{pmatrix}$, and the reflection B_3 in the median through vertex 3 corresponds to the permutation mapping

$$B_3 = \begin{pmatrix} 1 & 2 & 3 \\ 2 & 1 & 3 \end{pmatrix}.$$

It follows that the equilateral triangle has the set of six symmetries $\{I, A, A^2, B, B_2, B_3\}$, each corresponding to a unique permutation of the set of vertices $\{1, 2, 3\}$. But there are exactly $3! = 6$ such permutations. Hence the equilateral triangle has the symmetry group $\{I, A, A^2, B, B_2, B_3\}$ under composition of mappings.

Now $\quad AB(123) = A(132) \; = 213 = B_3(123) \quad$, so $\quad AB = B_3$.

Also $\quad A^2B(123) = A^2(132) = 321 = B_2(123) \quad$, so $\quad A^2B = B_2$.

Hence the equilateral triangle has symmetry group

$$G = \{I, A, A^2, B, AB, A^2B\}, \quad \text{where} \quad A^3 = I, \; B^2 = I.$$

This is clearly a representation of the non-abelian group of order 6 described under IV in Section **3**.

EXERCISE 9

1. Show that the following are abelian groups:
    (i) the set $\{2^k : k \in \mathbf{Z}\}$ under multiplication;
    (ii) the set $\{a + b\sqrt{3} : a$ and b rational numbers$\}$ under addition;
    (iii) the set $\{a + b\sqrt{2} : a$ and b rational numbers not both zero$\}$ under multiplication;
    (iv) the set of all complex numbers of modulus 1 under multiplication.

2. Draw up the following multiplication tables:
    (i) $\{1, 3, 5, 7\}$ under multiplication (mod 8);
    (ii) $\{1, 3, 7, 9\}$ under multiplication (mod 10);
    (iii) $\{1, 5, 7, 11\}$ under multiplication (mod 12);
    (iv) $\{2, 4, 6, 8\}$ under multiplication (mod 10).
    Verify that each is an abelian group, and for each determine whether it is isomorphic to the cyclic group C_4 or the four-group.

3. Show that the set $\{1, \omega, \omega^2, \omega^3, \omega^4\}$, where $\omega = \cos \frac{2}{5}\pi + i \sin \frac{2}{5}\pi$, forms a cyclic group of order 5 under multiplication.

4. A complex number z is called a **root of unity** if there is a positive integer n such that $z^n = 1$. Show that, if $z_1^{n_1} = 1$ and $z_2^{n_2} = 1$, then $(z_1 z_2)^{n_1 n_2} = 1$. Hence show that the set of all roots of unity forms a group under multiplication.

5. Show that the following sets of matrices form groups under the matrix operations given, and determine in each case whether the group is abelian or non-abelian.

    (i) The set $\left\{ \begin{bmatrix} a-b & b \\ -b & a \end{bmatrix} : a$ and b real numbers$\right\}$ under addition.

    (ii) The set $\left\{ \begin{bmatrix} a+2b & 3b \\ -b & a \end{bmatrix} : a$ and b real numbers, not both zero$\right\}$ under multiplication.

    (iii) The set $\left\{ \begin{bmatrix} a & b \\ 3b & a \end{bmatrix} : a$ and b rational numbers, not both zero$\right\}$ under multiplication.

    (iv) The set $\left\{ \begin{bmatrix} \alpha & \beta \\ -\bar{\beta} & \bar{\alpha} \end{bmatrix} : \alpha, \beta$ complex numbers, not both zero, with complex conjugates $\bar{\alpha}$ and $\bar{\beta} \right\}$ under multiplication.

    (v) The set $\left\{ \begin{bmatrix} 1 & 1 \\ 0 & 0 \end{bmatrix}, \begin{bmatrix} \omega & \omega \\ 0 & 0 \end{bmatrix}, \begin{bmatrix} \omega^2 & \omega^2 \\ 0 & 0 \end{bmatrix}, \begin{bmatrix} \omega^3 & \omega^3 \\ 0 & 0 \end{bmatrix}, \begin{bmatrix} \omega^4 & \omega^4 \\ 0 & 0 \end{bmatrix} \right\}$
    under multiplication, where $\omega = \cos \frac{2}{5}\pi + i \sin \frac{2}{5}\pi$.

    (vi) The set $\left\{ \begin{bmatrix} 1 & 0 \\ a & b \end{bmatrix} : a$ and b real numbers with $b > 0 \right\}$ under multiplication.

6. Show that the set $S = \left\{ \begin{bmatrix} x & y \\ -y & x \end{bmatrix} : x, y$ real numbers, not both zero$\right\}$ forms a group under matrix multiplication.
    Determine which of the following subsets of S are subgroups: (i) the elements of S with $x = 0$, (ii) the elements of S with $y = 0$, and (iii) the elements of S with $x^2 + y^2 = 1$.

7. Show that the set

$$G = \left\{ \begin{bmatrix} 1-n & n \\ -n & 1+n \end{bmatrix} : n \in \mathbf{Z} \right\}$$

forms an abelian group under matrix multiplication.
   Determine whether or not the following subsets of G are subgroups: (i) the elements of G for which $n \geqslant 0$, (ii) the elements of G for which n is even.

8. Show that the set $S = \left\{ \begin{bmatrix} a & b \\ 2b & a \end{bmatrix} : a, b \text{ rational numbers, not both zero} \right\}$ forms a group under matrix multiplication.
   Determine which of the following subsets of S are subgroups:
   (i) the elements of S with $a = 0, b \neq 0$;
   (ii) the elements of S with $a \neq 0, b = 0$;
   (iii) the elements of S with $a^2 - 2b^2 = 1$.

9. A square $n \times n$ matrix A is given. Show that the set of all non-singular $n \times n$ matrices X such that $AX = XA$ forms a group under matrix multiplication.

Prove that, if $\quad A = \begin{bmatrix} 0 & 0 & 0 \\ 1 & 0 & 0 \\ 0 & 1 & 0 \end{bmatrix}$,

then each matrix X is of the form $\begin{bmatrix} x & 0 & 0 \\ y & x & 0 \\ z & y & x \end{bmatrix}$, and deduce that, in this case, the group is abelian.

10. Show that the set $G = \{(a, b) : a, b \text{ complex numbers}, a \neq 0\}$ forms an abelian group under the operation $\circ$ defined by

$$(a, b) \circ (c, d) = (ac, ad + bc).$$

11. Show that the set $G = \mathbf{R} - \{1\}$ forms a group under the operation $\circ$ defined by

$$x \circ y = x + y - xy,$$

for x, y in G, where $+$ and $-$ are the usual operations for $\mathbf{R}$.

12. If, in a group, the element a is such that $a^2 = a$, show that $a = e$, the identity element.

13. If, in a group (written multiplicatively), the elements a and b are such that $a^2 = e$, $b^2 = e$ and $(ab)^2 = e$, where e is the identity element, show that $ab = ba$. Hence, or otherwise, show that, if every element a of a group G satisfies $a^2 = e$, then G is abelian.

14. Prove that, if $(ab)^2 = a^2b^2$ for all a and b in a group G (written multiplicatively), then G is abelian.

15. A group G (written multiplicatively) has precisely four elements $\{e, a, b, ab\}$, with e as identity element. Prove that either (i) $a^2 = e$, or (ii) $a^2 = b$, and show that in both cases G is abelian. Show also that, in case (i) when $b^2 \neq a$, G is the four-group, and, in case (ii), G is C_4, the cyclic group of order 4.

16. Show that, for a group $\{G, \circ\}$, $a^m \circ a^n = a^n \circ a^m = a^{m+n}$, and $(a^m)^n = a^{mn}$ for all integers m and n, positive, negative or zero. [**Note**: $a^0 = e$, the identity element, and, if n is a positive integer, $a^{-n} = (a^{-1})^n$.]

17. G is a group and a is a fixed element of G. Show that the set
$$G_a = \{x : x \in G, x = a^n \text{ for some integer } n\}$$
is a subgroup of G.

18. If G is a group and if C is the subset of G defined by
$$C = \{x \in G : x \text{ commutes with every element of } G, \text{ i.e. } \forall a \in G, ax = xa\},$$
show that C is a subgroup of G. [C is called the **centre** of G.]

19. G_1 and G_2 are groups written multiplicatively and $G = G_1 \times G_2$. Show that G forms a group under the operation $\circ$ defined by

$$(x_1, y_1) \circ (x_2, y_2) = (x_1 x_2, y_1 y_2),$$

for $(x_i, y_i) \in G_1 \times G_2 \quad (i = 1, 2)$.

20. Describe the symmetry group of each of the following figures:

    (i) an isosceles triangle,   (ii) a rhombus,

  (iii) a kite,   (iv) a parallelogram,

   (v) a scalene (i.e. non-isosceles) triangle,   (vi) a square,

 (vii) a parabola,   (viii) an ellipse,

  (ix) a circle,   (x) the curve $y = x^3$.

    In the case of (i), (ii), (iii), (iv) and (vi), describe the corresponding group of permutation mappings of the set of vertices.

Algebraic Structures 2
rings
integral domains
fields

1. Definition of a ring, examples of rings

In Chapter **9** we considered algebraic structures with a single binary operation and indicated how an abstract mathematical model such as C_4, the cyclic group of order 4, can be represented in many different ways. Many of the sets which arose in Chapter **9** possess two important binary operations and, in fact, a great deal of mathematics is concerned with such sets. For example, all the basic number systems have an addition and a multiplication; also the set of all 2×2 real matrices has an addition and a multiplication, and so has the set of all vectors in $\mathbf{R}^3$. It is useful and important to have several abstract models which cover the types of structure with two binary operations that most frequently appear. We shall define three such types of structure, namely, rings, integral domains and fields.

Definition of a ring. A **ring** is a structure $\{R, -, .\}$ consisting of a non-empty set R on which are defined two binary operations $+$ and $.$, called **addition** and **multiplication**, satisfying the following conditions:

  (i) R forms an abelian group under addition,
  (ii) R is closed under multiplication,
  (iii) multiplication is associative,
  (iv) the left and right distributive laws hold in R, i.e.

$$\forall a, b, c \in R, \quad \text{we have} \quad \begin{cases} a(b+c) = ab+ac, \\ \text{and } (a+b)c = ac+bc. \end{cases}$$

Note. For simplicity of notation we have omitted the multiplication symbol $.$ when writing products.

If we write out in detail the implications of (i), (ii), (iii) and (iv) we obtain the following list of conditions for a ring $\{R, +, .\}$. We shall use the letter

A for those associated with addition, **M** for properties of multiplication and **D** for distributive properties.

Addition properties	*Multiplication properties*
A_1. R is closed under $+$ (Cl.)	M_1. R is closed under .

A_2. $\forall a, b, c \in R$,
$$a+(b+c) = (a+b)+c$$
    (Ass.) M_2. $\forall a, b, c \in R, a(bc) = (ab)c$

A_3. $\exists 0 \in R$ such that, $\forall a \in R$,
$$a+0 = 0+a = a$$
    (Id.)

A_4. For each $a \in R, \exists -a \in R$
such that   (In.)
$$a+(-a) = (-a)+a = 0$$

A_5. $\forall a, b \in R, a+b = b+a$ (Comm.)

Distributive properties

D_1. $\forall a, b, c \in R, a(b+c) = ab+ac$,
D_2. $\forall a, b, c \in R, (a+b)c = ac+bc$.

Notes. 1. Since $\{R, +\}$ is a group, the elements 0 and $-a$ are unique. Also we write, as usual, $b-c$ for $b+(-c)$; in this notation we can check that $a(b-c) = ab-ac$, and $(a-b)c = ac-bc$.

2. If $\{R, +, .\}$ is a ring, we usually say that the set R forms a ring under $+$ and ., and often simply that R is a ring when the operations $+$ and . are clearly understood.

3. If, in addition to the above axioms, the ring $\{R, +, .\}$ satisfies also the following axiom:

$$M_3. \quad \exists e \in R \text{ such that, } \forall a \in R, ae = ea = a, \tag{1.1}$$

then we say that the ring has a **unity element** (or **identity element**) e. This element, if it exists, is often denoted by 1 and called a **one**. If it exists, then it is unique; for, if e_1 is also a unity element for the ring, then,

$$\forall a \in R, ae_1 = e_1a = a. \tag{1.2}$$

$a = e_1$ in (1.1) gives $e_1 = e_1e$, and $a = e$ in (1.2) gives $e = e_1e$; thus $e_1 = e$.

4. If, in addition to the axioms listed for a ring $\{R, +, .\}$, the ring also satisfies the following axiom:

$$M_5. \quad \forall a, b \in R, ab = ba,$$

then the ring is called **commutative**.

For a commutative ring, $(a+b)c = c(a+b)$, and, for such rings, the second distributive law can be deduced from the first.

Consequently for a commutative ring we need list only one of the distributive laws and we normally list $\mathbf{D}_1$.

When it is known that a ring does not satisfy $\mathbf{M}_5$, we say that the ring is **non-commutative**.

5. A ring $\{R, +, .\}$ which satisfies both $\mathbf{M}_3$ and $\mathbf{M}_5$ is called a **commutative ring with a unity element**.

6. If S is a subset of R where $\{R, +, .\}$ is a ring, and if $\{S, +, .\}$ is also a ring, then S is called a **subring** of the ring R.

7. For completeness we include a mention of **isomorphism of rings**.

If $\{R, +, .\}$ and $\{R', \oplus, \odot\}$ are rings, then they are said to be **isomorphic to each other** if $\exists$ a bijection f from R to R' such that, if f maps a to a' and f maps b to b', then f maps $a+b$ to $a'\oplus b'$ and ab to $a'\odot b'$, so that

$$f(a+b) = f(a)\oplus f(b) \quad \text{and} \quad f(a.b) = f(a)\odot f(b).$$

Some examples of rings.

(1) If $\mathbf{Z}$ is the set of all integers, then $\{\mathbf{Z}, +, .\}$, where $+$ and $.$ are the usual addition and multiplication for $\mathbf{Z}$, is clearly a ring; it is a commutative ring with a unity element, namely the integer 1.

(2) If E is the set of all *even* integers, then $\{E, +, .\}$, where $+$ and $.$ are the usual addition and multiplication, is a ring; it is a commutative ring, but does *not* have a unity element.

Check first that $\{E, +\}$ is an abelian group. Then check that the axioms $\mathbf{M}_1$, $\mathbf{M}_2$, $\mathbf{M}_5$ and $\mathbf{D}_1$ are satisfied.

Note that this ring is a subring of the ring given in (1).

Exercise. Explain why the set S of all *odd* integers does *not* form a ring under the usual addition and multiplication.

(3) If $S = \{0, 1, 2, 3\}$, then $\{S, +, .\}$, where $+$ and $.$ are addition and multiplication (mod 4), is a ring; it is a commutative ring with a unity element, namely the integer 1.

Check first that $\{S, + \pmod 4\}$ is an abelian group. Then check that the axioms $\mathbf{M}_1$, $\mathbf{M}_2$, $\mathbf{M}_3$, $\mathbf{M}_5$ and $\mathbf{D}_1$ are satisfied with 1 for e in the case of $\mathbf{M}_3$.

(4) If M is the set of all 2×2 real matrices, then $\{M, +, .\}$, where $+$ and $.$ are matrix addition and multiplication, is a ring; it has a unity element, namely the identity matrix I, but is non-commutative.

Check these facts; the details have all been covered by earlier work on matrices and groups.

(5) If $S = \left\{ \begin{bmatrix} x & y \\ 0 & z \end{bmatrix} : x, y \text{ and } z \text{ even integers} \right\}$, then $\{S, +, .\}$,

where $+$ and $.$ are matrix addition and multiplication, is a ring; it is non-commutative and does *not* have a unity element.

We check first that $\{S, +\}$ is an abelian group.

$\mathbf{A}_1$. If $A = \begin{bmatrix} x_1 & y_1 \\ 0 & z_1 \end{bmatrix}$ and $B = \begin{bmatrix} x_2 & y_2 \\ 0 & z_2 \end{bmatrix}$ are elements of S,

then $A + B = \begin{bmatrix} x_1 + x_2 & y_1 + y_2 \\ 0 & z_1 + z_2 \end{bmatrix}$; since $x_1 + x_2$, $y_1 + y_2$ and $z_1 + z_2$

are even integers, it follows that $A + B \in S$ and so that S is closed under matrix addition.

$\mathbf{A}_2$. Addition in S is associative since addition is associative for all 2×2 matrices.

$\mathbf{A}_3$. The matrix $O = \begin{bmatrix} 0 & 0 \\ 0 & 0 \end{bmatrix}$ is in S and acts as additive identity for S.

$\mathbf{A}_4$. If $A = \begin{bmatrix} x & y \\ 0 & z \end{bmatrix}$ is in S, then the matrix $-A = \begin{bmatrix} -x & -y \\ 0 & -z \end{bmatrix}$ is also

in S, and acts as additive inverse for A in S.

$\mathbf{A}_5$. From the symmetry of $A + B$ in $\mathbf{A}_1$, it is clear that, if $A, B \in S$, then

$$A + B = B + A.$$

Hence $\{S, +\}$ is an abelian group.
We now check properties of multiplication.

$\mathbf{M}_1$. In the notation used for $\mathbf{A}_1$,

$$AB = \begin{bmatrix} x_1 & y_1 \\ 0 & z_1 \end{bmatrix} \begin{bmatrix} x_2 & y_2 \\ 0 & z_2 \end{bmatrix} = \begin{bmatrix} x_1 x_2 & x_1 y_2 + y_1 z_2 \\ 0 & z_1 z_2 \end{bmatrix},$$

and this is in S since the entries are even integers.
Hence S is closed under matrix multiplication.

$\mathbf{M}_2$. Multiplication in S is associative since multiplication is associative for all 2×2 matrices.

Finally, the distributive laws $\mathbf{D}_1$ and $\mathbf{D}_2$ both hold in S since they hold for all 2×2 matrices.
Hence $\{S, +, .\}$ is a ring.

The ring is non-commutative since, in the notation used for $\mathbf{A}_1$,

$$BA = \begin{bmatrix} x_1x_2 & x_2y_1+y_2z_1 \\ 0 & z_1z_2 \end{bmatrix},$$

and we can easily choose matrices A, B in S for which $x_1y_2+y_1z_2 \neq x_2y_1+y_2z_1$, and so $AB \neq BA$, e.g.

$$A = \begin{bmatrix} 2 & 0 \\ 0 & 0 \end{bmatrix}, \quad B = \begin{bmatrix} 2 & 2 \\ 0 & 2 \end{bmatrix}.$$

Also the ring has no unity element. This can be shown as follows. If $E = \begin{bmatrix} a & b \\ 0 & c \end{bmatrix}$ in S $(a, b, c$ even integers) is a unity element for S, then,

for each $A = \begin{bmatrix} x & y \\ 0 & z \end{bmatrix}$ in S, we have $AE = A$, and so

$$\begin{bmatrix} xa & xb+yc \\ 0 & zc \end{bmatrix} = \begin{bmatrix} x & y \\ 0 & z \end{bmatrix}.$$

Consequently, $xa = x$, $zc = z$ and $xb+yc = y$ for all even integers x, y, z. These conditions imply that $a = 1$, $c = 1$ and $b = 0$. Since a and c must be even, it follows that the ring has no unity element.

(6) Let X be a fixed set and let S be the power set of X, i.e. the set of all subsets of X; define $+$ on S as usual by

$$A+B = (A \cap B') \cup (B \cap A'), \quad \forall A, B \in S,$$

and write $A.B$ for $A \cap B$.

Then $\{S, +, .\}$ is a commutative ring with a unity element.

Check first that $\{S, +\}$ is a group [see example in Section 7 of Chapter 2]. Then note that $\mathbf{M}_1, \mathbf{M}_2, \mathbf{M}_5$ are easily established, and show that $\mathbf{D}_1$ holds [see the example in Chapter 2, just mentioned].

Finally, we note that,

$$\forall A \in S, \ S.A = A.S = A \cap S = A,$$

so that S is a unity element for the ring.

The study of rings of this type is part of **Boolean Algebra**.

Exercise. If V is the set of all vectors in $\mathbf{R}^3$, show that $\{V, +, \times\}$ is *not* a ring, where $+$ is addition of vectors and $\times$ is the usual vector product multiplication.

2. Some properties of rings

Let $\{R, +, .\}$ be any ring; we prove first the following result:

(1) $\forall a \in R$, $a0 = 0 = 0a$.

Proof. From $0+0 = 0$, we deduce that $a(0+0) = a0$. Thus

$$a0 + a0 = a0 = a0 + 0,$$

and so $a0 = 0$, by the left cancellation property of the additive group of the ring.

The other part of (1), namely $0a = 0$, is proved similarly by using

$$(0+0)a = 0a.$$

(2) $\forall a, b \in R$, we have

(i) $(-a)b = -(ab)$, (ii) $a(-b) = -(ab)$, (iii) $(-a)(-b) = ab$.

Proof of (i). We start with:

$$(a+(-a))b = 0b = ((-a)+a)b.$$

Thus, using property (1), $ab+(-a)b = 0 = (-a)b + ab$.

Hence $(-a)b$ is an additive inverse of ab. But ab has a unique additive inverse, namely $-(ab)$. Hence $(-a)b = -(ab)$.

Proof of (ii). We proceed similarly, starting with:

$$b(a+(-a)) = b0 = b((-a)+a).$$

Proof of (iii).

$$\begin{aligned}(-a)(-b) &= -(a(-b)), \quad \text{by (i)},\\ &= -(-(ab)), \quad \text{by (ii)},\\ &= ab,\end{aligned}$$

since $-(-x) = x$ for any x in the ring.

Example. Let c be a given element of a ring $\{R, +, .\}$, and let

$$N(c) = \{x \in R : cx = 0\}.$$

Show that $N(c)$ is a subring of R (it is a subset of R).
We first show that $\{N(c), +\}$ is an abelian group.

A_1. If $x, y \in N(c)$, then $cx = 0$ and $cy = 0$, and so

$$cx + cy = 0 + 0.$$

Thus $c(x+y) = 0$, showing that $x + y \in N(c)$.

Hence $N(c)$ is closed under addition.

$\mathbf{A_2}$. The associative property of addition holds in $N(c)$ since it holds in R.

$\mathbf{A_3}$. Since $c0 = 0$, it follows that $0 \in N(c)$ and is an additive identity for $N(c)$.

$\mathbf{A_4}$. If $cx = 0$, then $c(-x) = -(cx) = -0 = 0$; thus $-x \in N(c)$ and is an additive inverse for x in $N(c)$.

$\mathbf{A_5}$. The commutative property of addition holds in $N(c)$ since it holds in R.

Hence $N(c)$ is an abelian group under $+$.
We now consider properties of multiplication.

$\mathbf{M_1}$. If $x, y \in N(c)$, then $cx = 0$ and $cy = 0$; thus

$$c(xy) = (cx)y = 0y = 0, \text{ and so } xy \in N(c).$$

Hence $N(c)$ is closed under multiplication.
[Note that, if $x \in N(c)$, then $xr \in N(c) \ \forall r \in R$.]

$\mathbf{M_2}$. The associative property of multiplication holds in $N(c)$ since it holds in R.

Finally, the distributive laws $\mathbf{D_1}$ and $\mathbf{D_2}$ hold in $N(c)$ since they hold in R.

Hence $N(c)$ forms a ring under the addition and multiplication of R, and so is a subring of the ring R.

3. Two important types of commutative ring with a unity element

I. Integral domains. An **integral domain** is a *commutative ring R with a unity element* which has *at least two elements* and satisfies the condition:

$$\text{If } a, b \in R \quad \text{and if } ab = 0, \quad \text{then either } a = 0 \text{ or } b = 0. \quad (3.1)$$

An alternative way of stating condition (3.1) is:
If a and b are non-zero elements of R, then $ab \neq 0$.

Note. The restriction "at least two elements" ensures that the zero element and the unity element are distinct.

The ring consisting of set $\{0\}$, with $+$ and $.$ defined by $0+0 = 0$ and $0.0 = 0$, is called the **zero ring**.

An obvious example of an integral domain is the set $\mathbf{Z}$ of all integers under the usual addition and multiplication. In fact this explains the use of the term "integral domain".

Cancellation property of an integral domain R.

If $a, b, c \in R$ and if $a \neq 0$, then

$$ab = ac \Rightarrow b = c.$$

Proof.
$$ab = ac \Rightarrow ab - ac = 0$$
$$\Rightarrow a(b-c) = 0$$
$$\Rightarrow a = 0 \text{ or } b - c = 0$$
$$\Rightarrow b = c, \text{ since } a \neq 0.$$

Note. It is important to note that this cancellation property does *not* hold for a general ring.

Example 1. Show that the set $S = \{0, 1, 2, 3, 4, 5\}$ forms a commutative ring with a unity element under addition and multiplication (mod 6), but that this ring is *not* an integral domain.

For an example of this type it is helpful to draw up the addition and multiplication tables. Then closure, commutativity, existence of identity and inverse can be easily checked from these tables. The associative properties of addition and multiplication and also the distributive property, as noted after Example 3 in Section 6 of Chapter 6, automatically hold since they hold in **Z**.

Check, by the process described, that $\{S, +(\text{mod } 6), .(\text{mod } 6)\}$ is a commutative ring with a unity element.

Now $2.3 = 6 \equiv 0 \,(\text{mod } 6)$, but $2 \not\equiv 0 \,(\text{mod } 6)$ and $3 \not\equiv 0 \,(\text{mod } 6)$. Hence, in this ring, we have two elements which differ from the additive identity, but whose product is the identity for addition (mod 6). Thus the ring is not an integral domain.

II. Fields. A **field** is a *commutative ring F with a unity element* which has *at least two elements* and is such that:

every non-zero element of F has a multiplicative inverse in F.

The complete list of properties for a field $\{F, +, .\}$ is as follows:

Addition properties		*Multiplication properties*
A_1. F is closed under $+$	(Cl.)	M_1. F is closed under .
A_2. $\forall a, b, c \in F,$ $a+(b+c) = (a+b)+c$	(Ass.)	M_2. $\forall a, b, c \in F, a(bc) = (ab)c$
A_3. $\exists 0 \in F$ such that, $\forall a \in F,$ $a+0 = 0+a = a$	(Id.)	M_3. $\exists e \in F$ such that, $\forall a \in F,$ $ae = ea = a$

A₄. For each $a \in F$, $\exists -a \in F$ such that $\qquad$ (In.)
$$a + (-a) = (-a) + a = 0$$

M₄. For each $a \in F - \{0\}$, $\exists a^{-1} \in F - \{0\}$ such that
$$aa^{-1} = a^{-1}a = e$$

A₅. $\forall a, b, \in F, a + b = b + a \qquad$ (Comm.)

M₅. $\forall a, b \in F, ab = ba$

Distributive property

D₁. $\forall a, b, c \in F, a(b + c) = ab + ac$

It follows that, if $\{F, +, .\}$ is a field, then

$$\{F, +\} \text{ is an abelian group,}$$
$$\{F - \{0\}, .\} \text{ is an abelian group,}$$

and the operations $+$ and $.$ are tied together by the distributive law **D₁**. We shall usually say simply that F is a field.

The unity element of an abstract field is usually denoted by 1.

Obvious examples of fields are **Q**, the set of rational numbers, **R**, the set of real numbers, and **C**, the set of complex numbers, each under the usual addition and multiplication.

Each of **Q** and **R** is a **subfield** of the field **C**, since each forms a field under the addition and multiplication of **C**.

An obvious question, following the definitions **I** and **II** is: How are fields and integral domains related, apart from the fact that they are both commutative rings with a unity element? The answer is as follows:

Every field is an integral domain, but not conversely.

Proof. Let F be a field, and suppose that the elements a, b in F are such that $ab = 0$; we show that either $a = 0$ or $b = 0$.

If $a \neq 0$, then a has a multiplicative inverse a^{-1} in F. Then

$$ab = 0 \Rightarrow a^{-1}(ab) = a^{-1}0$$
$$\Rightarrow (a^{-1}a)b = 0$$
$$\Rightarrow 1b = 0 \quad (1 \text{ denoting the unity element of } F)$$
$$\Rightarrow b = 0.$$

Hence F is an integral domain.

That the converse is not true is obvious, since e.g. **Z**, the set of all integers, is an integral domain but is not a field under the usual addition and multiplication.

On the other hand it is true that:

every finite integral domain is a field.

For interest we include a proof of this result.

Proof. Let R be a finite integral domain, and let $R - \{0\} = \{a_1, a_2, \ldots, a_n\}$ be the non-zero elements of R. Let a be any element in $R - \{0\}$. Since $a \neq 0$ and $a_j \neq 0$ $(j = 1, \ldots, n)$, it follows that $aa_j \neq 0$ $(j = 1, \ldots, n)$. Hence, if $S = \{aa_1, aa_2, \ldots, aa_n\}$, then $S \subseteq R - \{0\}$. But, since $a \neq 0$,

$$aa_j = aa_k \Rightarrow a_j = a_k, \quad \text{using cancellation.}$$

Hence the n displayed members of S are distinct, and so $S = R - \{0\}$.

Since the unity element e of R belongs to $R - \{0\}$, it follows that $e \in S$. Thus there is an integer i such that $aa_i = e$, and so, by commutativity, such that $aa_i = a_i a = e$. Therefore a_i is an inverse for a in R.

Thus every non-zero element in R has an inverse in R, and so R is a field.

Example 2. Show that the set $S = \{0, 1, 2, 3, 4, 5, 6\}$ forms a field under addition and multiplication (mod 7).

Draw up the addition and multiplication tables for S (mod 7) and verify that the field axioms A_1, A_3, A_4, A_5, M_1, M_3, M_4 and M_5 are all satisfied. The remaining axioms A_2, M_2 and D_1 are satisfied since addition and multiplication on $\mathbf{Z}$ are associative and the distributive property holds in $\mathbf{Z}$.

Example 3. Show that the set

$$S = \left\{ \begin{bmatrix} x & y \\ -y & x \end{bmatrix} : x, y \in \mathbf{R} \right\}$$

forms a field under matrix addition and multiplication.

First check that $\{S, +\}$ is an abelian group:

A_1. If $A = \begin{bmatrix} x_1 & y_1 \\ -y_1 & x_1 \end{bmatrix}$ and $B = \begin{bmatrix} x_2 & y_2 \\ -y_2 & x_2 \end{bmatrix}$ are in S, then

$$A + B = \begin{bmatrix} x_1 + x_2 & y_1 + y_2 \\ -(y_1 + y_2) & x_1 + x_2 \end{bmatrix}, \text{ and this is clearly in } S.$$

Hence S is closed under matrix addition.

A_2. The associative property of addition holds since it holds for all 2×2 matrices.

A_3. The matrix $O = \begin{bmatrix} 0 & 0 \\ 0 & 0 \end{bmatrix}$ is in S, and so is an additive identity for S.

A₄. If $A = \begin{bmatrix} x & y \\ -y & x \end{bmatrix}$ is in S, then $-A = \begin{bmatrix} -x & -y \\ -(-y) & -x \end{bmatrix}$ is in S, and is an additive inverse for A in S.

A₅. The commutative property of addition holds in S since it holds for all 2×2 matrices.

Hence $\{S, +\}$ is an abelian group.
We now deal with multiplication.

M₁. In the notation used for **A₁**,

$$AB = \begin{bmatrix} x_1 & y_1 \\ -y_1 & x_1 \end{bmatrix} \begin{bmatrix} x_2 & y_2 \\ -y_2 & x_2 \end{bmatrix} = \begin{bmatrix} x_1 x_2 - y_1 y_2 & x_1 y_2 + x_2 y_1 \\ -(x_1 y_2 + x_2 y_1) & x_1 x_2 - y_1 y_2 \end{bmatrix},$$

and clearly AB is in S.

Hence S is closed under matrix multiplication.

M₂. The associative property of multiplication holds in S since it holds for all 2×2 matrices.

M₃. The matrix $I = \begin{bmatrix} 1 & 0 \\ 0 & 1 \end{bmatrix}$ is in S, and so is a multiplicative identity for S.

M₄. If $A = \begin{bmatrix} x & y \\ -y & x \end{bmatrix}$ is in S and $A \neq O$, then x and y are not both zero.

It follows that $|A| = x^2 + y^2 \neq 0$ and consequently that the inverse matrix A^{-1} exists. Now

$$A^{-1} = \begin{bmatrix} \dfrac{x}{x^2 + y^2} & \dfrac{-y}{x^2 + y^2} \\ -\left(\dfrac{-y}{x^2 + y^2}\right) & \dfrac{x}{x^2 + y^2} \end{bmatrix}, \text{ and this clearly is in } S \text{ and is an}$$

inverse for A in S.

M₅. By symmetry of the product formed in **M₁**, $AB = BA$, so that multiplication is commutative.

Finally we have:

D₁. The distributive property holds since it holds for all 2×2 matrices. Hence S forms a field under matrix addition and multiplication.

Note. This field S is isomorphic to the field $\mathbf{C}$ of complex numbers. If f is the mapping from S to $\mathbf{C}$ which maps

$$A = \begin{bmatrix} x & y \\ -y & x \end{bmatrix} \quad \text{to} \quad \alpha = x+iy,$$

then f is a bijection and, if f maps A to α and B to β, then f maps $A+B$ to $\alpha+\beta$ and AB to $\alpha\beta$, i.e.

$$f(A+B) = f(A)+f(B) \quad \text{and} \quad f(AB) = f(A)f(B).$$

Check these facts.

EXERCISE 10

1. Determine which of the following subsets of $\mathbf{R}$, the set of real numbers, form rings under the usual addition and multiplication of $\mathbf{R}$. If a subset does form a ring, determine whether it is an integral domain or a field.
   (i) $S_1 = \{3m : m \in \mathbf{Z}\}$;
   (ii) $S_2 = \{4k+1 : k \in \mathbf{Z}\}$;
   (iii) $S_3 = \{x+y\sqrt{2} : x, y \in \mathbf{Z}\}$;
   (iv) $S_4 = \{x+y\sqrt{2} : x, y \in \mathbf{Q}\}$;
   (v) $S_5 = \{x+y\sqrt[3]{2} : x, y \in \mathbf{Z}\}$.

2. Determine whether the following subset of $\mathbf{Q}$ (the set of rational numbers),

$$S = \left\{ \frac{a}{7} : a \in \mathbf{Z} \right\}$$

   forms a ring under the usual addition but with multiplication defined by:

$$\frac{a}{7} \cdot \frac{b}{7} = \frac{ab}{7}, \quad \forall a, b \in \mathbf{Z}.$$

3. (i) Show that the set $S = \{0, 2, 4, 6, 8\}$ forms a field under addition and multiplication (mod 10).
   (ii) Show that the set $S = \{0, 1, 2, 3, 4, 5, 6, 7\}$ forms a commutative ring with a unity element under addition and multiplication (mod 8). Is this ring an integral domain?

4. Determine which of the following sets of 2×2 matrices form rings under matrix addition and multiplication. If a set does form a ring, determine whether it is an integral domain or a field.

   (i) $S_1 = \left\{ \begin{bmatrix} x & 0 \\ 0 & x \end{bmatrix} : x \in \mathbf{Z} \right\}$;

   (ii) $S_2 = \left\{ \begin{bmatrix} x & x \\ 0 & 0 \end{bmatrix} : x \in \mathbf{Q} \right\}$;

   (iii) $S_3 = \left\{ \begin{bmatrix} x & x \\ 0 & x \end{bmatrix} : x \in \mathbf{Z} \right\}$;

   (iv) $S_4 = \left\{ \begin{bmatrix} x & y \\ 3y & x \end{bmatrix} : x, y \in \mathbf{Q} \right\}$;

   (v) $S_5 = \left\{ \begin{bmatrix} x & y \\ 4y & x \end{bmatrix} : x, y \in \mathbf{Q} \right\}$;

(vi) $S_6 = \left\{ \begin{bmatrix} x+y & y \\ -y & x \end{bmatrix} : x, y \in \mathbf{R} \right\}$;

(vii) $S_7 = \left\{ \begin{bmatrix} x+y & y \\ y & x \end{bmatrix} : x, y \in \mathbf{R} \right\}$;

(viii) $S_8 = \left\{ \begin{bmatrix} x+y\sqrt{2} & y \\ 3y & x \end{bmatrix} : x, y \in \mathbf{Q} \right\}$.

5. If $R = \mathbf{Q} \times \mathbf{Q} = \{(a, b) : a, b \in \mathbf{Q}\}$, and if addition and multiplication are defined on R by:

$$(a_1, b_1) + (a_2, b_2) = (a_1 + a_2, b_1 + b_2),$$
$$(a_1, b_1)(a_2, b_2) = (a_1 a_2, b_1 b_2),$$

show that R forms a commutative ring with a unity element under these operations, and that this is *not* an integral domain.

6. Operations $\oplus$ and $\odot$ are defined on $\mathbf{Z}$, the set of integers, by:

$$m \oplus n = m + n - 1,$$
$$m \odot n = m + n - mn,$$

$\forall m, n \in \mathbf{Z}$, where the $+$, $-$ and multiplication used on the right-hand side are the usual addition, subtraction and multiplication for $\mathbf{Z}$.

Show that $\{\mathbf{Z}, \oplus, \odot\}$ is a commutative ring with a unity element, the zero element and unity element being the integers 1 and 0, respectively.

Show that, if $\mathbf{Z}$ is replaced by $\mathbf{R}$, the set of real numbers, then $\{\mathbf{R}, \oplus, \odot\}$ is a field.

7. R is a ring with a unity element e (R is not necessarily commutative). An element a of R is such that $ax = 0 \Rightarrow x = 0$.

Prove that, if $ay = az$, then $y = z$.

Hence show that, if there is an element b in R such that $ab = e$, then $ba = e$.

[**Hint.** Consider aba.]

8. An element e of a ring R is called a **left identity** if

$$ex = x \quad \forall x \in R,$$

and an element f of R is called a **right identity** if

$$xf = x \quad \forall x \in R.$$

Given that a ring R has a left identity e, show that the element $e + ae - a$ is also a left identity, where a is *any* element of R. Hence show that, if e is the *only* left identity of R, then it is also a right identity (and so is a unity element for R).

Additional Examples 1

The following examples are listed under the number of the chapter in which the appropriate topic is discussed. The number **I** is intended to imply that on the average (except for certain problems for Chapters **9** and **10**) the problems are easier than those listed in the succeeding collection numbered **II**.

Chapter 1: Introduction to Mathematical Logic

1. For each of the following statements, state whether it is true or false, and give a reason for your answer.
   (i) There are exactly seven odd integers between 2 and 20.
   (ii) For each natural number n, either $n = 1$ or $n^2 \geqslant 4$.
   (iii) There is a unique line in a plane perpendicular to a given line in the plane.
   (iv) For each 2×2 matrix A, the 2×2 matrix $I + A$, where I is the identity 2×2 matrix, is nonsingular.
   (v) Given that **x**, **y** are non-zero, non-parallel vectors, there are exactly two unit vectors through a point O perpendicular to both **x** and **y**.
      Write down the negation of each of the statements (i), (ii), (iii), (iv) and (v).

2. Say whether each of the following statements is true or false, and whether its converse is true or false.
   (i) If a quadrilateral is a rectangle, then its diagonals are equal.
   (ii) If a quadrilateral has more than one axis of symmetry, then it is a rectangle.
   (iii) If a triangle is right-angled, then its area is half the product of its two shortest sides.
   (iv) If a hexagon has all its angles equal, then it is regular.

3. Write down the converse of each of the following statements. In each case where the statement or its converse is false, give a counter-example.
   (i) $a + b > c \Rightarrow a^2 + b^2 > c^2$, $(a, b, c \in \mathbf{R})$.
   (ii) $a > b \wedge b = c \Rightarrow a > c$, $(a, b, c \in \mathbf{R})$.
   (iii) $a^2 > b^2 \Rightarrow a > b$, $(a, b \in \mathbf{R})$.
   (iv) All integers which are multiples of 15 are multiples of 5.
   (v) $(x + y)(x - y) = x - y \Rightarrow x + y = 1$, $(x, y \in \mathbf{R})$.
   (vi) $na \equiv nb \pmod{m} \Rightarrow a \equiv b \pmod{m}$, $(n, a, b, m \in \mathbf{N})$.

248

4. Use proof by contradiction to prove the following results.

    (i) If the composite integer n is expressed as $n_1 n_2$, where $n_1 \geqslant 2$ and $n_2 \geqslant 2$, then at least one of n_1, n_2 is $\leqslant \sqrt{n}$.

    (ii) If the product of three positive numbers exceeds 1000, then at least one of the numbers is > 10.

5. Examine the validity of the following arguments.

    (i) All Ruritarians are three-legged.

        All three-legged people have two heads.

        Hence all Ruritarians have two heads.

    (ii) Suppose that $a^x = -1$, where $a, x \in \mathbf{Z}$.

$$a^x = -1 \Rightarrow a^{2x} = 1 = a^0$$
$$\Rightarrow 2x = 0$$
$$\Rightarrow x = 0 \Rightarrow a^0 = -1 \Rightarrow 1 = -1.$$

    (iii) $\log(\sin \tfrac{1}{4}\pi) = \log(\sin \tfrac{1}{4}\pi)$

$$\Rightarrow 2\log(\sin \tfrac{1}{4}\pi) > \log(\sin \tfrac{1}{4}\pi)$$
$$\Rightarrow \log(\sin^2 \tfrac{1}{4}\pi) > \log(\sin \tfrac{1}{4}\pi)$$
$$\Rightarrow \sin^2 \tfrac{1}{4}\pi > \sin \tfrac{1}{4}\pi$$
$$\Rightarrow \frac{1}{2} > \frac{1}{\sqrt{2}}.$$

6. Which of the following conditions are (a) necessary, (b) sufficient, (c) necessary and sufficient, for $x \leqslant y$, where x and y are rational numbers:

        (i) $y = x+1$,   (ii) $x^3 \leqslant y^3$,

        (iii) $y \geqslant x-1$,   (iv) $2^x \leqslant 2^y$,

        (v) $y = x^2 + \tfrac{1}{4}$,   (vi) $[x] \leqslant [y]$?

[In (vi), $[x]$ means the integral part of x.]

Chapter 2: Sets and Relations

7. Which of the following sentences are meaningful?

        (i) $3 \subseteq \{1, 3, 5\}$,   (ii) $3 \in \{1, 3, 5\}$,   (iii) $\{3\} \subseteq \{1, 3, 5\}$,

        (iv) $\{3\} \in \{1, 3, 5\}$,   (v) $\emptyset \in \{\emptyset\}$,   (vi) $\emptyset \subseteq \{\emptyset\}$.

8. Pick out five pairs of equal sets from the following ten sets.

$A = \{1, 2, 3, 4, 5\}$, $F = \{\text{prime numbers} < 10\}$,

$B = \{2, 3, 5, 7\}$, $G = \{\text{positive integers} < 100$

$C = \{2n-1 : n \in \mathbf{N}, 2 \leqslant n \leqslant 5\}$,   which are perfect cubes$\}$,

$D = \{1, 8, 27, 64\}$, $H = \{\text{the first five natural numbers}\}$,

$E = \{x \in \mathbf{R} : x > 1 \text{ and } x < 0\}$, $K = \emptyset$,

                               $L = \{3, 5, 7, 9\}$.

9. From the set $\{1, 2, 3, 4, 5, 6, 7, 8, 9, 10\}$ list the following subsets:

    (i) the subset of even numbers,

    (ii) the subset of prime numbers,

    (iii) the complement of the subset of perfect squares,

    (iv) the subset of members of the form $4n-1$, $n \in \mathbf{N}$,

    (v) the subset consisting of numbers that are square roots of members of the set.

10. $S = \{1, 2, 3, 4, 5, 6, 7, 8, 9, 10\}$, $A = \{1, 3, 5\}$, $B = \{2, 4, 6, 8\}$, $C = \{2, 5, 10\}$ and a dash (') is used to denote the complement of a subset in S.

By writing out the elements of the appropriate sets verify that

(i) $(A \cup B)' = A' \cap B'$,   (ii) $(A \cap B)' = A' \cup B'$,

(iii) $A \cup (B \cap C) = (A \cup B) \cap (A \cup C)$,   (iv) $A \cap (B \cup C) = (A \cap B) \cup (A \cap C)$.

11. If $A = \{(x, y) \in \mathbf{R}^2 : x^2 + y^2 = 4\}$, list the sets of points obtained by taking the intersections with A of the lines   (i) $x = 2$,   (ii) $y = 0$,   (iii) $y = 1$,   (iv) $x = 3$.

12. Express as intervals the solutions of the following inequations on the real number line:

(i) $2x - 1 \geqslant 3$,   (ii) $3x + 2 < 14$,   (iii) $2x - 1 \geqslant 3 \wedge 3x + 2 < 14$.

13. If $A = \{x \in \mathbf{R} : 0 < x \leqslant 3\}$ and $B = \{x \in \mathbf{R} : x \leqslant 1$ or $x > 4\}$, express $A \cap B$ and $A \cup B$ in simplest form.

14. If $A = \{x \in \mathbf{R} : 0 < x < 2\}$ and $B = \{x \in \mathbf{R} : 1 \leqslant x < 4\}$, find, in simplest form, $A \cap B$, $A' \cap B$, $A' \cup B'$.

If $C = \{x \in \mathbf{R} : 0 \leqslant x \leqslant 2$ or $x \geqslant 3\}$, find, in simplest form, $(A' \cup B') \cap C$.

15. Show, using a Venn diagram, that, for all subsets A, B, C of a set S,

$$(A \cup C) + (B \cup C) \subseteq C'.$$

Show also that equality holds if and only if

$$(A + B') - C = \emptyset.$$

16. By using a Venn diagram, show that, if A, B, C are subsets of a set S such that $A \cap (B \cup C) = B \cap (C \cup A)$, then $B \cap C = C \cap A$. Is the converse true?

17. Prove that, for any three sets,

$$(A \cup B) \cap (B \cup C) \cap (C \cup A) = (A \cap B) \cup (B \cap C) \cup (C \cap A).$$

18. In a survey of 1000 households, washing machines, vacuum cleaners and refrigerators were counted. Each house had at least one of these appliances. 400 had no refrigerator, 380 had no vacuum cleaner and 542 no washing machine. 294 had both a vacuum cleaner and a washing machine, 277 both a refrigerator and a vacuum cleaner, 190 both a refrigerator and a washing machine. How many households had all three appliances? How many had only a vacuum cleaner?

19. For each of the following relations on $\mathbf{Z}$ determine whether it is (a) reflexive, (b) symmetric or (c) transitive:

aRb means   (i) $a > b$,

(ii) $a \geqslant b$,

(iii) $a = b$,

(iv) $|a - b| = 1$,

(v) $a + b$ is an odd integer,

(vi) $a + b$ is exactly divisible by 4.

20. A relation $\sim$ is defined on $\mathbf{Z}$ by: $a \sim b$ means that $a + b$ is exactly divisible by 2. Show that $\sim$ is an equivalence relation on $\mathbf{Z}$ and find the equivalence classes.

21. A, B denote points in a plane which contains a fixed point O. A relation $\sim$ is defined on the set of points in the plane by: $A \sim B$ means that $OA = OB$. Show that $\sim$ is an equivalence relation on the set of points in the plane and describe the equivalence classes.

22. l is a fixed line in a plane; a relation $\sim$ is defined on the set of points in the plane by: $P\sim Q$ means that the perpendicular distances from the points P and Q to l are equal. Show that $\sim$ is an equivalence relation on the set of points in the plane and describe the equivalence classes.

23. For x, $y \in \mathbf{Z}$ write xRy when $x^2 - y^2$ is divisible by 3. Prove that R is an equivalence relation on $\mathbf{Z}$ which partitions $\mathbf{Z}$ into two equivalence classes.

24. Let S be the set of all points in Scotland whose height above sea level is not less than 1000 metres. For x, $y \in S$ let $x \sim y$ mean: "y can be reached from x by travelling along the ground without descending below 1000 metres". Show that $\sim$ is an equivalence relation on S.

Chapter 3: Mappings

25. A is the set $\{0, 1, 2, 3, 4, 5\}$ and a mapping $g : A \to \mathbf{Z}$ is defined by $g(x) = 3x - 2$. Find the image of g.

26. If $f : \mathbf{R} \to \mathbf{R}$ is defined by $f(x) = \cos x + \sin x$, find $f(0)$, $f(\pi/4)$, $f(\pi/2)$, $f(3\pi/4)$, $f(\pi)$; determine the set $\{x \in \mathbf{R} : f(x) = 0\}$.

27. A is the set $\{-3, -2, -1, 0, 1, 2, 3\}$ and B is the interval $\{x \in \mathbf{R} : -3 \leqslant x \leqslant 3\}$. If $f : A \to \mathbf{R}$ and $g : B \to \mathbf{R}$ are defined by $f(x) = x^2$ and $g(x) = x^2$, sketch the graphs of f and g, and state the image in each case.

28. For each of the following mappings, find the given subset of the appropriate domain:
    (i) $f : \mathbf{R} \to \mathbf{R}$ defined by $f(x) = 2x + 3$; subset $\{x \in \mathbf{R} : f(x) = -4\}$;
    (ii) $g : \mathbf{R} \to \mathbf{R}^+$ defined by $g(x) = 2x^2 + 1$; subset $\{x \in \mathbf{R} : g(x) = 9\}$;
    (iii) $h : [0, \pi] \to \mathbf{R}$ defined by $h(x) = \cos x$; subset $\{x \in [0, \pi] : h(x) \leqslant \tfrac{1}{2}\}$.

29. A mapping $f : \mathbf{R} \to \mathbf{R}$ is defined by
$$f(x) = \begin{cases} x, & \text{when } x \in \mathbf{Z}, \\ 1/x, & \text{when } x \notin \mathbf{Z}. \end{cases}$$
Write down the numbers $f(1), f(0), f(\tfrac{1}{2}), f(2), f(-\tfrac{4}{3})$.
Which of the following numbers are in the image of f: $3, \tfrac{4}{3}, -\tfrac{5}{2}, \tfrac{1}{2}, \tfrac{1}{3}, \tfrac{1}{4}$?
Determine the image of f.

30. Sketch the graph of each of the following mappings and state whether the mapping is (a) injective, (b) surjective. In a case where a mapping is bijective, give a formula for the inverse mapping.
    (i) $f : \mathbf{R} \to \mathbf{R}, f(x) = 5x - 3$;   (ii) $g : [0, 2\pi] \to \mathbf{R}, g(x) = \sin x$;
    (iii) $h : \mathbf{R} \to \mathbf{R}^+, h(x) = x^4$;   (iv) $k : \mathbf{R}^+ \to \mathbf{R}^+, k(x) = x^4$.

31. The mappings f, g, h, k are defined as in question 30.
    Find the numbers   (i) $(f \circ g)(\tfrac{1}{2}\pi)$,   (ii) $(h \circ g)(\tfrac{1}{4}\pi)$,   (iii) $(h \circ f)(0)$,
    (iv) $(k \circ h)(2)$,   (v) $(k \circ k)(2)$,   (vi) $(f \circ f)(-1)$.

32. The mappings $f : \mathbf{R} \to \mathbf{R}$ and $g : \mathbf{R} \to \mathbf{R}$ are defined by
$$f(x) = \begin{cases} x^2 & \text{when } x \geqslant 0 \\ 1/x & \text{when } x < 0 \end{cases}, \quad g(x) = \begin{cases} x & \text{when } x \geqslant 0 \\ -x & \text{when } x < 0 \end{cases}.$$
    (i) Find formulae for $f \circ g$ and $g \circ f$, and sketch the graphs of f and $g \circ f$.
    (ii) Show that f is bijective and find its inverse.

33. The mappings $f:\mathbf{R}\to\mathbf{R}$ and $g:\mathbf{R}\to\mathbf{R}$ are defined by

$$f(x) = \begin{cases} x+3 \text{ when } x\geqslant 0 \\ x \text{ when } x<0 \end{cases} \text{ and } g(x) = \begin{cases} x-2 \text{ when } x\geqslant 0 \\ 2-x \text{ when } x<0. \end{cases}$$

For each mapping, sketch its graph and state whether it is (a) injective, (b) surjective. If it is not injective, specify an element of the codomain that is the image of more than one element of the domain; if it is not surjective, state its image.

Find formulae for $g\circ f$ and $f\circ g$.

Show that the restriction h of $f\circ g$ which has $\mathbf{R}^+$ as domain and the image of $f\circ g$ as codomain is bijective, and find the inverse of h.

34. State the (maximal) domain of definition and the image of the real function defined by each of the following formulae.

    (i) $f(x) = x^2+2$,   (ii) $g(x) = 2x-5$,

    (iii) $h(x) = \sqrt{(9-x^2)}$,   (iv) $k(x) = \sqrt{(x-3)}$,

    (v) $u(x) = \sqrt{(1-x)}+\sqrt{(x-3)}$,   (vi) $v(x) = \sin x^2$.

35. If $A = \mathbf{R} - \{0, 1\}$, mappings $f_i:A\to A$ $(i = 1, 2, 3, 4, 5, 6)$ are defined by the formulae:

$$f_1(x) = x, \qquad f_2(x) = 1-x, \qquad f_3(x) = \frac{x-1}{x},$$

$$f_4(x) = \frac{1}{x}, \qquad f_5(x) = \frac{1}{1-x}, \qquad f_6(x) = \frac{x}{x-1}.$$

Show that the set $\{f_1, f_2, f_3, f_4, f_5, f_6\}$ is closed under composition of mappings.

Chapter 4: Number Systems

36. Express in simplest form the rational numbers
$$2^3, \quad 3^2, \quad 2^{-3}, \quad (-3)^2, \quad (-2)^{-3}, \quad (-3)^{-2}, \quad 3^{-2}, \quad (-2)^3.$$

37. Simplify (i) $(4\sqrt{2})^{-4/5}$,   (ii) $(12)^{5/6}\times(\frac{2}{3})^{1/3}$.

38. Determine whether the following numbers are rational, irrational or neither.

    (i) $(15\frac{5}{8})^{1/3}$,   (ii) $\sqrt[3]{9}$,   (iii) roots of $x^2+x-1 = 0$,

    (iv) roots of $x^2+x+1 = 0$,   (v) $\sqrt{(4-x^2)}$ when $x = \sqrt{3}$,

    (vi) $\sqrt{(4-x^2)}$ when $x = 3$,   (vii) $\sqrt{(4-x^2)}$ when $x = \sqrt{2}$.

39. Prove that $3^{1/3}$ is irrational.

40. Express as a rational number in its lowest terms $0\cdot90\dot{7}\dot{4}$ $(= 0\cdot9074074074\ldots)$.

41. For each of the following irrational numbers find a polynomial equation with rational coefficients which has that number as a root:

    (i) $1+\sqrt{2}$,   (ii) $1+2^{1/3}$,   (iii) $-5+\sqrt{3}$,

    (iv) $\sqrt{3}-\sqrt{2}$,   (v) $1+\sqrt{2}+\sqrt{3}$,   (vi) $2^{1/3}+2^{2/3}$.

42. Prove by the method of contradiction that, if x and y are unequal integers, then $\dfrac{x+\sqrt{2}}{y+\sqrt{2}}$ is irrational.

43. Let $S = \{a+b\sqrt{2}:a, b \in \mathbf{Q}\}$. Prove that S is closed under both addition and multiplication and also that the reciprocal of each non-zero member of S is again a member of S.

44. Find all solutions of the simultaneous equations
$$\begin{cases} |x+y| = 3, \\ x-y = 1. \end{cases}$$

45. By considering $\left(x+\dfrac{1}{x}\right)^2 - 4$, prove that, $\forall x \in \mathbf{R} - \{0\}$, $\left|x+\dfrac{1}{x}\right| \geqslant 2$.

46. For which values of $x \in [0, \infty)$ is it true that $\sqrt{x} < x$?

47. If the number pairs (a, b) and (c, d) are multiplied by the rule:
$$(a, b) \times (c, d) = (ac+bd, bc+ad),$$
find $(5, 3) \times (10, 2)$. If $(6, 3) \times (p, q) = (60, 48)$, find p and q.

48. Express the set
$$\left\{x \in \mathbf{R}: \frac{x}{x-1} \geqslant \frac{x+2}{x}\right\}$$
as a union of intervals.

49. Express as an interval or union of intervals the set
$$\left\{y \in \mathbf{R}: \frac{y-3}{y-1} < 0\right\}.$$
Deduce a similar expression for $\left\{x \in \mathbf{R}: \dfrac{|x|-3}{|x|-1} < 0\right\}$.

50. Express the set
$$S = \left\{x \in \mathbf{R}: \frac{x-1}{x+1} \leqslant \frac{x+1}{x-1}\right\}$$
as a union of intervals, and find $S \cap [0, 2]$.

Chapter 5: Induction, Finite Summations, Permutations and Selections

51. Prove the following results by induction:

(i) $\forall n \in \mathbf{N}$, $1+3+5+\ldots+(2n-1) = n^2$;

(ii) $\forall n \in \mathbf{N}$, $1+3+3^2+\ldots+3^{n-1} = \frac{1}{2}(3^n-1)$;

(iii) $\forall n \in \mathbf{N}$, $\displaystyle\sum_{r=1}^{n} 2^{r-1}(r+1) = 2^n n$;

(iv) $\forall n \in \mathbf{N}$, $\displaystyle\sum_{r=1}^{2n} (-1)^r r^2 = n(2n+1)$;

(v) $\forall n \in \mathbf{N}$, $\displaystyle\sum_{r=1}^{2n} (-1)^{r-1}\frac{1}{r} = \sum_{r=n+1}^{2n} \frac{1}{r}$;

(vi) $\forall n \in \mathbf{N}$, $\displaystyle\prod_{r=1}^{n} \frac{2r-1}{r} = \frac{1}{2^n}\binom{2n}{n}$.

52. (i) By noting that $(r+1)^3 - r^3 = 3r^2 + 3r + 1$ (for $r = 1, 2, \ldots, n$), prove that

$$\sum_{r=1}^{n} r^2 = \tfrac{1}{6}n(n+1)(2n+1), \text{ assuming that } \sum_{r=1}^{n} r = \tfrac{1}{2}n(n+1).$$

(ii) By noting that $(r+1)^4 - r^4 = 4r^3 + 6r^2 + 4r + 1$, prove that

$$\sum_{r=1}^{n} r^3 = \tfrac{1}{4}n^2(n+1)^2.$$

53. Show that, $\forall n \in \mathbf{N}$,

    (i) $\displaystyle\sum_{r=1}^{n} (n+2r) = n(2n+1);$

    (ii) $\displaystyle\sum_{r=1}^{n} r(n^2 - r^2) = \tfrac{1}{4}n^2(n^2 - 1);$

    (iii) $\displaystyle\sum_{r=1}^{2n+1} (n^2 + r) = n^3 + (n+1)^3.$

    [Hints: For (i), $\displaystyle\sum_{r=1}^{n} (n+2r) = n\sum_{r=1}^{n} 1 + 2\sum_{r=1}^{n} r$

                            $= n.n + 2.\tfrac{1}{2}n(n+1),$

and proceed similarly for (ii) and (iii).]

54. (i) Evaluate ${}^9P_3$, $\dbinom{12}{8}$ and $\dbinom{13}{4}$.

(ii) Find r so that ${}^7P_r = 840$.

(iii) Find the natural number $n > 2$ for which

$$\binom{n}{1} + \binom{n}{3} = 2\binom{n}{2}.$$

55. In the expansion of $(1+x)^{10}$ in ascending powers of x, what is the ratio of
    (i) the $(r+1)$th coefficient to the rth coefficient,
    (ii) the $(r+1)$th term to the rth term?
    Find the largest coefficient in the expansion, and the largest term when $x = \tfrac{1}{4}$.

56. Use the binomial theorem to prove that
$$(1 + t + t^2)^n \geqslant 1 + nt + \tfrac{1}{2}n(n+1)t^2$$
for all $n \in \mathbf{N}$ and $t > 0$.

57. Show that the term independent of x in the expansion of $(x + \tfrac{1}{2}x^{-1})^{2n}$, where $n \in \mathbf{N}$, can be expressed as

$$\prod_{r=1}^{n} \left(2 - \frac{1}{r}\right).$$

58. Find the coefficient of x^{-1} in the expansion of

$$(x+2)^3\left(1 - \frac{1}{x}\right)^{10}.$$

59. Find, without attempting to simplify, the coefficient of x^5 in the expansion of $(x+3)^2(x-1)^n$, where n is an integer $\geqslant 5$.

60. Show that in the expansion of
$$\{1+x-\tfrac{1}{2}(n-1)x^2\}^n$$
in ascending powers of x the coefficient of x^2 is zero. Find the coefficient of x^3.

61. (i) By using the identity $\binom{m}{r-1}+\binom{m}{r} = \binom{m+1}{r}$, show that, for all $n \in \{0, 1, 2, \ldots, m\}$,
$$\sum_{r=0}^{n} (-1)^r \binom{m+1}{r} = (-1)^n \binom{m}{n}.$$

(ii) By using the identity $(1+x)^n(1-x)^n = (1-x^2)^n$, evaluate $\sum_{r=0}^{n} (-1)^r \binom{n}{r}^2$ when (a) n is even, (b) n is odd.

62. Using the digits 1, 2, 3, 4, 5, how many
(i) five-digit numbers can be formed without repetitions,
(ii) five-digit numbers can be formed allowing repetitions,
(iii) five-digit even numbers can be formed without repetitions?
In (i), how many of the numbers are $> 30\,000$?

63. How many different permutations are there of all the letters of the word *ARRANGEMENT*? How many of these permutations begin and end with a vowel?

64. How many different selections of 3 letters can be made from A, A, B, C, D, E, F?

65. A set S has 7 elements. How many subsets of S have 2 elements? In how many ways can one choose 2 disjoint subsets of S, each containing 2 elements?

66. An ordinary pack of 52 cards is to be dealt out equally among 4 players. Show that the number of ways in which this can be done is $(52!)/(13!)^4$ and that the fraction of these that result in all four aces going to the same player is $44/4165$.

Chapter 6: Elementary Number Theory

67. Find the quotient and principal remainder if
(i) 259 is divided by 13,
(ii) -362 is divided by 23,
(iii) 28745 is divided by 57.

68. (i) Express 4031 in the scale of 5 and $(4031)_5$ in decimal notation.
(ii) Divide 110110111 by 1011, where the numbers are in binary notation.
(iii) If 64 is expressed in the scale of g and the last digit is 3, find the value of g.
(iv) If $(3341)_g = 1205$ in decimal notation, find the base g.

69. For the pairs of integers a_1, a_2 given in the following table use the euclidean algorithm to determine $d = (a_1, a_2)$ and hence find integers x, y such that $d = xa_1 + ya_2$. Write down the positive common divisors of each pair of integers.

a_1	a_2
201	320
3142	1592
29 638	18 542

70. Use the results of problem **69** to determine
    (i) (201, 320, 101),   (ii) (3142, 1592, 44),
    (iii) (29 638, 18 542, 3650).

71. Find the prime decompositions of the following integers:
    (i) 1022,   (ii) 1023,   (iii) 1024,   (iv) 1025,   (v) 4567,   (vi) 999 999.

72. (i) Prove that, if $m > 1$ and m divides l, then m is not a factor of $l+1$. Is $l(l+1)$ ever a square for $l > 0$?
    (ii) Prove that the difference between cubes of consecutive integers is always odd.

73. (i) Find the smallest positive integer n such that $n+1$, $n+2$ and $n+3$ are all composite.
    (ii) Find also the smallest positive integer n such that $n+1, n+2, n+3$ and $n+4$ are all composite.

74. Examine the following statements involving integers. Prove those statements which you claim to be true and give a counter-example to each statement which you claim to be false.
    (i) If p is a prime and $p|a$ and $p|(a^2 + 6b^2)$, then $p|b$.
    (ii) If $a|m$ and $b|m$ and if $(a, b) = 1$, then $ab|m$.
    (iii) If $m|(a-1)$, then $m|(a^3-1)$.
    (iv) If $2a \equiv 2b(\bmod m)$, then $a \equiv b(\bmod m)$.

75. List all the integers x in the set
$$\{x \in \mathbf{Z} : 1 \leqslant x \leqslant 50, x \equiv 3(\bmod 7)\}.$$

76. Extend the set $\{5, 8, 21, 35, 50, -6, -24\}$ to form a CSR (mod 17).

77. Which of the following statements involving integers are true and which false? If a statement is true give a proof; if false give a counter-example.
    (i) $a \equiv b(\bmod m) \Rightarrow a^2 \equiv b^2(\bmod m)$;
    (ii) $a^2 \equiv b^2(\bmod m) \Rightarrow a \equiv b(\bmod m)$;
    (iii) $a \equiv b(\bmod m) \Rightarrow a^2 \equiv b^2(\bmod m^2)$;
    (iv) $a^2 \equiv b^2(\bmod m^2) \Rightarrow a \equiv b(\bmod m)$;
    (v) $a \equiv b(\bmod 3) \wedge a \equiv b(\bmod 5) \Leftrightarrow a \equiv b(\bmod 15)$.

78. Solve the congruence equations
$$\text{(i) } 2x \equiv 5(\bmod 9), \qquad \text{(ii) } x^3 \equiv 6(\bmod 7),$$
$$\text{(iii) } x^4 \equiv 1(\bmod 13), \qquad \text{(iv) } x^2 + 2x + 3 \equiv 0(\bmod 6).$$

79. Show that any integer which is a fourth power is congruent (mod 7) to 0, 1, 2 or 4.

80. Find the general solution of the simultaneous congruence equations
$$x \equiv 1(\bmod 2), \qquad x \equiv 0(\bmod 3), \qquad x \equiv 0(\bmod 5).$$

81. A quadratic polynomial $ax^2 + bx + c$ is said to be *irreducible* (mod 3) if the congruence equation $ax^2 + bx + c \equiv 0$ (mod 3) has no root for x. Find all the irreducible (mod 3) quadratic polynomials which arise from the 18 such polynomials obtained by allowing a, b, c to take values from the set $\{0, 1, 2\}$ with $a \neq 0$.

Chapter 7: Complex Numbers

82. Express in the form $a + ib$ with a and b real

$$\text{(i)} \quad \frac{2-i}{3+4i}, \qquad \text{(ii)} \quad \frac{(1-i)(1+2i)}{2+i}, \qquad \text{(iii)} \quad \frac{1}{1+i} - \frac{2}{1+3i}.$$

83. Find the modulus and principal value of the argument of each of the complex numbers

$$\text{(i)} \quad -1+i, \qquad \text{(ii)} \quad \sqrt{3}-i, \qquad \text{(iii)} \quad 3+4i, \qquad \text{(iv)} \quad (-2+\sqrt{3})i.$$

84. Simplify $\dfrac{(\cos \alpha + i \sin \alpha)^5}{(\cos \beta + i \sin \beta)^3}$ and $\dfrac{(\cos \alpha + i \sin \alpha)^2}{\cos \beta - i \sin \beta}$.

85. If $\dfrac{1}{w} = \dfrac{1}{2-i} - \dfrac{1}{1-3i}$, express w in the form $a + ib$ with a and b real numbers.

86. If $z = x + iy$, where x and y are real, express $w = \dfrac{z+8i}{z+6}$ in the form $u + iv$, where u and v are real. Deduce that the points on the Argand diagram that represent complex numbers z for which w is a purely imaginary complex number all lie on a circle through the origin.

87. Find the complex numbers z for which $z^2 + 2\bar{z}^2 + z - \bar{z} + 9 = 0$, where $\bar{z}$ is the complex conjugate of z.

88. Show that $\dfrac{1+\sin \theta + i \cos \theta}{1+\sin \theta - i \cos \theta} = \cos(\tfrac{1}{2}\pi - \theta) + i \sin(\tfrac{1}{2}\pi - \theta)$, and deduce that, for a positive integer n,

$$\left(1+\sin\frac{\pi}{n}+i\cos\frac{\pi}{n}\right)^n + i\left(1+\sin\frac{\pi}{n}-i\cos\frac{\pi}{n}\right)^n = 0$$

if and only if $n \equiv 1 \pmod 4$.

89. If $z = \cos \theta + i \sin \theta$, find the modulus and argument of $1 + z^5$ when $0 < \theta < \tfrac{1}{5}\pi$.

90. Find the 5 fifth roots of unity. Hence write down the 5 roots of the equation $z^5 = 32$.

91. (i) Find the 3 cube roots of -1.
(ii) Find the 4 fourth roots of $1 - i\sqrt{3}$.

92. (i) Express $128 \sin^2 \theta \cos^6 \theta$ in the form

$$a \cos 8\theta + b \cos 6\theta + c \cos 4\theta + d \cos 2\theta + e,$$

where a, b, c, d, e are constants.
(ii) Express $128 \sin^7 \theta \cos \theta$ in the form

$$a \sin 8\theta + b \sin 6\theta + c \sin 4\theta + d \sin 2\theta,$$

where a, b, c, d are constants.

93. Prove that, if r and θ are real and $|r| < 1$, then

$$\sin 2\theta + r \sin 5\theta + r^2 \sin 8\theta + \ldots = \frac{\sin 2\theta + r \sin \theta}{1 - 2r \cos 3\theta + r^2}.$$

94. Find the 3 cube roots of unity.
Show that, if ω is either of the non-real cube roots of unity, then $1 + \omega + \omega^2 = 0$. Hence show that

    (i) $(1+t)^5 + (1+\omega t)^5 + (1+\omega^2 t)^5 = 3 + 30t^3$,

    (ii) $(a+b+c)(a+\omega b + \omega^2 c)(a+\omega^2 b + \omega c) = a^3 + b^3 + c^3 - 3abc$.

95. (i) Given that the equation $z^4 + 7z^2 - 12z + 130 = 0$ has a root $-2+3i$, find all the roots of the equation.

    (ii) Given that the equation $2z^4 + 6z^3 + 43z^2 + 44z + 85 = 0$ has a root $-1+4i$, find all the roots of the equation.

        [Note that a real polynomial equation has non-real roots in complex conjugate pairs.]

96. Solve the equation $x^7 + 1 = 0$, and hence find real quadratic factors for the polynomial $x^6 - x^5 + x^4 - x^3 + x^2 - x + 1$.

97. Explain the geometrical significance of the following mapping of the Argand diagram into itself:

$$f : z \rightarrow \bar{z}(\cos \alpha + i \sin \alpha),$$

where $\bar{z}$ is the complex conjugate of z and α is the radian measure of a fixed angle.

    Show that $f \circ f = I$, where I is the identity mapping.

Chapter 8: Matrices

98. If $3 \begin{bmatrix} 2 & -1 & 3 \\ 1 & 2 & -1 \end{bmatrix} + X = 4 \begin{bmatrix} -2 & 1 & 4 \\ 1 & 3 & 0 \end{bmatrix}$, find X.

99. If $A = \begin{bmatrix} 1 & 2 \\ 3 & 4 \end{bmatrix}$ and $X = \begin{bmatrix} x & y \\ z & t \end{bmatrix}$, show that $AX = XA$ if and only if $2z = 3y$ and $2t = 2x + 3y$.

    Show that the only symmetric matrices X (i.e. $X' = X$) for which $AX = XA$ are the scalar multiples of I, the identity 2×2 matrix.

100. If $A = \begin{bmatrix} 0 & 1 \\ -1 & 0 \end{bmatrix}$ and the real matrix $X = \begin{bmatrix} x & y \\ z & t \end{bmatrix}$ is such that $XAX' = A$, show that $|X| = 1$.

101. If $A = \begin{bmatrix} 2 & 1 \\ 5 & 4 \end{bmatrix}$, find A^{-1} and show that A^{-1} can be expressed in the form $pA + qI$, where p and q are scalars.

102. Find all the 2×2 matrices X for which $X^2 = \begin{bmatrix} 4 & 0 \\ 0 & 1 \end{bmatrix}$.

103. Show that, if $A = \begin{bmatrix} 0 & 1 \\ 0 & 0 \end{bmatrix}$, there is no matrix $X = \begin{bmatrix} x & y \\ z & t \end{bmatrix}$ such that $X^2 = A$.

    State the only real number k for which the equation $X^2 = kA$ has a solution, and find a non-zero solution X for the equation with this value of k.

104. The square matrix B satisfies the equation $B^2 = I$. Prove by induction that, for all $n \in \mathbf{N}$,

$$(I + B)^n = 2^{n-1}(I + B).$$

105. If $A = \begin{bmatrix} 3 & -1 \\ 1 & 1 \end{bmatrix}$, show by induction that, for all $n \in \mathbf{N}$,

$$A^n = 2^{n-1} \begin{bmatrix} 2+n & -n \\ n & 2-n \end{bmatrix}.$$

106. If $A = \begin{bmatrix} 1 & 3 & 2 \\ 4 & 0 & 6 \\ -1 & 2 & -3 \end{bmatrix}$ and $B = \begin{bmatrix} 2 & 1 & 0 \\ 0 & 2 & 1 \\ 1 & 0 & 2 \end{bmatrix}$, verify that $|A||B| = |AB|$.

107. Evaluate the determinants

(i) $\begin{vmatrix} 3 & -2 \\ 4 & -1 \end{vmatrix}$, (ii) $\begin{vmatrix} 1 & 3 & 2 \\ 3 & 0 & 5 \\ -2 & 1 & 4 \end{vmatrix}$, (iii) $\begin{vmatrix} 2 & 0 & 1 \\ -1 & 2 & 1 \\ 2 & 4 & 4 \end{vmatrix}$,

(iv) $\begin{vmatrix} 1 & 2 & 3 & 4 \\ 2 & 1 & 2 & 1 \\ 0 & 0 & 1 & 1 \\ 3 & 4 & 1 & 2 \end{vmatrix}$, (v) $\begin{vmatrix} 1 & 2 & 1 & 2 & 1 \\ 0 & 0 & 1 & 1 & 1 \\ 1 & 1 & 0 & 0 & 0 \\ 0 & 0 & 1 & 1 & 2 \\ 1 & 2 & 2 & 1 & 1 \end{vmatrix}$.

108. (i) Show that $\begin{vmatrix} 1 & 1 & 1 \\ a & b & c \\ a^2 & b^2 & c^2 \end{vmatrix} = -(a-b)(a-c)(b-c)$,

and deduce the factors of $\begin{vmatrix} 1 & 1 & 1 \\ a^2 & b^2 & c^2 \\ a^4 & b^4 & c^4 \end{vmatrix}$.

(ii) Show that $\begin{vmatrix} 1 & 1 & 1 \\ a & b & c \\ a^3 & b^3 & c^3 \end{vmatrix} = -(a-b)(a-c)(b-c)(a+b+c)$.

109. Factorise $\begin{vmatrix} 1 & 1 & 1 & 1 \\ a & b & c & d \\ a^2 & b^2 & c^2 & d^2 \\ a^3 & b^3 & c^3 & d^3 \end{vmatrix}$.

110. (i) Find the inverse if it exists of

(a) $\begin{bmatrix} 1 & -1 & 0 \\ 3 & 1 & 1 \\ 2 & 1 & 1 \end{bmatrix}$, (b) $\begin{bmatrix} 1 & 2 & -1 \\ 3 & 4 & 6 \\ 1 & 0 & 8 \end{bmatrix}$.

(ii) What condition must x satisfy to ensure that the matrix $\begin{bmatrix} 2 & 3 & 1 \\ -4 & x & 0 \\ 1 & 5 & 2 \end{bmatrix}$ has an inverse?

111. Find the adjugate and inverse of the matrix

$$A = \begin{bmatrix} 1 & 2 & 3 \\ 2 & 3 & 0 \\ 0 & 1 & 2 \end{bmatrix},$$

and hence solve the system of equations

$$\begin{aligned} x + 2y + 3z &= 1, \\ 2x + 3y \quad &= 1, \\ y + 2z &= 1. \end{aligned}$$

112. Show that, if $a \neq \pm 2$, the system of linear equations

$$\begin{aligned} x+2y-z &= b, \\ (a+1)x+3y+z &= b, \\ 2x+y+az &= 0 \end{aligned}$$

has a unique solution for each given value of b and find this solution.

Prove that, if $a = 2$, then the system is consistent for all values of b.

Consider the consistency of the system in the case $a = -2$, determining the general solution when it exists.

113. Show, using echelon reduction, that the system of linear equations

$$\begin{aligned} x+3y+z-2t &= 3, \\ x+4y+3z-t &= 4, \\ 2x+3y-4z-7t &= 3, \\ 3x+8y+z-7t &= 8 \end{aligned}$$

is consistent and find the general solution.

Chapter 9: Algebraic Structures I—Groups

114. Show that the set $\left\{ \begin{bmatrix} 1 & 0 \\ 0 & 1 \end{bmatrix}, \begin{bmatrix} -2 & -3 \\ 1 & 1 \end{bmatrix}, \begin{bmatrix} 1 & 3 \\ -1 & -2 \end{bmatrix} \right\}$ forms a group under matrix multiplication.

115. Let $\omega = \frac{1}{2}(-1+i\sqrt{3})$, so that $\omega^3 = 1$.

Draw up the multiplication table for the set of matrices

$$S = \left\{ I = \begin{bmatrix} 1 & 0 \\ 0 & 1 \end{bmatrix}, A = \begin{bmatrix} -1 & 0 \\ 0 & 1 \end{bmatrix}, B = \begin{bmatrix} 1 & 0 \\ 0 & \omega \end{bmatrix}, \right.$$
$$\left. C = \begin{bmatrix} 1 & 0 \\ 0 & \omega^2 \end{bmatrix}, D = \begin{bmatrix} -1 & 0 \\ 0 & \omega \end{bmatrix}, E = \begin{bmatrix} -1 & 0 \\ 0 & \omega^2 \end{bmatrix} \right\},$$

and show that S forms a group under matrix multiplication. (Assume that multiplication of 2×2 matrices is associative.)

116. Show that the set

$$G = \left\{ \begin{bmatrix} 1 & a \\ 0 & 1 \end{bmatrix} : a \in \mathbf{R} \right\}$$

forms an abelian group under matrix multiplication.

If H is the subset of G consisting of the matrices with non-negative a, explain why H does not form a group under matrix multiplication.

117. Show that the following sets of matrices form groups under matrix multiplication, and determine in each case whether the group is abelian or non-abelian.

(i) $G_1 = \left\{ \begin{bmatrix} a & a \\ 0 & 0 \end{bmatrix} : a \in \mathbf{R} - \{0\} \right\}$;

(ii) $G_2 = \left\{ \begin{bmatrix} p+q\sqrt{2} & r \\ 0 & p-q\sqrt{2} \end{bmatrix} : \begin{array}{l} p, q \in \mathbf{Q}, p \text{ and } q \text{ not} \\ \text{both zero}; r \in \mathbf{R} \end{array} \right\}$;

(iii) $G_3 = \left\{ \begin{bmatrix} \cos x & \sin x \\ -\sin x & \cos x \end{bmatrix} : x \in \mathbf{R} \right\};$

(iv) $G_4 = \left\{ \begin{bmatrix} p & 1-p \\ 1-p & p \end{bmatrix} : p \in \mathbf{R} - \{\frac{1}{2}\} \right\}.$

118. Show that the set of matrices

$$G = \left\{ \begin{bmatrix} a & b \\ 0 & c \end{bmatrix} : a, c \in \mathbf{C} - \{0\}; b \in \mathbf{C} \right\}$$

forms a group under matrix multiplication.

Demonstrate that this group is non-abelian by finding two elements of G that do not commute with one another.

Show that the subset of G consisting of those members of G for which b is real does not form a subgroup of G.

119. Show that the set of matrices

$$G = \left\{ \begin{bmatrix} a & 0 & 0 \\ 0 & b & 0 \\ 0 & 0 & c \end{bmatrix} : a, b, c \in \mathbf{R} - \{0\} \right\}$$

forms a group under matrix multiplication. Is this group abelian?

Show that the set of eight matrices obtained from $\begin{bmatrix} \pm 1 & 0 & 0 \\ 0 & \pm 1 & 0 \\ 0 & 0 & \pm 1 \end{bmatrix}$ by

taking all possible choices of the signs forms a subgroup of the group G.

120. Mappings f_1, f_2, f_3, f_4 from $\mathbf{R} - \{0\}$ to $\mathbf{R} - \{0\}$ are defined by: $f_1(x) = x$, $f_2(x) = -x, f_3(x) = \dfrac{1}{x}, f_4(x) = -\dfrac{1}{x}.$

Show that $\{f_1, f_2, f_3, f_4\}$ forms a group under composition of mappings.

121. Show that the set

$$G = \{(a, b): a, b \in \mathbf{R}, a \neq 0\}$$

forms a group under the operation on G defined by:

$$(a, b)(c, d) = (ac, ad + b).$$

Show that $H = \{(1, b): b \in \mathbf{R}\}$ forms a subgroup of G.

122. An element g of a group G satisfies the equation $g^{2n+1} = e$, where e is the identity element and n is an integer. Prove that $\exists\, h \in G$ such that $g = h^2$.

123. Elements g and h of a group (written multiplicatively) satisfy $(gh)^3 = e$ (i.e. $ghghgh = e$), where e is the identity element. Prove that $(hg)^3 = e$.

Chapter 10: Algebraic Structures II—Rings, Integral Domains, Fields

124. Determine which of the following subsets of $\mathbf{R}$ form rings under the usual addition and multiplication of $\mathbf{R}$. If a subset does form a ring, determine whether it is an integral domain or a field.

    (i) $S_1 = \mathbf{Q} - \mathbf{Z}$;   (ii) $S_2 = \{5m : m \in \mathbf{Z}\}$,

    (iii) $S_3 = \{x + y\sqrt{3} : x, y \in \mathbf{Z}\}$,

    (iv) $S_4 = \{x + y\sqrt{3} : x, y \in \mathbf{Q}\}$,

(v) $S_5 = \{x+y2^{1/3}:x, y\in\mathbf{Q}\}$,

(vi) $S_6 =$ set of all rational numbers whose denominators are 1 or a power of 2,

(vii) $S_7 = \{x+y5^{1/4}:x, y\in\mathbf{Z}\}$,

(viii) $S_8 = \{x+y5^{1/3}:x, y\in\mathbf{Q}\}$,

(ix) $S_9 = \{x+y5^{1/3}+z5^{2/3}:x, y, z\in\mathbf{Q}\}$.

125. Which of the following subsets of $\mathbf{C}$ form rings under the usual addition and multiplication of $\mathbf{C}$. If a subset does form a ring, determine whether it is an integral domain or a field.

(i) $S_1 = \mathbf{C}-\mathbf{R}$ (ii) $S_2 = \{xi:x\in\mathbf{Z}\}$,

(iii) $S_3 = \{xi:x\in\mathbf{R}\}$,   (iv) $S_4 = \{x+iy:x, y\in\mathbf{Z}\}$,

(v) $S_5 = \{x+iy:x, y\in\mathbf{Q}\}$,

(vi) $S_6 = \{x+i\sqrt{5}y:x, y\in\mathbf{Z}\}$,

(vii) $S_7 = \{x+i\sqrt{5}y:x, y\in\mathbf{Q}\}$.

126. Show that the set

$$\left\{\begin{bmatrix} x & 2y \\ -y & x \end{bmatrix}:x, y\in\mathbf{R}\right\}$$

forms a field under matrix addition and multiplication, but that under the same addition and multiplication, the set

$$\left\{\begin{bmatrix} x & 2y \\ -y & x \end{bmatrix}:x, y\in\mathbf{C}\right\}$$

is a commutative ring with a unity element which is not an integral domain.

127. The following are the addition table and part of the multiplication table for a ring having four elements a, b, c and d:

+	a	b	c	d
a	a	b	c	d
b	b	c	d	a
c	c	d	a	b
d	d	a	b	c

.	a	b	c	d
a	a	a	a	a
b	a	—	c	d
c	a	c	—	—
d	a	—	—	b

Use the distributive laws to complete the multiplication table. [Note that a is the zero element.]

[For example, $d^2+db = d(d+b) = da = a$, and so $b+db = a$; thus $db = (-b)+a = d+a = d$. Or, $db = d(-d) = -d^2 = -b = d$.]

128. Show that, in any field,

$$ab^{-1}+ac^{-1} = a(b+c)^{-1} \Rightarrow a = 0 \text{ or } b^2+bc+c^2 = 0.$$

129. If $\mathbf{R}[x]$ denotes the set of all polynomials in x with coefficients in $\mathbf{R}$, show that $\mathbf{R}[x]$ forms a commutative ring with a unity element under addition and multiplication of polynomials. Is this ring an integral domain or a field?

130. An operation $\oplus$ is defined on **R** by
$$x \oplus y = (x^3 + y^3)^{1/3}, \quad \forall x, y \in \mathbf{R}.$$
Show that $\{\mathbf{R}, \oplus, .\}$ is a field, . being the usual multiplication of **R**.

131. Operations $\oplus$ and $\otimes$ are defined on **R** in terms of the usual addition and multiplication on **R** by:
$$a \oplus b = a+b+1,$$
$$a \otimes b = a+b+ab.$$
Show that $\{\mathbf{R}, \oplus, \otimes\}$ is a field.
Show that $\{\mathbf{Z}, \oplus, \otimes\}$ is a commutative ring with a unity element. Is it an integral domain? Is $\{\mathbf{Q}, \oplus, \otimes\}$ a field?

132. Let S be a ring and let subset X of S form a field under the addition and multiplication of S; let m be a fixed element in S with a multiplicative inverse m^{-1} in S. Show that the subset
$$F = \{mxm^{-1} : x \in X\}$$
of S forms a field under the addition and multiplication of S.

Taking S as the ring of 2×2 real matrices and assuming that
$$X = \left\{\begin{bmatrix} x & y \\ -y & x \end{bmatrix} : x, y \in \mathbf{R}\right\}$$
forms a field under matrix addition and multiplication, use the matrix
$M = \begin{bmatrix} 1 & 1 \\ 0 & 1 \end{bmatrix}$ (corresponding to the element m above) to obtain another subset
of the ring of all 2×2 real matrices which also forms a field under matrix addition and multiplication.

133. In a ring R every element x satisfies $x^2 = x$. By considering $x+y$, show that, $\forall x, y \in R$, $xy + yx = 0$. Deduce, by taking $y = x$, that
$$x + x = 0 \quad \forall x \in R.$$
Hence show that R is commutative (i.e. $xy = yx$, $\forall x, y \in R$).
[*Hint*: $yx = -(-yx) = -(xy)$, etc.]

134. R is a *commutative* ring and
$$S = \{(x, n) : x \in R, n \in \mathbf{Z}\}.$$
Addition and multiplication are defined on S by:
$$(x, n) + (y, m) = (x+y, n+m),$$
and $\qquad (x, n).(y, m) = (xy + ny + mx, nm).$

Show that S forms a commutative ring with a unity element under these operations.
[Note that $nx = x+x+\ldots+x$ (n entries) for $n \in \mathbf{N}$,
$\qquad 0x = 0,$
and $nx = (-x)+(-x)+\ldots+(-x)$ ($|n|$ entries) for negative n.]

Additional Examples 2

Chapter 1: Introduction to Mathematical Logic

1. Are any of the following statements equivalent to the negation of $p \Rightarrow q$:
$$p \Rightarrow \sim q, (\sim p) \Rightarrow q, (\sim p) \Rightarrow \sim q, q \Rightarrow p?$$

2. By writing out their truth tables show that the statements
$$(p \wedge (\sim q)) \vee ((\sim p) \wedge q)$$
and
$$(p \vee q) \wedge \sim (p \wedge q)$$
are logically equivalent. Convince yourself that both correspond to the idea of "p or q but not both". This is the *exclusive* "or" common in non-mathematical usage.

3. A compound statement which has the truth value T whatever the truth values of its constituent statements $p, q, \ldots$ may be is called a *tautology*. Prove that the following are tautologies:
$$p \vee (\sim p), (p \wedge q) \Rightarrow p, p \Rightarrow (p \vee q), p \vee (p \Rightarrow q),$$
$$[(p \Rightarrow q) \wedge (q \Rightarrow r)] \Rightarrow (p \Rightarrow r).$$

4. Let p denote the statement "$m = 1$" and q denote the statement "$n = 2$". Which of the following statements concerning *natural numbers* m and n are equivalent to $p \wedge q$, and which are equivalent to $p \vee q$?
   (i) $(m-1)^2 + (n-2)^2 = 0$,   (ii) $mn - 2m - n + 2 = 0$,
   (iii) $(m-1)^2 (n-2)^2 = 0$,   (iv) $m(n-1) = 1$.

5. In the following statements x denotes a real number. Say whether each statement is true or false. Write down the converse statements and say whether they are true or false.
   (i) If $x > 1$ then $x/(x+1) > \frac{1}{2}$.
   (ii) If $x > -1$ then $x^2 > 1$.
   (iii) If $x^3 > x$ then $x^5 > x$.

6. Determine which of the following statements about natural numbers are true and which are false.
   (i) $\forall n, 2n^2 - 8n + 7 > 0$;   (ii) $\exists n$ such that $11n - n^2 > 29$;
   (iii) $\forall n, n^2 - n$ is even;   (iv) $\exists n$ such that $n^3 - n$ is odd.

7. Which of the following statements are true and which false?
   (i) A necessary condition for the even integer n to be a perfect square is that n be divisible by 4.
   (ii) A necessary condition for the odd integer n to be a perfect square is that n be divisible by 9.

264

(iii) A necessary condition for the integer n to be a perfect square is that n be odd or divisible by 4.

(iv) A necessary condition for the integer n to be divisible by 5 is that the last (decimal) digit of n be 5.

8. The statement

$$(\forall n)(\exists m)(m > 2n),$$

referring to natural numbers m and n, asserts that for each number n there exists a number m such that $m > 2n$. Is this statement true?

Express in words the statements

$$(\exists m)(\forall n)(m > 2n),$$
$$(\exists n)(\forall m)(m > 2n).$$

Are they true?

9. Let p_x and q_x be statements about the objects x of some collection. Then the compound statements

$$(\forall x)(p_x \vee q_x)$$

and

$$[(\forall x)p_x] \vee [(\forall x)q_x]$$

are not equivalent. Convince yourself of this by observing that when the objects are the natural numbers, p_x means "x is even" and q_x means "x is odd", then one of the statements is true, the other false.

Chapter 2: Sets and Relations

10. S is the set of natural numbers between 1 and 12 inclusive. A, B, C, D are the subsets of S consisting of the multiples of 2, 3, 4, 6 respectively; E is the set of squares in S. List the elements of these subsets. Verify that $A \cap B = D$, $C \subseteq A$, $D \subseteq B$, $E \subseteq D'$. List the elements of B' and C' and verify the relation $(B \cup C)' = B' \cap C'$. Find the symmetric difference $C + E$.

11. Express as an interval or a union of intervals each of the sets
  (i) $[-1, 2] \cap (1, 3)$;   (ii) $[0, 3] - [1, 2]$;   (iii) $[-1, 2] + [1, 3]$;
  (iv) $\{x \in \mathbf{R} : x(x-1)(x-2)(x-3) < 0\}$;
  (v) $\{x \in \mathbf{R} : \exists \theta \in \mathbf{R} \text{ such that } x = \sin \theta\}$.

12. List the elements of the sets
  (i) $\{x \in \mathbf{R} : x^3 = x\} \cap \{x \in \mathbf{R} : x^3 + 3x^2 + 2x = 0\}$;
  (ii) $\{x \in \mathbf{Q} : x^2 < 1\} \cap \{x \in \mathbf{Q} : 4x \in \mathbf{Z}\}$;
  (iii) $\{x \in \mathbf{Z} : x^2 < 50\} \cap \{x \in \mathbf{Z} : \frac{1}{2}(x-1) \in \mathbf{Z}\}$.

13. P, Q and R are sets such that $P \subseteq Q$, $Q \subseteq R$ and $R \subseteq P$. Prove that $P = Q = R$.

14. Simplify the following expressions, in which A and B are subsets of a set S.
  (i) $(A \cap B') \cup A' \cup B$;   (ii) $(A \cup B) \cap A' \cap B'$;
  (iii) $A \cap B \cap (A \cup B)$.

15. (i) Prove that $A \subseteq B$ if and only if $A \cap B = A$.
  (ii) Prove that $A \subseteq B$ if and only if $A \cup B = B$.
  (iii) Prove that $A = B$ if and only if $A \cap B = A \cup B$.

16. Prove that for subsets A, B, C of a set S,

$$A \cap B \subseteq C \Rightarrow A \cap C' \subseteq B'.$$

17. Prove by means of a Venn diagram, or otherwise, that
$$A \cup (B+C) \neq (A \cup B)+(A \cup C)$$
if A is non-empty.

18. Prove that for any three sets A, B, C
$$A \cap (B \cup C) \subseteq (A \cap B) \cup C.$$
Show further that
$$A \cap (B \cup C) = (A \cap B) \cup C$$
if and only if $C \subseteq A$.

19. If $P = \{a, b, c\}$, $Q = \{x, y, z\}$ list the elements of $P \times Q$. Let $A = \{a, b\}$, $B = \{b, c\}$, $X = \{x, y\}$, $Y = \{y, z\}$. List the elements of $A \times X$, $B \times Y$, $(A \cap B) \times (X \cap Y)$.
Is $(A \cap B) \times (X \cap Y) = (A \times X) \cap (B \times Y)$?
Is $(A \cup B) \times (X \cup Y) = (A \times X) \cup (B \times Y)$?

20. If X, Y, Z are subsets of S and $n(S) = 80$, $n(X) = 32$, $n(Y) = 27$, $n(Z) = 29$, $n(X \cap Y) = 12$, $n(X \cap Z) = 13$, $n(Y \cap Z) = 10$, $n(X \cap Y \cap Z) = 3$, find $n(X' \cap Y' \cap Z')$.

21. It is known that in a group of people, each of whom speaks at least one of the languages English, German and Russian, 31 speak English, 36 speak German and 27 speak Russian. 10 speak both English and German, 9 both English and Russian, 11 both German and Russian. Prove that the group contains at least 64 people and not more than 73.

22. Let S be the subset of $\mathbf{R} \times \mathbf{R}$ consisting of those pairs (x, y) such that $xy \neq 0$. Define relations $\sim$ and $\approx$ on S by
$$(x, y) \sim (x', y') \text{ if and only if } x'/x = y'/y,$$
$$(x, y) \approx (x', y') \text{ if and only if } x'/x = y/y'.$$
Prove that $\sim$ and $\approx$ are equivalence relations. Sketch some of the corresponding equivalence classes in diagrams of the x, y-plane.

23. For m, $n \in \mathbf{N}$ write mRn if m/n can be expressed in the form p/q where the natural numbers p, q are both odd. Prove that R is an equivalence relation on $\mathbf{N}$ and describe the equivalence classes into which R partitions $\mathbf{N}$.

24. Let R and S be two equivalence relations on the same set E. We can define new relations T and U on E as follows:
$$xTy \text{ means } \text{``}xRy \text{ and } xSy\text{''},$$
$$xUy \text{ means } \text{``}xRy \text{ or } xSy\text{''}.$$
Prove that T is an equivalence relation. By considering the example in which $E = \mathbf{Z}$, xRy means "$x-y$ is even" and xSy means "$x-y$ is divisible by 3" show that U is not necessarily an equivalence relation.
Describe T in this case.

Chapter 3: Mappings

25. Mappings f, $g : \mathbf{R} \to \mathbf{R}$ are defined by
$$f(x) = 2x+3, \qquad g(x) = 4x+9.$$
Prove that $f \circ g = g \circ f$.

26. S is the set $\{a, b, c\}$ and mappings $f, g : S \to S$ are defined by
$$f(a) = b, \quad f(b) = b, \quad f(c) = a;$$
$$g(a) = c, \quad g(b) = a, \quad g(c) = b.$$
Find the mappings $f \circ g, g \circ f, g^{-1}$ and $g^{-1} \circ f \circ g$.

27. A mapping $h : S \to S$, where $S = \mathbf{R} - \{2\}$, is defined by
$$h(x) = (2x + 3)/(x - 2).$$
Prove that h is bijective and determine the inverse mapping h^{-1}.

28. A mapping $f : \mathbf{R} \to \mathbf{R}$ is defined by
$$f(x) = (3x + 2)/(x - 1) \text{ for } x \neq 1, f(1) = 3.$$
Prove that f is bijective and find f^{-1}.

29. A mapping $g : \mathbf{R} \to \mathbf{R}$ is defined by
$$g(x) = \begin{cases} 2x - 1 & \text{for } x < 1, \\ 3x - 2 & \text{for } x \geq 1. \end{cases}$$
Prove that g is bijective and determine g^{-1}.

30. Mappings $f, g, h : \mathbf{R} \to \mathbf{R}$ are defined by
$$f(x) = x^2 + 1, \quad g(x) = x^3 + 1, \quad h(x) = |x|.$$
Prove that $f \circ h = h \circ f$ but $f \circ g \neq g \circ f, g \circ h \neq h \circ g$.

31. For each non-zero real number c a mapping $f_c : \mathbf{R} \to \mathbf{R}$ is defined by
$$f_c(x) = \frac{x}{c(x - 1)} \text{ for } x \neq 1, \quad f_c(1) = c.$$

Prove that f_2 is neither injective nor surjective. Show that there are two numbers c such that f_c is a bijection.

32. Prove that the equation
$$ax^2 - x + a = 0 \qquad (a \in \mathbf{R})$$
has a real root if and only if $4a^2 \leq 1$. Determine the image of the mapping $f : \mathbf{R} \to \mathbf{R}$ given by $f(x) = x/(x^2 + 1)$. Is f injective?

33. If $S = \{a, b\}, T = \{c, d\}$ write down (or represent by diagrams) the *four* different mappings from S to T. How many of them are surjections?

How many different mappings are there from S to $U = \{x, y, z\}$? How many are injective? How many mappings are there from U to S and how many are injective?

34. For mappings $f : Y \to Z, g : X \to Y$ prove that
   (i) if f and g are injective so is $f \circ g$;
   (ii) if $f \circ g$ is injective so is g;
   (iii) if $f \circ g$ is surjective so is f.

35. If $U = \{a, b, c\}$ how many different mappings from U to U are there? How many of these have at least one fixed element? [*Note.* A *fixed element* of a mapping f from a set to itself is an element x of the set such that $f(x) = x$.]

36. f and g are mappings of a set X into itself, g being a bijection. Prove that if c is a fixed element of f then $g(c)$ is a fixed element of $g \circ f \circ g^{-1}$.
Illustrate this in the case $X = \mathbf{R}$, with f and g defined by
$$f(x) = 3x - 2, \quad g(x) = 2x + 1.$$

37. Two mappings f, $g:S \to S$ commute, i.e. $f \circ g = g \circ f$. The mapping f has a fixed element which is not fixed for g. Show that f has at least one further fixed element.

38. If $f:S \to T$ is a mapping and A is a subset of S then $f(A)$ denotes the set of all elements $f(x)$ with $x \in A$; it is a subset of T.

Let $S = T = \{0, 1, 2, 3, 4\}$ and define f by $f(x) = (x-2)^2$. If $A = \{1, 2, 4\}$, $B = \{2, 3\}$ find $f(A)$, $f(B)$, $f(A \cap B)$ and $f(A \cup B)$. Verify that $f(A \cup B) = f(A) \cup f(B)$, $f(A \cap B) \subseteq f(A) \cap f(B)$. Show also that neither of the sets $f(B')$ and $[f(B)]'$ is a subset of the other.

39. Let $X = \mathbf{R} - \{0, 1\}$. Prove that, if $f:X \to X$ is defined by $f(x) = 1 - x^{-1}$, then $f^3 = i_X$. [*Note.* $f^3 = f \circ f \circ f$.]

40. Mappings $h:S \to T$, $f:T \to U$ and $g:T \to U$ are such that
$$f \circ h = g \circ h.$$
If h is surjective (but not necessarily bijective), show that $f = g$.

Chapter 4: Number Systems

41. Knowing that $\sqrt{2}$ is irrational prove that $(\sqrt{2}+3)/(\sqrt{2}-1)$ is irrational.

42. The rational numbers x and y satisfy $x^2 - 2xy - 2y^2 = 0$. Prove that $x = y = 0$.

43. Express as a union of intervals the set of real numbers x such that
$$\frac{x-2}{x+3} > \frac{2(2x-1)}{x+1}.$$

44. Express as a union of intervals the set
$$\left\{ x \in \mathbf{R} : |x| < \frac{3}{x+4} \right\}.$$

45. Express as an interval the set
$$\{x \in \mathbf{R} : |x+1| + |2-x| = 3\}.$$

46. Find the set $\{x \in \mathbf{R} : |x+1| = |2x+1|\}$.

47. Use the triangle inequality to prove that for all θ, $\phi \in \mathbf{R}$
$$|\sin(\theta + \phi)| \leq |\sin \theta| + |\sin \phi|.$$

48. Prove that for all non-zero real numbers x,
$$\left| x + \frac{1}{x} \right| = |x| + \frac{1}{|x|}.$$
Is it true that $\left| x - \frac{1}{x} \right| = |x| - \frac{1}{|x|}$ for all such x?

49. Define a relation $\sim$ on $\mathbf{R} - \{0\}$ by
$$x \sim y \text{ if and only if } x/y \text{ is rational.}$$
Prove that $\sim$ is an equivalence relation on $\mathbf{R} - \{0\}$.
What is the equivalence class determined by 1?

50. Prove that if a, b, $c \in \mathbf{R}$ then
$$a^2 + b^2 + c^2 \geq bc + ca + ab,$$
equality holding if and only if $a = b = c$.

Chapter 5: Induction, Finite Summations, Permutations and Selections

51. Prove by induction that, for all $n \in \mathbf{N}$,

$$\sum_{r=1}^{n} (-1)^r r = \tfrac{1}{4}[(-1)^n(2n+1)-1].$$

52. Prove by induction that, for all $n \in \mathbf{N}$,

$$\sum_{r=1}^{n} (-1)^{r-1} \frac{2r+1}{r(r+1)} = 1 + \frac{(-1)^{n-1}}{n+1}.$$

53. Prove that, for $n \geq 2$,

$$\prod_{r=2}^{n} \frac{r^2}{r^2-1} = \frac{2n}{n+1}.$$

54. Prove by induction that, if θ is a given acute angle then, for all $n \in \mathbf{N}$, $\sin n\theta \leq n \sin \theta$.

55. A sequence $u_1, u_2, u_3, \ldots$ of integers is defined by $u_1 = 1$ and for all $n \in \mathbf{N}$

$$u_{n+1} = 2u_n + 3.$$

Write down the values of u_2, u_3, u_4. Prove by induction that for all $n \in \mathbf{N}$

$$u_n = 2^{n+1} - 3.$$

56. A sequence $v_1 = 1, v_2 = \tfrac{1}{3}, v_3 = \tfrac{1}{7}, \ldots$ is defined by $v_1 = 1$ and $v_{n+1} = v_n/(v_n+2)$, $n \in \mathbf{N}$. Show by induction that for all $n \in \mathbf{N}$, $v_n = 1/(2^n - 1)$.

57. Let $f_1 = 1, f_2 = 1, f_3 = 2, f_4 = 3, f_5 = 5, \ldots$ be the *Fibonacci numbers* defined by

$$f_1 = f_2 = 1, \qquad f_{n+2} = f_{n+1} + f_n \quad \text{for all } n \in \mathbf{N}$$

[cf. Chapter 6, Exercise 6, Q. 5]. Prove by induction that, for all $n \in \mathbf{N}$, f_{3n} is even. [*Hint.* $f_{3n+3} = 2f_{3n+1} + f_{3n}$.]

Write out some more terms of the sequence, guess a similar result with "divisible by 3" in place of "even", and prove your conjecture.

58. How many numbers between 100 and 999 contain 3 different digits?

59. How many seven-digit numbers can be formed by permuting the digits 2, 2, 2, 3, 4, 5, 7? How many of them are divisible by 25?

60. The set S contains 8 elements; a given subset B has 3 elements. How many subsets of S contain B? [The sets B and S are to be included.]

For how many subsets C of S containing 4 elements does $B \cap C$ consist of 2 elements?

61. A and B are disjoint subsets of a set E. A, B and E contain 3, 4 and 10 elements respectively. For how many subsets C of E do $A \cap C$ and $B \cap C$ both contain just one element?

62. The sets A and B have 4 and 7 elements respectively. Find the number of injective mappings from A into B.

63. Prove that $(r+1)\dbinom{n+1}{r+1} = (n+1)\dbinom{n}{r}$ for $0 \leq r \leq n$.

64. Prove that

$$\binom{n+2}{r} = \binom{n}{r} + 2\binom{n}{r-1} + \binom{n}{r-2} \qquad (2 \le r \le n),$$

$$\binom{n+3}{r} = \binom{n}{r} + 3\binom{n}{r-1} + 3\binom{n}{r-2} + \binom{n}{r-3} \qquad (3 \le r \le n)$$

and guess a similar result for $\binom{n+4}{r}$.

65. Prove that the coefficient of x^4 in the expansion of $(1+x+x^3)^n$ is

$$\tfrac{1}{24}n(n-1)(n^2 - 5n + 30).$$

66. If $(1+3x+x^2)^n = c_0 + c_1 x + c_2 x^2 + \ldots + c_{2n}x^{2n}$

find $c_0 + c_1 + c_2 + \ldots + c_{2n}, \quad c_0 - c_1 + c_2 - \ldots + c_{2n}$

and $c_1 + 2c_2 + 3c_3 + \ldots + 2nc_{2n}.$

67. Use the binomial theorem to show that, for all $n \in \mathbf{N}$,

$$\left(1 + \frac{1}{n+1}\right)^{n+1} > 2.$$

Hence prove by induction that, for all $n \in \mathbf{N}$, $n! \le \{\tfrac{1}{2}(n+1)\}^n$.

68. Prove that $\binom{n}{r} \le \dfrac{n^r}{2^{r-1}}$ for $1 \le r \le n$.

Hence (*harder*) use the binomial theorem to show that for all $n \in \mathbf{N}$

$$\left(1 + \frac{1}{n^2}\right)^n < \frac{2n+1}{2n-1}.$$

Chapter 6: Elementary Number Theory

69. Prove that in the square of an odd integer (expressed to base 10) the tens digit is always even, e.g. $39^2 = 1521$.

70. Find the two 2-digit numbers (in decimal notation) whose representations in the scale of 7 consist of the same digits in reverse order.

71. Find *two* solutions in integers of $779x + 529y = 1$, one in which $x < 0$ and one in which $y < 0$.

72. Show that if a, b, x, y are integers such that $ax + by = (a, b)$ then $(x, y) = 1$.

73. Prove that, for all $n \in \mathbf{N}$, $(4n+3, 7n+5) = 1$.

74. A student worked out the greatest common divisors of pairs of the natural numbers a, b, c and found that

$$(a, b) = 9, \quad (b, c) = 30, \quad (c, a) = 12.$$

Why can you be sure that he made a mistake?

75. Show that if p and q are two distinct primes then the pq numbers $px + qy$, where $x, y \in \mathbf{N}$ and $1 \le x \le q, 1 \le y \le p$, are all different. [*Hint*. If $px + qy = px' + qy'$ then $q|p(x - x')$; use Th. 5.3.]

76. Prove that $(n+1)|(2n+1)\binom{2n}{n}$ by identifying $\dfrac{2n+1}{n+1}\binom{2n}{n}$ with a binomial coefficient. Show that $(2n+1, n+1) = 1$ for all $n \in \mathbf{N}$ and deduce that $\binom{2n}{n} \equiv 0$ (mod $n+1$).

77. Prove that, for all $n \in \mathbf{N}$,
$$3^{2n} \equiv 2^n \text{ (mod 7) and } 2^{2n+1} + 3^{2n+1} \equiv 0 \text{ (mod 5)}.$$

78. Prove that for every natural number n (expressed to base 10) n and n^5 end in the same digit.

79. Factorize 1001. When n is the six-digit number $abcdef$ in the scale of 10, $F(n)$ denotes the number $abc - def$. For instance, when $n = 571361$, $F(n) = 571 - 361 = 210$. Prove that $n \equiv 0$ (mod 7) if and only if $F(n) \equiv 0$ (mod 7). What other primes could replace 7 in this statement?

Chapter 7: Complex Numbers

80. Evaluate: (i) $(1+2i)^2(1-i) + (1-2i)^2(1+i)$,

(ii) $\dfrac{1+3i}{3+i} - \dfrac{3-i}{1-3i}$.

81. Simplify $(a+ib)^{10} + (b-ia)^{10}$.

82. Prove that for any $z \in \mathbf{C}$, $\mathcal{R}(iz) = -\mathcal{I}z$, $\mathcal{I}(iz) = \mathcal{R}z$.

83. Prove that if $z = \tan \theta \, (\cos \theta + i \sin \theta)$ then
$$\mathcal{R}(z + z^{-1}) = \operatorname{cosec} \theta.$$

84. Show that for $z, w \in \mathbf{C}$, $\mathcal{R}(zw) = \mathcal{R}z.\mathcal{R}w$ if and only if at least one of z and w is real.

85. Find the modulus and the principal value of the argument of each of the complex numbers $-3+3i$, $-\dfrac{1}{\sqrt{3}} - i$, $3 - i\sqrt{3}$ and $1 - i \tan \alpha$ $(-\frac{1}{2}\pi < \alpha < \frac{1}{2}\pi)$.

86. Evaluate $(\cos \frac{1}{7}\pi + i \sin \frac{1}{7}\pi)(\cos \frac{2}{7}\pi + i \sin \frac{2}{7}\pi)(\cos \frac{4}{7}\pi + i \sin \frac{4}{7}\pi)$.

87. Use de Moivre's theorem to find the square roots of $1 + i\sqrt{3}$.

88. Find positive real numbers x and y such that
$$(x+iy)^2 = 1+i.$$
Hence write down the values of $\cos \frac{1}{8}\pi$ and $\sin \frac{1}{8}\pi$.

89. The complex number z has modulus 1. Prove that
$$\frac{(2+z)(1+2z)}{(3+z)(1+3z)} \in \mathbf{R}.$$

90. Points A, B, C in the complex plane represent complex numbers a, b, c such that $(a-b)^2 + 3(b-c)^2 = 0$. Describe the triangle ABC.

91. Prove that, if z, a, b are complex numbers such that
$$(2z - a - b)^2 + (a-b)^2 = 0,$$
then $|z - a| = |z - b|$.

92. Express $\sin^8 \theta$ in terms of cosines of integral multiples of θ.

93. If p, q, r are the roots of the equation $z^3 - 7z - 6 = 0$ calculate $p^2 + q^2 + r^2$ and $p^4 + q^4 + r^4$. Check by finding p, q, r.

94. Prove that, if z satisfies

$$z^3 + az^2 + bz + c = 0, \qquad (*)$$

then
$$z^2(z^2 + b)^2 = (az^2 + c)^2.$$

Hence write down a cubic equation whose roots are the squares of those of ($*$). Before trying questions **95** and **96** consider Topic to Explore II, p. 143.

95. Find the sum of the squares of the roots of the equation

$$z^4 + 3z^3 + 2z^2 - 2z - 4 = 0.$$

Check by solving the equation completely.

96. Show that the roots of $(z+1)^5 = z^5$ are the numbers $-\frac{1}{2}(1 + i \cot \frac{1}{5}r\pi)$, $r = 1, 2, 3, 4$. Remembering that $\cot \theta = -\cot (\pi - \theta)$, check the sum of these roots.

97. Explain why a cubic equation with real coefficients cannot have a non-real double root.

98. Write down all the 8th roots of unity in the form $a + ib$ $(a, b \in \mathbf{R})$.

99. Simplify $\sqrt{2}(1+i)(\sqrt{3}+i)^{-1}$ and show that this is a 24th root of unity.

100. Mappings $f, g : \mathbf{C} \to \mathbf{C}$ are defined by

$$f(z) = \bar{z}, \qquad g(z) = cz,$$

where $c \in \mathbf{C}$ is constant. Prove that $f \circ g = g \circ f$ if and only if c is real.

101. Let $S = \mathbf{C} - \{0\}$ and define mappings $f, g : S \to S$ by

$$f(z) = z|z|^2, \qquad g(z) = z/|z|.$$

Determine whether f and g are (i) injective, (ii) surjective.

102. Define mappings $p, q : \mathbf{C} \to \mathbf{C}$ by $p(z) = i\bar{z} + 1, q(z) = \bar{z} - i$. Find $p \circ q$ and $q \circ p$.

103. A mapping $f : \mathbf{C} \to \mathbf{C}$ is defined by $f(z) = a\bar{z} + b$ where $a, b \in \mathbf{C}$ are constants and $a \neq 0$. Prove that f is a bijection, determining f^{-1}. Prove that if $|a| \neq 1$ then f has exactly one fixed element [see Additional Examples II, Q. **35**.]

104. A, B, C are three consecutive vertices of a regular polygon inscribed in a circle with its centre at the origin of the complex plane. The complex numbers corresponding to A, B are a, b respectively. Find the number corresponding to C.

105. Show that $z^2 - z\sqrt{3} + 1$ is a factor of $z^{12} - 1$.

106. Find $\sum\limits_{n=1}^{\infty} \sin^n \theta \sin n\theta$, where $\theta \neq \frac{1}{2}(2r+1)\pi, r \in \mathbf{Z}$.

107. Write down the expansion of $(1+i)^n$ by the binomial theorem. Deduce that

$$1 - \binom{2m}{2} + \binom{2m}{4} - \binom{2m}{6} + \ldots + (-1)^m \binom{2m}{2m} = 2^m \cos \tfrac{1}{2}m\pi.$$

Chapter 8: Matrices

108. Prove that all the matrices X such that $XA = A'X$, where $A = \begin{bmatrix} 1 & 2 \\ 3 & 4 \end{bmatrix}$, are symmetric.

109. The 2×2 matrix A satisfies $|I + A| = 1 + |A|$, where I is the unity 2×2 matrix. Prove that $|I - A| = 1 + |A|$.

110. Verify that if $A = \begin{bmatrix} 2 & 3 \\ 3 & 5 \end{bmatrix}$ then $A^2 = 7A - I$, where I is the unity 2×2 matrix.

Express A^3 in the form $A^3 = pA + qI$ where p, q are integers.

111. A matrix of the form $\begin{bmatrix} a & b & c \\ 0 & d & e \\ 0 & 0 & f \end{bmatrix}$ is called *upper triangular*. Prove that the product of two matrices of this kind is itself upper triangular.

112. Show that for all a, b, c the matrix $T = \begin{bmatrix} 1 & a & b \\ 0 & 1 & c \\ 0 & 0 & 1 \end{bmatrix}$ is invertible, and find T^{-1}.

113. Find the general solution of the system of equations
$$2x + 3y - 5z = 0, \qquad 3x - y + 4z = 0.$$

114. Solve the system of equations
$$x + 2y - 3z = 5, \qquad 3x + 7y - 2z = 0, \qquad 4x - y + 2z = 1.$$

115. Solve the system of equations
$$x - y + 4z = 3, \qquad 2x + 3y - 6z = 7, \qquad x + 9y - 24z = 5.$$

116. Solve the system of equations
$$ax + (a+1)y = a, \qquad (a+1)x + (a+3)y = 2$$
for x and y, distinguishing between the cases $a \neq 1$ and $a = 1$.

117. Prove that if $a \neq -5$ the system of equations
$$x + y + z = 0, \qquad x + ay - z = a + 3, \qquad 2x - y + z = -1$$
have a unique solution which is independent of a. Is this still a solution when $a = -5$? If so, is it the only solution?

118. Show that if
$$ax + by = 0, \qquad cx + dy = 0 \qquad\qquad (*)$$
with x, y not both zero, then $ad = bc$. Prove also the converse that, if $ad = bc$, then the equations $(*)$ have a non-zero solution.

119. If $A = \begin{bmatrix} 1 & -2 & 5 \\ 2 & -3 & 8 \\ -1 & 4 & -10 \end{bmatrix}$, $B = \begin{bmatrix} 2 & -2 & -1 \\ 3 & -2 & -1 \\ -5 & 7 & 3 \end{bmatrix}$ use row operations to find the matrix C such that $AC = B$. [*Note*. It is not necessary to find A^{-1} independently.]

120. Prove that if A is a square matrix such that $A + I$ is invertible, I being the unity matrix, then
$$A(A+I)^{-1} = (A+I)^{-1}A.$$

121. Prove that if A is a non-singular matrix then $|A^{-1}| = |A|^{-1}$. Show that if all the elements of A and of A^{-1} are integers, then $|A| = \pm 1$.

122. Let B be a non-singular 3×3 matrix. By taking determinants on both sides of the relation adj $B = |B|B^{-1}$ prove that
$$|\text{adj } B| = |B|^2.$$
Verify this for $B = \begin{bmatrix} 1 & 1 & 2 \\ 2 & 1 & -1 \\ 3 & 1 & 1 \end{bmatrix}$. [*Note.* It can be shown that $|\text{adj } B| = |B|^2$

also holds when $|B| = 0$; thus if B is singular so is adj B.]

123. Show that if B is a skew-symmetric $m \times m$ matrix and A is any matrix of order $m \times n$, then $A'BA$ is skew-symmetric.

124. Prove that if A is a square matrix such that A and $\dfrac{1}{\sqrt{2}}(I+A)$ are both ortho-

gonal then A is skew-symmetric and $\dfrac{1}{\sqrt{2}}(I-A)$ is orthogonal. Verify that this

holds when $A = \begin{bmatrix} 0 & 1 \\ -1 & 0 \end{bmatrix}$.

125. Verify that if $A = \begin{bmatrix} -1 & -1 \\ 2 & -2 \end{bmatrix}$ then $A^2 = -3A - 4I$ where I is the 2×2 unity

matrix. Find a matrix B of the form $B = xA + yI$, where x, y are real, such that $B^2 = A$.

126. Let M_n be the set of $n \times n$ matrices with real elements. For $A, B \in M_n$ write $A \sim B$ when $\exists P \in M_n$ such that P is invertible and $A = PBP^{-1}$. Prove that $\sim$ is an

equivalence relation on M_n. Taking $n = 2$ prove that $\begin{bmatrix} 1 & 0 \\ 0 & 2 \end{bmatrix} \sim \begin{bmatrix} 2 & 0 \\ 0 & 1 \end{bmatrix}$.

127. Define mappings f, g, h from M_n [see Q. **126**] to itself by
$$f(A) = A', \qquad g(A) = A^2, \qquad h(A) = I + A,$$
where I is the unity matrix in M_n.
(a) Which of f, g, h are (i) injective, (ii) surjective?
(b) Do f and g commute—i.e. is $f \circ g = g \circ f$? Do f and h commute? Do g and h commute?

128. Mappings $p, q : M_n \to M_n$ [see Q. **126**] are defined by
$$p(A) = A + A', \qquad q(A) = A + 2A'.$$
Prove that (i) the image of p is the set of all symmetric matrices in M_n; (ii) q is a bijection. Is p injective?

Chapter 9: Algebraic Structures I—Groups

129. Prove that the set of rational numbers expressible in the form m/n where m and

n are odd integers forms an abelian group under multiplication. Is this set closed under addition?

130. Prove that the set $\{x \in \mathbf{Q} : 2^n x \in \mathbf{Z} \text{ for some } n \in \mathbf{N}\}$ is an abelian group under addition.

131. An operation o on $\mathbf{Z}$ is defined by $x \circ y = xy + 2(x + y) + 2$.
   (i) Show that o is commutative.
   (ii) Verify that $(1 \circ 2) \circ 3 = 1 \circ (2 \circ 3)$.
   (iii) Show that o is associative.
   (iv) Does o have an identity element?

132. An operation $*$ is defined on $\mathbf{Z}$ by $x * y = xy + 2(x + y) + 1$. By considering a numerical example, or otherwise, show that $*$ is not associative.

133. Prove that matrix multiplication is a binary operation on the set

$$S = \left\{ \begin{bmatrix} 1 & r \\ 0 & 2^n \end{bmatrix} : r \in \mathbf{Q}, n \in \mathbf{Z} \right\}.$$

Is S a group under this operation?

134. Prove by induction that if a, b are elements of a multiplicative group G then, $\forall n \in \mathbf{N}$,

$$(bab^{-1})^n = ba^n b^{-1}.$$

[*Note.* Any element of the form bab^{-1} is said to be *conjugate* to a in G.]

135. Prove that if $A = \begin{bmatrix} 1 & 1 \\ -2 & -1 \end{bmatrix}$, $I = \begin{bmatrix} 1 & 0 \\ 0 & 1 \end{bmatrix}$ then the matrices $I, -I, A, -A$
form a group isomorphic to C_4 under multiplication.

136. Prove that the set O_n of all orthogonal $n \times n$ matrices with real elements is a group under matrix multiplication.
   Show that if $A \in O_n$ then $|A| = \pm 1$. Which of the sets $\{A \in O_n : |A| = 1\}$, $\{A \in O_n : |A| = -1\}$ is a subgroup of O_n?

137. An operation o on the set $S = \{(0, 0), (1, 0), (0, 1), (1, 1)\}$ is defined by $(a, b) \circ (c, d) = (a +_2 c, b +_2 d)$, where $+_2$ denotes addition (mod 2). Prove that S is a group under o and determine whether it is isomorphic to C_4 or to the four-group.

138. Prove that if H and K are subgroups of a group G then $H \cap K$ is a subgroup of G.

139. G is a group with its operation written multiplicatively. A new operation $*$ is defined on G by

$$x * y = xgy$$

for all $x, y \in G$, where g is a given fixed element of G. Prove that G is also a group under $*$.

140. The elements a, b of a multiplicative group G satisfy $a^6 = b^2 = e$, $ab = ba^5$, where e is the identity element. Prove that $(ab)^2 = e$, $a^2 b = ba^4$ and $ba = a^5 b$.

141. Elements x, y, z, t of a group G satisfy $xy = zt$. By considering suitable elements of the symmetry group of an equilateral triangle show that for G non-abelian it is not necessarily true that $yx = tz$.

142. a, b are elements of a group with identity e such that $ba = a^2b$, $b^3 = e$. Prove that (i) $bab^{-1} = a^2$, (ii) $b^2ab^{-2} = a^4$, (iii) $a^8 = a$, (iv) $a^7 = e$.

143. Prove that $\mathbf{Z}$ is a group under the operation $\times$ defined by

$$m \times n = m + (-1)^m n.$$

[Note that $(-1)^k = (-1)^{-k}$ for all $k \in \mathbf{Z}$.]

144. S is a set on which is defined an associative operation $\circ$. S contains an element e such that $x \circ e = x$ for all $x \in S$. Moreover, for each $x \in S$ there exists $\bar{x} \in S$ such that $\bar{x} \circ x = x \circ \bar{x} = e$. Prove that $e \circ x = x$ for all $x \in S$. Thus $\{S, \circ\}$ is a group. [*Hint.* $e \circ x = (x \circ \bar{x}) \circ x$.]

Chapter 10: Algebraic Structures II—Rings, Integral Domains, Fields

145. Prove that the set of all matrices of the form $\begin{bmatrix} a & b & c \\ c & a & b \\ b & c & a \end{bmatrix}$, where $a, b, c \in \mathbf{R}$, forms

a ring under matrix addition and multiplication. Is the ring an integral domain?

146. Define an operation $\circ$ on $\mathbf{Z}$ by

$$x \circ y = \begin{cases} x & \text{if } x \geq y; \\ y & \text{if } x < y. \end{cases}$$

Is $\circ$ associative? If $+$ has its usual meaning is $\{\mathbf{Z}, +, \circ\}$ a ring?

147. The complex numbers of the form $a + bi$ where $a, b \in \mathbf{Z}$ are called *Gaussian integers*. Prove that under the usual addition and multiplication the set G of Gaussian integers is an integral domain, but not a field.

148. The elements $0, 1, 2, 3, 4$ form a field under addition and multiplication (mod 5). Solve the equations $x^2 + 1 = 0$ and $x^2 + x - 1 = 0$ in this field.

149. Let S be the ring formed by $0, 1, 2, 3, 4, 5, 6, 7$ under addition and multiplication (mod 8) [Exercise **10**, No. 3(ii)]. Show that the equation $x^2 - 1 = 0$ has *four* roots in S. Find quadratic equations with coefficients in S having (i) precisely two roots, (ii) precisely one root, (iii) no root, in S.

150. Let F be the field consisting of the elements $0, 1, 2$ with addition and multiplication (mod 3). Define addition and multiplication on $F \times F$ by

$$(a, b) + (c, d) = (a + c, b + d),$$
$$(a, b)(c, d) = (ac - bd, ad + bc).$$

Prove that $F \times F$ then becomes a field.
Prove that if x and y are any two elements of $F \times F$ then

$$(x + y)^3 = x^3 + y^3.$$

151. A field K has the property that for all $x \in K$, $x + x = 0$. Prove that if $y, z \in K$ satisfy $y^2 = z^2$ then $y = z$.

152. A field F contains q elements.

(i) How many different 2×2 matrices $\begin{bmatrix} a & b \\ c & d \end{bmatrix}$ are there with elements in F?

    (ii) Show that among these matrices there are $(2q-1)^2$ for which
$$ad = bc = 0.$$
    (iii) How many matrices are there with $ad = bc \neq 0$?

    (iv) Deduce that there are $q(q+1)(q-1)^2$ non-singular matrices [those for which $ad \neq bc$].

153. It is given that the elements $0, 1, j, j^2$ form a field F when addition and multiplication are defined as shown in the tables

+	0	1	j	j^2
0	0	1	j	j^2
1	1	0	j^2	j
j	j	j^2	0	1
j^2	j^2	j	1	0

.	0	1	j	j^2
0	0	0	0	0
1	0	1	j	j^2
j	0	j	j^2	1
j^2	0	j^2	1	j

    (i) What type of group is $\{F, +\}$?

    (ii) What type of group is $\{F-\{0\}, .\}$?

    (iii) Working in the field F simplify the expression $(1+j)(1+j^2)$.

    (iv) Solve the equation $x^2 + jx + j^2 = 0$ in F. Find a quadratic equation with coefficients in F which has no root in F.

    (v) Solve the equation $x^4 = x$ in F.

154. Let $S = \{2n : n \in \mathbf{Z}\}$, $T = \{3n : n \in \mathbf{Z}\}$. Then, under the customary addition and multiplication, S and T are rings. Prove that the mapping from S to T defined by $2n \to 3n$ is not an isomorphism between them. [*Note.* This does *not* show that (as is in fact the case) S and T are non-isomorphic. (Why not?) This is a little harder to prove.]

Answers

Exercise 1
2. (i) F, (ii) T, (iii) T, (iv) F, (v) T, (vi) F, (vii) T, (viii) T, (ix) F, (x) T, (xi) F, (xii) F.
3. Truth values of given statements: (i) F, (ii) T, (iii) T, (iv) F, (v) F, (vi) T, (vii) T, (viii) F, (ix) T, (x) T, (xi) F, (xii) F.
5. (a): (i) (iv) (vi); (b): (iii) (iv) (v) (vi) (vii); (c): (iv) (vi).

Exercise 2
1. $\{1, 6\}, \{3, 5, 8\}$.
2. (i), (ii), (v).
3. $\{1, 2\}, \{1, 2, 3, \{1, 2\}, \{1, 2, 3\}\}$.
4. $\{3\}, (0, 3), (-\infty, 3] \cup [5, \infty)$.
5. $(1, 6), \emptyset, [3, 4], (-\infty, 1] \cup [3, 4]$.
6. $(-\frac{1}{2}, 0] \cup [\frac{1}{2}, 1]$.
13. 11.
15. (i), (iv).
16. (i), (ii).
17. $\{O\}$ and circles with centre O, where O is $(0, 0)$; $\{(x, y): x^2 + y^2 = 2\}$.

Exercise, p. 43. Those of Examples **6, 7.**

Exercise, p. 45. (i) Those of Examples **1, 2, 7, 9.**
(ii) Surjective, not injective.

Exercise, p. 46. $g^{-1}(y) = \frac{1}{3}(y+2); g^{-1}(y) = \begin{cases} \sqrt{y} & \text{when} \quad y \geqslant 0, \\ y & \text{when} \quad y < 0. \end{cases}$

Exercise 3
1. (i) $-1, 3, 8; [-1, \infty)$. (ii) $[-5, 10]$. (iii) $3, 0, 0; [0, \infty) = \mathbf{R}^+$. (iv) $[-1, 1]$.
2. (i) Injective, not surjective, image $[1, \infty)$.
(ii) Injective, not surjective, image $(-\infty, -1) \cup [1, \infty)$.
(iii) Surjective, not injective, $h(1) = h(-1) = 1$.
(iv) Neither injective nor surjective, $k(1) = k(-1) = 2$, image $[1, \infty)$.

279

3. (i) (a), (b), $f^{-1}(y) = y - 2$. (ii) (b).

(iii) (a), (b), $h^{-1}(y) = \begin{cases} \sqrt{(1+y)} & \text{when} \quad y \geqslant -1, \\ -\sqrt{(-1-y)} & \text{when} \quad y < -1. \end{cases}$

(iv) (a). (v) (a). (vi) (a), (b), $v^{-1}(m) = \begin{cases} m-2 & (m \text{ even}), \\ m & (m \text{ odd}). \end{cases}$

4. Possible answers: (i) $f_1(s) = s + 1$; (ii) $f_2(s) = 2$, all s; (iii) $f_3(s) = 2s$; (iv) $f_4(s) = |s - 2| + 1$.

5. $f \circ f(x) = x^4$; $f \circ g(x) = \begin{cases} (x-1)^2 & \text{when} \quad x \geqslant 0, \\ x^2 & \text{when} \quad x < 0 \end{cases}$

$g \circ f(x) = x^2 - 1$; $g \circ g(x) = \begin{cases} x - 2 & \text{when} \quad x \geqslant 1, \\ 1 - x & \text{when} \quad 0 \leqslant x < 1, \\ -x - 1 & \text{when} \quad x < 0. \end{cases}$

6. $f \circ g(x) = \begin{cases} 4x+2 & \text{when} \quad x \geqslant 0, \\ x+2 & \text{when} \quad x < 0, \end{cases}$ $g \circ f(x) = \begin{cases} 4x+1 & \text{when} & x \geqslant 0, \\ 2x+5 & \text{when} & -2 \leqslant x < 0, \\ x+2 & \text{when} & x < -2. \end{cases}$

$f \circ g$ is bijective, with $(f \circ g)^{-1}(y) = \begin{cases} \frac{1}{4}(y-2) & \text{when} \quad y \geqslant 2, \\ y - 2 & \text{when} \quad y < 2. \end{cases}$

7. $g \circ f(x) = |\cos x| : \mathbf{R} \to \mathbf{R}$, image $[0, 1]$.
$f \circ g(x) = \sin \left[\sqrt{(1-x^2)}\right] : [-1, 1] \to [-1, 1]$, image $[0, \sin 1]$.

8. $g \circ f(x) = \begin{cases} 2x-1 & \text{when} & x \geqslant 0, \\ 1-x^2 & \text{when} & -1 < x < 0, \\ 2x^2 - 3 & \text{when} & x \leqslant -1. \end{cases}$ $f \circ g(x) = \begin{cases} 2x-2 & \text{when} & x \geqslant \frac{3}{2}, \\ (2x-3)^2 & \text{when} & 1 \leqslant x < \frac{3}{2}, \\ 2 - x & \text{when} & x < 1. \end{cases}$

13. (i) $[3, \infty)$, (iii) $g : \mathbf{R}^+ \to [3, \infty)$, $g(x) = x^2 + 3$, (iv) $g^{-1}(y) = \sqrt{(y-3)}$.
14. $g : [0, \pi] \to [-1, 1]$, $g(x) = \cos x$.
15. (i) $\mathbf{R}^+$, (ii) $\mathbf{R} - \{(2k+1)\pi : k \in \mathbf{Z}\}$, (iii) $\{0\}$, (iv) $[-2, 2]$, (v) $(-1, 1]$, (vi) $\emptyset$, (vii) $\mathbf{R}$, (viii) $\mathbf{R}$.

Exercise, p. 68. $(-2, -1] \cup (2, 4]$, $\{-1\} \cup (2, 3]$.

Exercise 4
6. (i) F [e.g. $a = \sqrt{2}$, $b = -\sqrt{2}$], (ii) T, (iii) F [e.g. $a = \pi$, $b = 1/\pi$], (iv) T, (v) T.
8. Rational ($= 2$).
9. (i) $(0, 1]$, (ii) $(-\infty, \frac{7}{3}) \cup (3, \infty)$, (iii) $(-1, 2) \cup (4, \infty)$, (iv) $(-2, 0)$, (v) $(0, 1) \cup (2, 4)$, (vi) $[-2, 1] \cup \{4\}$, (vii) $(-\infty, 4)$, (viii) $[-4, -2]$.

Exercise 5

3. (i) 63·05, (ii) 1·80, (iii) 1·02.
4. (ii) 10084.

5. (i) $-\binom{12}{3}.2^9.3^3$, (ii) 84, (iii) 70, (iv) $-\binom{18}{3} = -816, \binom{18}{6} = 18564$, (v) 1792,

(vi) $1 + 12x + 54x^2 + 100x^3$,

(vii) $\dfrac{(-1)^n.(2n)!}{n!(n+2)!}(9n^2 + 7n + 2)$.

6. (i) 55, (ii) $4845x^4/8$ and $4845x^8/8$, (iii) 14, (iv) The coefficients of $x^{(n-1)/2}$ and $x^{(n+1)/2}$ are equal largest.
8. 1062.
9. 300, 140.
10. $13!/48, 11!/8$.
11. $10!/2 = 1814400, 15120$.
12. 3360, 96, 2160, 480.
13. $\frac{1}{6}m^2n(n-1)(mn+m-3)$.
14. 120, 72; 625, 108.
15. $(p+q+r)!/p!q!r!, 280$.

Exercise 6.1

1. 11, 6; 8, 89; $-26, 17; -22, 0; 130, 9$.
2. $a = b$.
4. No [e.g. $a = b_1 = b_2 = 2$].

Exercise 6.2

1. (i) $(1010001)_2 = (311)_5 = (144)_7 = (100)_9$.
   (ii) $(1001000011)_2 = (4304)_5 = (1455)_7 = (713)_9$.
   (iii) $(10000101000)_2 = (13224)_5 = (3050)_7 = (1412)_9$.
   (iv) $(11101110110111)_2 = (442122)_5 = (62366)_7 = (22865)_9$.
   (i) $(69)_{12}$, (ii) $(403)_{12}$, (iii) $(748)_{12}$, (iv) $(8t1e)_{12}$.
2. (i) 157, (ii) 578, (iii) 2766, (iv) 419184.
3. (i) 333041, (ii) 1121021, (iii) 111110.
4. 13.
5. 624.

Exercise 6.3

1. $d = 37, x = 3, y = -4$, divisors 1, 37;
   $d = 53, x = 3, y = -4$, divisors 1, 53;
   $d = 1, x = -952, y = 1731$, divisor 1;
   $d = 91, x = 2, y = -3$, divisors 1, 7, 13, 91.
   [*Note.* The values of x, y given are those obtained by direct application of the euclidean algorithm; in each case the equation $d = xa_1 + ya_2$ has other integral solutions.]

2. (i) $37 = 3.1147 - 4.851 + 0.407$,

  (ii) $1 = -849.1219 + 1132.901 + 15.1000$,

  (iii) $1 = -952.5213 + 1731.2867 + 0.a$,

  (iv) $13 = 56.4277 - 84.2821 - 3.845$.

3. $2(a, b)$ if $a = 2^n\alpha$, $b = 2^n\beta$ with $n \geqslant 0$ and α, β odd; (a, b) otherwise.

Exercise 6.4

1. (i) $2.5.7^2$, (ii) $3^2.5^3$, (iii) $2^3.11.31$, (iv) $2.3.5.11.13^{-2}$, (v) $2^{-4}.7^2.11^{-1}.17$, (vi) $3^2.11.101$.

2. For primes <100 see p. 98; 101, 103, 107, 109, 113, 127, 131, 137, 139, 149, 151, 157, 163, 167, 173, 179, 181, 191, 193, 197, 199.

3. $n = 6$, $30031 = 59.509$.

4. (1) T, (2) T, (3) T, (4) F [e.g. $p = 5$, $a = 1$, $b = 2$, $c = 1$], (5) T, (6) F [e.g. $a = 2$, $m = 15$].

5. E.g. $a = b = m = 2$.

7. p or p^2; p, p^2 or p^3; p^2 or p^3.

8. 2, 6, 10, 14, 18, 22; $36 = 2.18 = 6.6$, $60 = 2.30 = 6.10$.

Exercise 6.5

1. (i) 5, (ii) 4, (iii) 0, (iv) 2, (v) 2.

2. 6, 17, 28, 39, 50, 61, 72, 83, 94.

3. (E.g.) $\{3k : k \in \mathbf{Z}, -8 \leqslant k \leqslant 8\}$.

4. (i) $\{x : x = 12k + 5 \text{ for some } k \in \mathbf{Z}\}$.

  (ii) $\{x : x = 13k + 2 \text{ for some } k \in \mathbf{Z}\}$.

8. (i) $x \equiv 8 \pmod{13}$, (ii) $x \equiv 23 \pmod{37}$, (iii) $x \equiv 6 \pmod 9$, (iv) $x \equiv 4$ or 9 (mod 10), (v) $\emptyset$, (vi) $x \equiv 5$ or 8 (mod 13), (vii) $\emptyset$, (viii) $x \equiv 1 \pmod 8$, (ix) $x \equiv 1$ or 5 (mod 6).

11. Addition: Closed, identity 0, all elements have inverses. Multiplication: Closed, identity 1, only 1, 5 have inverses.

12. Closed, identity 1, each element is its own inverse.

Exercise 6

4. (E.g.) $x = 2$, $y = 0$, $z = -1$.

18. $x \equiv 5 \pmod 6$.

20. E.g. $p = 4$, $a = b = 1$.

Exercises, p. 120

(a) $5 + 15i$, -18, $-38 + 41i$.

(b) $1 - i$, $-1 + i$, $-1 - i$.

(c) $-2 + 7i$, $\frac{1}{4}(-3 \pm i\sqrt 7)$.

(d) $x^2 - y^2$, $3x^2y - y^3$, $[1 - x^2 - y^2]/[(1 - x)^2 + y^2]$.

(e) $(5 - i)/26$.

Exercise, p. 124. $\pm(1 + 4i)$.

Exercises, pp. 125–6

(a) A square.

(b) Multiplication of a vector by a scalar.

Exercise, p. 127. $\sqrt{2}$, $\frac{1}{4}\pi$; $3\sqrt{2}$, $\frac{3}{4}\pi$; 5, $-\frac{1}{2}\pi$; 10, π; 2, $\frac{2}{3}\pi$; 2, $-\frac{3}{4}\pi$; $\sqrt{13} \div 3 \cdot 606$, $-56°19' \div -0 \cdot 983$ radians.

Exercise, p. 130
(a) (ii) For $\pi < \theta < 2\pi$, PV arg $z = \frac{1}{2}\theta - \pi$.
(d) $r^2 + s^2 + 2rs \cos(\theta - \phi)$.
(e) (ii) $(-\pi, \pi]$, (iii) Yes, (iv) Yes, $\mathbf{R}^+$, no, no.

Exercises, p. 131
(a) (iii) arg $\bar{z} = -$arg z except for arg $z = \pi = $ arg $\bar{z}$.
(c) $\phi \circ \phi = $ identity mapping of $\mathbf{C}$.

Exercises, p. 135
(a) $\frac{1}{2}(i - \sqrt{3})$, $(1 + i)/\sqrt{2}$, $8(1 + i)$, $-512i$.
(b) $\cos \theta + i \sin \theta$, $\cos 4\theta + i \sin 4\theta$.

Exercises, p. 136
(a) $c^5 - 10c^3 s^2 + 5cs^4$, $5c^4 s - 10c^2 s^3 + s^5$, $16c^5 - 20c^3 + 5c$, $16c^4 - 12c^2 + 1$ $[c = \cos \theta$, $s = \sin \theta]$, 5, No.
(b) (i) $\frac{1}{16}(\cos 5\theta + 5 \cos 3\theta + 10 \cos \theta)$.
    (ii) $\frac{1}{64}(35 \sin \theta - 21 \sin 3\theta + 7 \sin 5\theta - \sin 7\theta)$.

Exercises, p. 139
(b) $(z + 1)^2 (z - 2)(z + 3)$.

Exercises, p. 142
(a) $2 \pm i$, $-1 \pm i$.
(b) $z^6 - 4z^5 + 7z^4 - 4z^3 - 4z^2 + 8z - 4 = 0$.

Exercises, p. 143
(a) -2, 14, -20; roots -1, 2, -3.
(b) $b^2 - 2ac$.
(d) Not all real.

Exercise, p. 144
(i) Equilateral triangle.
(iii) $(\alpha + \beta + \gamma)(\alpha + \omega\beta + \omega^2\gamma)(\alpha + \omega^2\beta + \omega\gamma)$.

Exercises, p. 146
(a) $2i$, $\pm\sqrt{3} - i$.
(b) The same roots appear in different order.
(c) Square.

Exercise, p. 147. $(z - 1)(z + 1)(z^2 + 1)(z^2 - z\sqrt{2} + 1)(z^2 + z\sqrt{2} + 1)$.

Exercise, p. 148. *na.*

Exercise, p. 150
(a) $2^n \cos^n \frac{1}{2}\theta \sin \frac{1}{2}n\theta$.
(b) No (see p. 130). No.
(c) $(-1)^{(n-1)/2} \sin n\theta \sin^n \theta$.

Exercise, p. 151
(a) $r \sin \theta/(1 - 2r \cos \theta + r^2)$.

Exercise 7
 1. (i) 125, (ii) $-\frac{1}{5}(4 + 7i)$, (iii) $-\frac{1}{4}(1 + i)$.
 2. (i) ± 1, (ii) 0.
 3. (i) 3, $n \equiv 3 \pmod 6$; (ii) 6, $n \equiv 0 \pmod 6$.
 4. $\frac{1}{3}(2z + w)$.
 5. (i) $\frac{1}{3}(z_1 + z_2 + z_3)$, (ii) parallelogram.
 6. Right-angled isosceles.
 7. $2\sqrt{2}$, $-\frac{1}{12}\pi$.
 8. $\frac{1}{2}(\theta + \phi)$.
 9. $1/\sqrt{3}$.
16. (i) $\cos \theta + i \sin \theta$, (iv) $n \equiv 1 \pmod 4$.
17. Approx. $7°7'$ or $0 \cdot 124$ radian, $n = 13$.
18. $c^8 - 28c^6s^2 + 70c^4s^4 - 28c^2s^6 + s^8$ $[c = \cos \theta, s = \sin \theta]$.
    (ii) $\frac{1}{128}(6 \sin 2\theta + 2 \sin 4\theta - 2 \sin 6\theta - \sin 8\theta)$.
19. (i) $-15 - 8i$, (ii) $-3i$, $-2 + 5i$.
20. $\pm 2(\sqrt{3} + i)$.
21. (i) 1, $-2 \pm 2i$, (ii) $\pm i$, $\pm 2i$, (iii) $-\frac{1}{2}$, $-\frac{1}{2}(1 \pm i)$.
22. 2.
23. $z^5 + z^4 - 2z^3 - 2z^2 + z + 1 = 0$.
25. $-1, 5, 4, z^3 + z^2 + 5z - 4 = 0$.
27. (i) $\theta = (8k + 1)\pi/12$, $k = 0, 1, 2$.
    (ii) $\theta = (6k + 1)\pi/12$, $k = 0, 1, 2, 3$.
28. ± 3, $\frac{3}{2}(1 \pm i\sqrt{3})$, $\frac{3}{2}(-1 \pm i\sqrt{3})$.
29. (i) T, (ii) F.
30. $(z^2 + z\sqrt{2} + 1)(z^2 - z\sqrt{2} + 1)$.
31. $a, b = \frac{1}{2}(1 \pm \sqrt{5})$.

32. $\displaystyle\prod_{k=1}^{4} \left(z^2 - 2z \cos \frac{2k\pi}{9} + 1\right)$, $\dfrac{1}{8}$.

37. $(1 - z)^{-2}$.

Exercise, p. 158

1. $A - B = \begin{bmatrix} 7 & -1 \\ -1 & -5 \end{bmatrix}$.

2. (b) $\begin{bmatrix} -5 & -1 & -6 \\ -6 & -7 & -14 \\ -9 & -5 & -8 \end{bmatrix}$, (c) $\begin{bmatrix} -5 & 0 & -4 \\ -3 & -3 & -9 \\ -3 & 2 & 0 \end{bmatrix}$, (d) $\begin{bmatrix} -5 & 1 & -2 \\ 0 & 1 & -4 \\ 3 & 9 & 8 \end{bmatrix}$.

Exercise, p. 159

2. $\begin{bmatrix} -14 & 14 & 12 \\ 34 & 6 & -4 \end{bmatrix}$.

Exercise, p. 166. A column matrix.

Exercise, p. 170. $y_1 = 11z_1 + 12z_2$, $y_2 = -19z_2$.

Exercise, pp. 172–3

1. $[x \quad y \quad z] \begin{bmatrix} 2 & 0 & 2 \\ 0 & -1 & \frac{5}{2} \\ 2 & \frac{5}{2} & 3 \end{bmatrix} \begin{bmatrix} x \\ y \\ z \end{bmatrix}$ 2. $\begin{bmatrix} 1 & 1 & 3 \\ 1 & 0 & 3 \\ 3 & 3 & 1 \end{bmatrix}$.

Exercise, p. 180

(i) $\dfrac{1}{2} \begin{bmatrix} 3 & 1 \\ 4 & 2 \end{bmatrix}$, (ii) no inverse, (iii) $\begin{bmatrix} \cos\theta & -\sin\theta \\ \sin\theta & \cos\theta \end{bmatrix}$.

Exercise 8

1. (i) $\begin{bmatrix} 4 & -2 \\ 3 & 5 \\ -11 & 3 \\ 5 & 2 \end{bmatrix}$, $\begin{bmatrix} 1 & 3 \\ 6 & -11 \\ -1 & 6 \\ -4 & 11 \end{bmatrix}$, none, $\begin{bmatrix} 0 & 2 \\ 1 & 3 \\ -3 & -5 \\ -4 & 1 \end{bmatrix}$, $\begin{bmatrix} 0 & 2 & 4 \\ 1 & 3 & 3 \\ -3 & -5 & -1 \\ -4 & 1 & 14 \end{bmatrix}$,

$\begin{bmatrix} -1 & 1 & 5 \\ 4 & 3 & -6 \\ -3 & -2 & 5 \\ -5 & -1 & 13 \end{bmatrix}$, none, $\begin{bmatrix} 6 & -19 \\ -7 & 1 \end{bmatrix}$, $\begin{bmatrix} -1 & 3 & 0 & -1 \\ -1 & 0 & 3 & -2 \\ 1 & -9 & 6 & -1 \end{bmatrix}$,

(ii) $E^2 = \begin{bmatrix} 0 & 0 & 1 & -2 \\ 0 & 0 & 0 & 1 \\ 0 & 0 & 0 & 0 \\ 0 & 0 & 0 & 0 \end{bmatrix}$, $E^3 = \begin{bmatrix} 0 & 0 & 0 & 1 \\ 0 & 0 & 0 & 0 \\ 0 & 0 & 0 & 0 \\ 0 & 0 & 0 & 0 \end{bmatrix}$,

$E^n = O$ for $n > 3$.

5. $X = \begin{bmatrix} x & y \\ z & -x \end{bmatrix}$ where (i) $x^2 = -yz$, (ii) $x^2 = 1 - yz$; also in (ii) $X = I, -I$.

6. E.g. $X = \begin{bmatrix} 1 & 1 \\ 0 & 1 \end{bmatrix}$, $\quad Y = \begin{bmatrix} 0 & 1 \\ 1 & 2 \end{bmatrix}$.

8. $\begin{bmatrix} -1 & 0 \\ y & -2 \end{bmatrix}$, $\begin{bmatrix} -2 & 0 \\ y & -1 \end{bmatrix}$ (y arbitrary), $-I$, $-2I$.

10. E.g. $P = \dfrac{1}{\sqrt{2}} \begin{bmatrix} 1 & 1 \\ 1 & -1 \end{bmatrix}$.

11. $n = 12$.

13. $P^2 = (a^2 + b^2 + c^2)I$, $P^{-1} = (a^2 + b^2 + c^2)^{-1} P$.

15. $U^2 = \begin{bmatrix} 0 & 0 & 1 \\ 0 & 0 & 0 \\ 0 & 0 & 0 \end{bmatrix}$, $U^3 = O$, $AB = a^3 I$, $A^{-1} = a^{-3} B$.

17. $A^{100} = 50A^2 - 49I = \begin{bmatrix} 1 & 0 & 0 \\ -50 & 1 & 0 \\ 150 & 0 & 1 \end{bmatrix}$.

18. (i) $\dfrac{1}{6} \begin{bmatrix} -3 & 3 & 3 \\ 7 & -5 & -3 \\ 5 & -1 & -3 \end{bmatrix}$, $\begin{matrix} x = 1 \\ y = 0 \\ z = -1 \end{matrix}$ (ii) $\begin{bmatrix} 5 & 2 & -3 \\ 1 & 1 & -1 \\ -3 & -1 & 2 \end{bmatrix}$, $\begin{matrix} x = -1 \\ y = -1 \\ z = 1. \end{matrix}$

21. $\begin{bmatrix} -3 & -2 & -7 \\ -1 & 0 & -2 \\ 3 & 1 & 6 \end{bmatrix}$.

23. (i) $A = \begin{bmatrix} 1 & -1 & \frac{3}{2} \\ -1 & 2 & 0 \\ \frac{3}{2} & 0 & -1 \end{bmatrix}$,   (ii) $A = \begin{bmatrix} 2 & 0 & -\frac{3}{2} \\ 0 & 4 & 3 \\ -\frac{3}{2} & 3 & 0 \end{bmatrix}$.

24. (i) $a = 1$, $b = c = -2$.

25. $a = 2$.

26. $x = 0$, $y = a/(a-1)$, $z = 1/(1-a)$; $a = 1$:no solution; $a = 2$:$y = 2-x$, $z = -1$, x arbitrary.

27. $x = (a-1)/(a+1)$, $\; y = 0$, $\; z = -2/(a+1)$; $\; a = 1$:$x = 3+3z$, $\; y = -2-2z$, $\; z$ arbitrary; $a = -1$:no solution.

28. $a = -2$, $x = -8t$, $y = 3t$, $z = 7t$ (t arbitrary).

29. $15x = 7p + 2q - 29y - 66t$, $15z = q - 4p + 23y + 12t$.

30. $x = -4 + 5z - t$, $y = 9 - 7z$, z and t arbitrary.

Exercise 9.1

1. (i) on, (ii) on, (iii) in, (iv) on, (v) in, (vi) in, (vii) neither, (viii) on.

2. Addition: no; multiplication: yes; the same.

3. Addition: no; multiplication: no; without $a \neq 0$: no, yes.
4. (mod 6) neither closed; (mod 5) both closed.

Exercise 9

2. (i) four-group, (ii) C_4, (iii) four-group, (iv) C_4 [identity $= 6$].
5. All except (iv), (vi) abelian.
6. (i) Not subgroup, (ii) subgroup, (iii) subgroup.
7. (i) Not subgroup, (ii) subgroup.
8. (i) Not subgroup, (ii) subgroup, (iii) subgroup.
20. (i) C_2, (ii) four-group, (iii) C_2, (iv) C_2, (v) $\{e\}$, (vi) A group of order 8 comprising e, rotations through $\pm 90°$ and $180°$ about the centre, reflections in the diagonals and in the lines joining the midpoints of opposite sides, (vii) C_2, (viii) four-group, (ix) an infinite group comprising all rotations (including e) about the centre O and all reflections in lines through O, (x)C_2.

Exercise 10

1. (i) ring, (ii) not ring, (iii) integral domain, (iv) field, (v) not ring.
2. Yes.
3. (ii) No.
4. (i) integral domain, (ii) field, (iii) not ring, (iv) field, (v) ring, (vi) field, (vii) ring, (viii) field.

Additional Examples I

1. (i) F, (ii) T, (iii) F, (iv) F, (v) T.
2. (i) T, converse F, (ii) F, T, (iii) T, T, (iv) F, T.
3. (i) F, converse F, (ii) T, F, (iii) F, F, (iv) T, F, (v) F, T, (vi) F, T.
5. (i) Valid; (ii), (iii) invalid.
6. (i) (b), (ii) (c), (iii) (a), (iv) (c), (v) (b), (vi) (a).
7. (ii), (iii), (v), (vi).
8. $A = H, B = F, C = L, D = G, E = K.$
9. (i) $\{2, 4, 6, 8, 10\}$, (ii) $\{2, 3, 5, 7\}$, (iii) $\{2, 3, 5, 6, 7, 8, 10\}$, (iv) $\{3, 7\}$, (v) $\{1, 2, 3\}$.
10. (i) $\{7, 9, 10\}$, (ii) S, (iii) $\{1, 2, 3, 5\}$, (iv) $\{5\}$.
11. (i) $\{(2, 0)\}$, (ii) $\{(2, 0), (-2, 0)\}$, (iii) $\{(\sqrt{3}, 1), (-\sqrt{3}, 1)\}$, (iv) $\emptyset$.
12. (i) $[2, \infty)$, (ii) $(-\infty, 4)$, (iii) $[2, 4)$.
13. $\{x \in \mathbf{R} : 0 < x \leqslant 1\}$, $\{x \in \mathbf{R} : x \leqslant 3 \text{ or } x > 4\}$.
14. $\{x \in \mathbf{R} : 1 \leqslant x < 2\}$, $\{x \in \mathbf{R} : 2 \leqslant x < 4\}$, $\{x \in \mathbf{R} : x < 1 \text{ or } x \geqslant 2\}$, $\{x \in \mathbf{R} : 0 \leqslant x < 1 \text{ or } x = 2 \text{ or } x \geqslant 3\}$.
16. Yes.
18. 83, 132.
19. (i) c, (ii) a, c, (iii) a, b, c, (iv) b, (v) b, (vi) b.
20. $\{a \in \mathbf{Z} : a \text{ odd}\}$, $\{a \in \mathbf{Z} : a \text{ even}\}$.
21. O and circles with centre O.
22. l and pairs of parallels equidistant from l.
23. $\{3k : k \in \mathbf{Z}\}$, $\{3k \pm 1 : k \in \mathbf{Z}\}$.
25. $\{-2, 1, 4, 7, 10, 13\}$.
26. $1, \sqrt{2}, 1, 0, -1, \{k\pi - \frac{1}{4}\pi : k \in \mathbf{Z}\}$.
27. $\{0, 1, 4, 9\}$, $\{x \in \mathbf{R} : 0 \leqslant x \leqslant 9\}$.
28. (i) $\{-\frac{7}{2}\}$, (ii) $\{-2, 2\}$, (iii) $[\frac{1}{3}\pi, \pi]$.
29. $1, 0, 2, 2, -\frac{3}{4}; 3, \frac{4}{3}, -\frac{5}{2}; \mathbf{R} - \{1/k : k \in \mathbf{Z} \text{ and } |k| > 1\}$.

30. (i) (a), (b), $f^{-1}(y) = \frac{1}{3}(y+3)$, (ii) neither, (iii) (b), (iv) (a), (b), $k^{-1}(y) = \sqrt[4]{y}$.

31. (i) 2, (ii) $\frac{1}{4}$, (iii) 81, (iv) $2^{16} = 65536$, (v) 2^{16}, (vi) -43.

32. (i) $f \circ g(x) = x^2$, $g \circ f(x) = \begin{cases} x^2 & \text{when} \quad x \geqslant 0 \\ -1/x & \text{when} \quad x < 0. \end{cases}$

(ii) $f^{-1}(y) = \sqrt{y}$ when $y \geqslant 0$, $1/y$ when $y < 0$.

33. f: (a), not (b), image $(-\infty, 0) \cup [3, \infty)$.

g: not (a) [e.g. $3 = g(-1) = g(5)$], not (b), image $[-2, \infty)$.

$g \circ f(x) = x + 1$ when $x \geqslant 0$; $= 2 - x$ when $x < 0$.

$$f \circ g(x) = \begin{cases} x+1 & \text{when} \quad x \geqslant 2, \\ x-2 & \text{when} \quad 0 \leqslant x < 2, \\ 5-x & \text{when} \quad x < 0. \end{cases} \qquad h^{-1}(y) = \begin{cases} y-1, & y \geqslant 3 \\ y+2, & -2 \leqslant y < 0. \end{cases}$$

34. (i) **R**, $[2, \infty)$; (ii) **R**, **R**; (iii) $[-3, 3]$, $[0, 3]$; (iv) $[3, \infty)$, $[0, \infty) = \mathbf{R}^+$; (v) $\emptyset$, $\emptyset$; (vi) **R**, $[-1, 1]$.

36. $8, 9, 1/8, 9, -1/8, 1/9, 1/9, -8$.

37. (i) $\frac{1}{4}$, (ii) $4\sqrt{3}$.

38. (i) Rational, (ii) irrational, (iii) irrational, (iv) neither, (v) rational, (vi) neither, (vii) irrational.

40. 49/54.

41. (i) $x^2 - 2x - 1 = 0$, (ii) $x^3 - 3x^2 + 3x - 3 = 0$, (iii) $x^2 + 10x + 22 = 0$, (iv) $x^4 - 10x^2 + 1 = 0$, (v) $x^4 - 4x^3 - 4x^2 + 16x - 8 = 0$, (vi) $x^3 - 6x - 6 = 0$.

44. $x = 2, y = 1; x = -1, y = -2$.

46. $x > 1$.

47. $(56, 40)$; $p = 8, q = 4$.

48. $(-\infty, 0) \cup (1, 2]$.

49. $(1, 3), (-3, -1) \cup (1, 3)$.

50. $(-1, 0] \cup (1, \infty), \{0\} \cup (1, 2]$.

54. (i) 504, 495, 715; (ii) 4; (iii) 7.

55. (i) $(11-r)/r$, (ii) $(11-r)x/r$; 252, 45/16.

58. -50.

59. $(-1)^n \left[-9\binom{n}{5} + 6\binom{n}{4} - \binom{n}{3} \right].$

60. $\frac{1}{6}n(n-1)(1-2n)$.

61. (ii) (a) $(-1)^m \binom{2m}{m}$ where $n = 2m$; (b) 0.

62. (i) 120, (ii) 3125, (iii) 48; 72.

63. $11!/16 = 2494800$, $\frac{3}{4}(9!) = 272160$.

64. 25.

65. 21, 105 [disregarding order of subsets].

67. (i) 19, 12; (ii) -16, 6; (iii) 504, 17.

68. (i) 112111, 516; (ii) 100111, remainder 1010; (iii) 61; (iv) 7.

69. $d = 1, x = 121, y = -76; d = 2, x = 379, y = -748; d = 146, x = -5, y = 8.$

70. (i) 1, (ii) 2, (iii) 146.

71. (i) $2.7.73$, (ii) $3.11.31$, (iii) 2^{10}, (iv) $5^2.41$, (v) 4567 (prime), (vi) $3^3.7.11.13.37$.

72. (i) No.

73. (i) 7, (ii) 23.

74. (i) F [e.g. $p = a = 3, b = 1$], (ii) T, (iii) T, (iv) F [e.g. $m = 2, a = 1, b = 2$].

75. $\{3, 10, 17, 24, 31, 38, 45\}$.

76. Adjoin (e.g.) 2, 3, 6, 7, 9, 12, 13, 14, 15, 17.

77. (i) T, (ii) F [e.g. $a = 1, b = 2, m = 3$], (iii) F [e.g. $a = 1, b = 4, m = 3$], (iv) F [e.g. $a = 5, b = 4, m = 3$], (v) T.

78. (i) $x \equiv 7 \pmod 9$, (ii) $x \equiv 3, 5$ or $6 \pmod 7$, (iii) $x \equiv 1, 5, 8$ or $12 \pmod{13}$, (iv) $x \equiv 1$ or $3 \pmod 6$.

80. $x \equiv 15 \pmod{30}$.

81. $x^2 + 1, 2x^2 + x + 1, 2x^2 + 2x + 1, x^2 + x + 2, x^2 + 2x + 2, 2x^2 + 2$.

82. (i) $(2 - 11i)/25$, (ii) $(7 - i)/5$, (iii) $(3 + i)/10$.

83. (i) $\sqrt 2, \frac{3}{4}\pi$; (ii) $2, -\frac{1}{6}\pi$; (iii) $5, \tan^{-1} \frac{4}{3} [= 53°8'$ or $0·927$ rad, app.], (iv) $2 - \sqrt 3, -\frac{1}{2}\pi$.

84. $\cos(5\alpha - 3\beta) + i \sin(5\alpha - 3\beta), \cos(2\alpha + \beta) + i \sin(2\alpha + \beta).$

85. $3 + i$.

86. $u = \dfrac{x^2 + y^2 + 6x + 8y}{x^2 + y^2 + 12x + 36}, \quad v = \dfrac{8x + 6y + 48}{x^2 + y^2 + 12x + 36}.$

87. $1 \pm 2i$.

89. $2 \cos \dfrac{5\theta}{2}, \dfrac{5\theta}{2}.$

90. $1, \cos \dfrac{2\pi}{5} \pm i \sin \dfrac{2\pi}{5}, \cos \dfrac{4\pi}{5} \pm i \sin \dfrac{4\pi}{5}$; products of these by 2.

91. (i) $-1, \frac{1}{2}(1 \pm i\sqrt 3)$; (ii) $2^{1/4}\left(\cos \dfrac{k\pi}{12} + i \sin \dfrac{k\pi}{12}\right)$, where $k = 5, 11, 17, 23$.

92. (i) $-\cos 8\theta - 4 \cos 6\theta - 4 \cos 4\theta + 4 \cos 2\theta + 5$,
(ii) $-\sin 8\theta + 6 \sin 6\theta - 14 \sin 4\theta + 14 \sin 2\theta$.

94. $1, \frac{1}{2}(-1 \pm i\sqrt 3)$.

95. (i) $-2 \pm 3i, 2 \pm \sqrt 6 i$; (ii) $-1 \pm 4i, \frac{1}{2}(-1 \pm 3i)$.

96. $-1, \cos \dfrac{k\pi}{7} \pm i \sin \dfrac{k\pi}{7}$ $(k = 1, 3, 5)$; $x^2 - 2x \cos \dfrac{k\pi}{7} + 1$ $(k = 1, 3, 5)$.

97. Reflection in the real axis followed by rotation through α about O; i.e. reflection in the line $\theta = \frac{1}{2}\alpha$.

98. $\begin{bmatrix} -14 & 7 & 7 \\ 1 & 6 & 3 \end{bmatrix}.$

101. $\dfrac{1}{3}\begin{bmatrix} 4 & -1 \\ -5 & 2 \end{bmatrix}$; $p = -\frac{1}{3}$, $q = 2$.

102. $\begin{bmatrix} \pm 2 & 0 \\ 0 & \pm 1 \end{bmatrix}$ (four matrices).

103. $k = 0$, any matrix of the form $\begin{bmatrix} x & y \\ z & -x \end{bmatrix}$ with $yz = -x^2$.

107. (i) 5, (ii) -65, (iii) 0, (iv) 0, (v) 2.
108. (i) $-(a-b)(a-c)(b-c)(a+b)(a+c)(b+c)$.
109. $(a-b)(a-c)(a-d)(b-c)(b-d)(c-d)$.

110. (i) (a) $\begin{bmatrix} 0 & 1 & -1 \\ -1 & 1 & -1 \\ 1 & -3 & 4 \end{bmatrix}$, (b) no inverse. (ii) $x \neq -\frac{4}{3}$.

111. adj $A = \begin{bmatrix} 6 & -1 & -9 \\ -4 & 2 & 6 \\ 2 & -1 & -1 \end{bmatrix}$, $A^{-1} = \frac{1}{4}$ adj A, $x = -1$, $y = 1$, $z = 0$.

112. $x = z = -b/2(a+2)$, $y = b/2$. When $a = -2$ consistent if and only if $b = 0$, solution $x = z$, $y = 0$.
113. $x = 5z + 5t$, $y = 1 - 2z - t$, z and t arbitrary.
116. Inverse law fails.
117. All abelian except (ii).
119. Yes.
124. (i) Not ring, (ii) ring, (iii) integral domain, (iv) field, (v) not ring, (vi) integral domain, (vii) not ring, (viii) not ring, (ix) field.
125. (i) Not ring, (ii) not ring, (iii) not ring, (iv) integral domain, (v) field, (vi) integral domain, (vii) field.
127. $b^2 = b$, $c^2 = a$, $cd = c$, $db = d$, $dc = c$.
129. Integral domain.
131. Yes, yes.

132. $\left\{ \begin{bmatrix} x-y & 2y \\ -y & x+y \end{bmatrix} : x, y \in \mathbf{R} \right\}$.

Additional Examples II
   1. No.
   4. (i), (iv) $p \wedge q$; (ii), (iii) $p \vee q$.
   5. (i) T, F; (ii) F, F; (iii) T, T.
   6. (i) F, (ii) T, (iii) T, (iv) F.
   7. (i) T, (ii) F, (iii) T, (iv) F.
   8. Yes, no, no.

10. $A = \{2, 4, 6, 8, 10, 12\}$, $B = \{3, 6, 9, 12\}$, $C = \{4, 8, 12\}$, $D = \{6, 12\}$,
   $E = \{1, 4, 9\}$, $B' = \{1, 2, 4, 5, 7, 8, 10, 11\}$,
   $C' = \{1, 2, 3, 5, 6, 7, 9, 10, 11\}$, $C + E = \{1, 8, 9, 12\}$.
11. (i) $(1, 2]$, (ii) $[0, 1) \cup (2, 3]$, (iii) $[-1, 1) \cup (2, 3]$, (iv) $(0, 1) \cup (2, 3)$, (v) $[-1, 1]$.
12. (i) $\{-1, 0\}$, (ii) $\{-\frac{3}{4}, -\frac{1}{2}, -\frac{1}{4}, 0, \frac{1}{4}, \frac{1}{2}, \frac{3}{4}\}$, (iii) $\{-7, -5, -3, -1, 1, 3, 5, 7\}$.
14. (i) S, (ii) $\emptyset$, (iii) $A \cap B$.
19. $\{(a, x), (a, y), (a, z), (b, x), (b, y), (b, z), (c, x), (c, y), (c, z)\}$, $\{(a, x), (a, y), (b, x), (b, y)\}$,
   $\{(b, y), (b, z), (c, y), (c, z)\}$, $\{(b, y)\}$, yes, no.
20. 24.
22. Straight lines through O (other than $x = 0$, $y = 0$) with O deleted; hyperbolas
   with $x = 0$, $y = 0$ as asymptotes.
23. The sets $S_h = \{2^h k : k \in \mathbf{N}, k \text{ odd}\}$, $h = 0, 1, 2, \ldots$; e.g. $S_2 = \{4, 12, 20, 28, \ldots\}$.
24. T means "$x - y$ is divisible by 6".
26. Images of a, b, c under $f \circ g$, $g \circ f$, g^{-1}, $g^{-1} \circ f \circ g$ are a, b, b; a, a, c; b, c, a;
   b, c, c respectively.
27. $h^{-1}(y) = (2y + 3)/(y - 2)$.
28. $f^{-1}(y) = (y + 2)/(y - 3)$ when $y \neq 3$; $f^{-1}(3) = 1$.
29. $g^{-1}(y) = \frac{1}{2}(y + 1)$ when $y < 1$; $= \frac{1}{3}(y + 2)$ when $y \geqslant 1$.
31. $c = \pm 1$.
32. $[-\frac{1}{2}, \frac{1}{2}]$, no.
33. Map a, b to c, c; c, d; d, c; d, d; 2 surjections. 9; 6 injective. 8; none injective.
35. 27, 19.
38. $\{0, 1, 4\}$, $\{0, 1\}$, $\{0\}$, $\{0, 1, 4\}$.
43. $(-4, -3) \cup (-1, \frac{1}{3})$.
44. $(-4, -3) \cup (-1, \sqrt{7} - 2)$.
45. $[-1, 2]$.
46. $\{-\frac{2}{3}, 0\}$.
48. No.
49. $\mathbf{Q} - \{0\}$.
57. f_{4n} is divisible by 3 for all $n \in \mathbf{N}$.
58. 648.
59. 840, 80.
60. 32, 30.
61. 96.
62. 840.
66. 5^n, $(-1)^n$, $5^n . n$.
70. 23, 46.
71. $x = -146$, $y = 215$; $x = 383$, $y = -564$ (not unique).
79. 11, 13.
80. (i) 2, (ii) 0.
81. 0.
85. $3\sqrt{2}, \frac{3}{4}\pi$; $2/\sqrt{3}, -\frac{2}{3}\pi$; $2\sqrt{3}, -\frac{1}{6}\pi$; $\sec \alpha, -\alpha$.
86. -1.
87. $\pm(\sqrt{3} + i)/\sqrt{2}$.
88. $x = \sqrt{[\frac{1}{2}(1 + \sqrt{2})]}$, $y = \sqrt{[\frac{1}{2}(\sqrt{2} - 1)]}$, $\cos \frac{1}{8}\pi = 2^{-1/4}x$, $\sin \frac{1}{8}\pi = 2^{-1/4}y$.
90. Half of an equilateral triangle, with B a right angle.
92. $\frac{1}{128}(\cos 8\theta - 8 \cos 6\theta + 28 \cos 4\theta - 56 \cos 2\theta + 35)$.
93. 14, 98; roots $-1, -2, 3$.
94. $w^3 + (2b - a^2)w^2 + (b^2 - 2ac)w - c^2 = 0$.

95. $5; 1, -2, -1\pm i$.

96. -2.

97. Non-real roots form conjugate pairs.

98. $\pm 1, \pm i, \pm(1+i)/\sqrt{2}, \pm(1-i)/\sqrt{2}$.

99. $[(\sqrt{3}+1)+i(\sqrt{3}-1)]/2\sqrt{2} = \cos\dfrac{\pi}{12}+i\sin\dfrac{\pi}{12}$.

101. (i) f yes, g no; (ii) f yes, g no.

102. $p \circ q(z) = iz, q \circ p(z) = 1-i-iz$.

103. $f^{-1}(w) = (\bar{w}-\bar{b})/\bar{a}$.

104. b^2/a.

106. $\dfrac{\sin^2\theta}{1-2\sin\theta\cos\theta+\sin^2\theta}$.

110. $48A-7I$.

112. $\begin{bmatrix} 1 & -a & ac-b \\ 0 & 1 & -c \\ 0 & 0 & 1 \end{bmatrix}$.

113. $x:y:z = -7:23:11$.

114. $x = 1, y = -1, z = -2$.

115. $x = \frac{2}{5}(8-3z), y = \frac{1}{5}(1+14z)$, z arbitrary.

116. $a \neq 1 : x = 2+a, y = -a; a = 1 : x = 1-2y$, y arbitrary.

117. $x = 1, y = 1, z = -2$; yes, no.

119. $\begin{bmatrix} -1 & 3 & 1 \\ 1 & 0 & 1 \\ 1 & -1 & 0 \end{bmatrix}$.

125. $\pm(A+2I)$.

127. (a) (i) f, h; (ii) f, h. (b) Yes, yes, no.

128. No.

129. No.

131. (iv) Yes: -1.

133. Yes.

136. The first, not the second.

137. Four-group.

145. No.

146. Yes. No [distributive law fails].

148. $2, 3; 2$ (double root).

149. $1, 3, 5, 7$. Possible answers: (i) $x^2-x = 0$, (ii) $2x^2+x-4 = 0$, (iii) $x^2-3 = 0$.

152. (i) q^4, (iii) $(q-1)^3$.

153. (i) Four-group, (ii) C_3, (iii) 1, (iv) $1, j^2$; (e.g.) $x^2+x+j = 0$; (iv) $0, 1, j, j^2$.

Index